Driven by Demand

D0558422

Energy plays a central role in shaping our society and infrastructure, making it increasingly important for today's leaders to understand the impact of energy decisions. Discussions about energy often neglect important historical lessons about previous energy transformations and provide inadequate consideration of context – *Driven by Demand* takes a fresh approach by exploring the emergence of energy systems, outcomes and priorities. It outlines select historical and current events, challenges, and developing energy trends using a range of case studies. Readers will gain foundational knowledge about energy flows and end-uses, helping them to become more conversant about energy outcomes and priorities. This accessible book paves the way for broader discussions about societal resilience, privacy, and security concerns associated with the move towards "smart" infrastructure. This is a must-read for business executives, policymakers and students working in energy policy, energy management, and sustainable business.

A link to the authors' companion website, which features additional commentary and upcoming events, can be found at www.cambridge.org/demand.

JIMMY Y. JIA leads the Sustainable Energy Solutions Certificate program at Pinchot University, a year-long course that educates MBA students on how to incorporate energy issues into the decision-making process. Together with Jason Crabtree, he founded Distributed Energy Management to help business owners transparently manage utility costs via a money management account, combining total lifecycle costs of operational and capital expenses. He has been passionate about promoting energy literacy and understanding how energy consumption affects our society since his time as a student at MIT.

JASON CRABTREE is the CEO of Rationem, a bespoke analysis, advisory, and technology firm, and Chairman of Distributed Energy Management. He is a risk management and complex adaptive systems aficionado with a strong focus on the nexus of information, energy, security, and technology. He maintains deep involvement in energy, cybersecurity, and national security strategy and policy. He is a graduate of the US Military Academy at West Point and a Rhodes Scholar.

Driven by Demand

How Energy Gets its Power

JIMMY Y. JIA AND JASON CRABTREE

CAMBRIDGE
UNIVERSITY PRESS

CAMBRIDGE
UNIVERSITY PRESS

University Printing House, Cambridge CB2 8BS, United Kingdom

Cambridge University Press is part of the University of Cambridge.

It furthers the University's mission by disseminating knowledge in the pursuit of education, learning and research at the highest international levels of excellence.

www.cambridge.org
Information on this title: www.cambridge.org/9781107507104

First published 2015

Printed in the United States of America by Sheridan Books, Inc.

A catalog record for this publication is available from the British Library.

Library of Congress Cataloging in Publication data
Jia, Jimmy Yi-Jie, 1980–
    Driven by demand : how energy gets its power / Jimmy Y. Jia and Jason Crabtree.
        pages   cm
    Includes bibliographical references and index.
    ISBN 978-1-107-10466-2 (hbk) – ISBN 978-1-107-50710-4 (pbk)
    1. Power resources.   2. Electric power production.   I. Crabtree, Jason.   II. Title.
    TJ163.2.J53   2014
    621.042–dc23   2014049354

ISBN 978-1-107-10466-2 Hardback
ISBN 978-1-107-50710-4 Paperback

Additional resources for this publication at www.cambridge.org/demand

To my mother, for her unwavering support, guidance, and acceptance that my fascination with energy is not a temporary one. — Jimmy

To Bridget, for her toleration of my unabashed love for electric power systems, simulations, and my unfailing desire to describe them to her ad infinitum. — Jason

Contents

Figures

Tables

Acknowledgments

Books are never the work of individuals; they are the result of large numbers of people who shape them in a myriad of ways. This effort was no exception.

We are profoundly grateful for the support and encouragement of a broad group of people whose insight and experience was formative. Special thanks goes to Pinchot University, ConocoPhillips, Mary Kay Chess, John Gardner, Lola Ball, Paula Parish, Gideon and Gordon Parke, Andrew Sellers, Stan Gent, Roel Hammerschlag, Phillip Schmidt-Pathmann, Steven Tobias, Philip Hahn, Matt Brown, Jake Beal, Kyle Capizzi, Olivia Kim, Rich Sonstelie, Frank Heidt, Ian Macleod, Jack Dillon, Oliver Venn, Brewster Earle, Stephanie Hare, David Johnson and Steve Flynn.

Your contributions of insights, support, and encouragement made this book possible.

Jimmy Y. Jia and Jason Crabtree

PART I

Introduction

1 | *You get what you ask for*

There is nothing either good or bad, but thinking makes it so.

William Shakespeare, *Hamlet*

Energy is full of paradox. Depending on the mix of resources used to generate power where you live, driving an "environmentally friendly" electric vehicle can increase carbon emissions relative to an equivalent petroleum-fueled mode of conveyance (Doucette and McCullough, 2011). Even if solar energy were ubiquitous, we would still require coal in the modern economy. Energy efficiency improvements do not reduce overall demand for energy products. Effectively reasoning about energy, despite numerous similar paradoxes and competing factors, is the primary function for this book.

A friend in the coal industry, with experience in pit mining, recently lamented about how much the environmental costs of energy extraction and the residual scarring on the landscape bothered him. At the same time, he quipped that "As long as people keep buying coal, I'll keep digging it up." People certainly do not keep lumps of coal sitting at home. Yet virtually all of humanity is dependent on coal in some form because of its role in electricity generation and in the process of steel-making.

Making good decisions about energy—despite these kinds of counterintuitive facts—requires context. Our modern society uses vast amounts of energy, even when compared to human civilization only a few generations ago. Our energy infrastructures are oft forgotten but are foundational elements of our society which are completely approachable if explored from the right perspective. Energy consumption is a part of everyday living, but energy issues are mistakenly treated as too complex or arcane to be considered by decision-makers in business or policy. Nothing could be further from the truth.

The networks which feed society's growing thirst for energy developed specific technologies and grew into their particular structure for specific historical reasons. Often, these reasons have more to do with the demand for energy—specifically the way in which we consume energy to do useful work—than with any particular attribute of how we locate, extract, transport, and transform the energy products being consumed. This book is structured as an explanation about *why* we consume energy to better understand *how* we supply it today and the implications for our collective future.

1.1 The energy we buy

So what exactly is the *energy* that we buy? What is this product that silently supports our entire economy and society? What form does our energy come in? What would happen if this product were interrupted or no longer available? What are the underlying services that the product enables? How can we start to determine ways to transition to a more sustainable energy future? Will the solution require a single replacement or a mix of many resources that balance the needs of the economy, society, and the environment?

Like virtually any other modern product or service, energy comes in many forms: electric energy, petroleum energy, solar energy, etc. Often, differing forms can deliver the same benefits, such as gasoline-powered vehicles or battery-powered vehicles. But sometimes only one form of energy is suitable for a particular task, such as using electric energy to charge a laptop. In many cases, one form has a competitive advantage over another based on the type of energy needed to do useful work. An example is using a natural gas boiler to run a central heating plant with electricity co-generation to heat and power maximally efficient commercial buildings.

Regardless of the form of energy selected for individual tasks, energy runs our society. The "energy problem" is not an isolated issue whose solution resides in an engineering or process optimization book. Rather, energy is a central component of modern issues on a global scale, including poverty, national security, health, commerce, and the environment. Energy underpins the creation and delivery of nearly all modern benefits to society. It feeds the processes that we deem necessary, as most visibly expressed by our wallets.

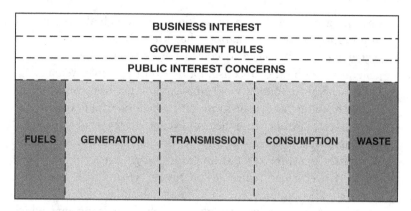

Figure 1.1 Major energy silos

1.2 A few guiding principles

Energy is more than just a supply and fuel source problem or an energy efficiency and consumption problem. The energy supply chain, from fuels to waste, is shown in Figure 1.1. Intersecting each of these silos are business interests, government rules, and public interest concerns. To truly understand and act on energy issues, one needs to understand the interactions between these foundational components of energy as an economic system which has significant overlap with the responsibilities of sovereign states. Figure 1.1 also shows a conceptual framework that depicts the major silos of the energy industry. The bottom silos in gray represent the infrastructure and supply chain. The top silos in white represent societal interaction with the infrastructure.

A few key points about energy help ground our exploration:

1. Energy itself is always conserved. Energy cannot be created or destroyed, but it can be transformed from one form to another to meet our specific needs. Each transformation incurs a loss in useful energy.[1] Generally speaking, the more times we convert forms of energy, the larger the losses.
2. Energy must always be in balance; what goes in must come out. Too frequently, the conversation about sustainable energy revolves exclusively around generation issues, such as promoting wind or

---

[1] Losses of *useful* energy can also be expressed as an increase in entropy, which is discussed in more detail later in the book.

protesting coal. Although these kinds of individual issue are an important part of the equation, they are only a small part of the bigger picture. We need to consider how to balance sustainable generation with rational consumption over the whole system.

3. Useful energy is foundational to society. At a basic level, useful energy from the sun is transformed into the food which we consume to survive. Beyond these needs, excess energy enables societal development and the emergence of specialized skills and services. Businesses generate revenue by consuming energy to do useful work. Governments consume energy to secure their resources, territory, and populations. Individuals consume energy to better enjoy their lives in increasingly controlled environments which provide light and thermal comfort, and are populated with a myriad of energy-consuming devices to enhance recreation and productivity. Each of these activities is enabled by society's growing ability to harness more useful energy.

1.3 Features versus benefits

Society buys energy for one simple reason: we value energy outcomes. We value the heat, light, or mobility that energy provides when transformed for our end uses. These outcomes, which are really aspects of consumer benefits, ultimately drive energy supply and infrastructure decisions. As a result, to truly understand key facets of today's energy issues, we must begin with an evaluation of end uses and work back towards the underlying infrastructure and resources available. Throughout the book, we will continually tie decisions and examples from each considered industry vertical back to the specific transformations that allow it to meet customers' demands.

Too much focus is often given to energy *features* without adequate attention to energy *benefits*. Energy features include topics such as fossil fuels, renewables, carbon sinks, or heat reclamation. These types of properties refer almost entirely to the energy sources we consume. Energy benefits are the services and outcomes we enjoy, ranging from an international cruise to a video call with family to modern refrigeration and food delivery networks.

As anyone with experience in sales knows, benefits sell products. Product benefits are usually driven by quantifiable demand. Benefits, not features, are ultimately what drive most consumers to open their wallets. Product features describe product attributes that may influence a decision but rarely close a deal. This holds true for the energy sector as well. Societal demands for energy in previous generations shaped the current system and drove the construction of our existing infrastructure. However, the problems that we faced in the late 1800s are different than the problems of today or those likely to manifest in the future. Just like law, regulations, and most other human endeavors, energy infrastructure and delivery systems lag behind what is technologically feasible.

Our energy infrastructure—physical, financial, regulatory, political, and others—is very good at solving many problems from a century ago. For most industrial powers, this meant providing low-cost electricity to the largest number of people. As our society has evolved since, our requirements have shifted substantially. Today's society has a greater need for reliability, security, resilience, and sustainability in its energy strategy. Cost is still dominant, but other factors have grown substantially in importance. These ongoing shifts in demand will dictate much about how energy infrastructure will continue to evolve. By understanding why we consume energy, we can better understand where both energy products and services are likely to be headed.

This book emphasizes priorities and problem formulations to reframe much of the existing energy debate around possible actions. It eschews a focus on tradeoffs, which give readers binary solutions to complex problems, and embraces decision matrices to help provide a more nuanced decision-making tool to align actions with priorities. It explores interactions between siloed industry sectors. Because the discussion revolves around the core value proposition of energy consumption—doing useful work—readers can re-examine real-world technologies and services based on energy outcomes.

1.4 Where we are headed

The rest of the book is on society and the demands we place on energy. It explores the questions of "What are we buying?" and "Can we procure the same benefit while consuming less energy?" We make powerful contributions to energy policy and strategy each day by voting with our wallets. By internalizing energy outcome-focused thinking into everyday business decisions, we can become more conscious consumers who are equipped with the proper tools and vocabulary. We can also better manage energy-related risks for our families, our organizations and communities, and our world by seeking a mix of efficiency and resilience.

A significant portion of the book is dedicated to electric energy infrastructure, as it is the most prolific and easily recognizable energy infrastructure in developed nations. However, many of the concepts applied to electric power systems are applicable to other energy infrastructures. The fundamental problem of delivering reliable energy services across a large, complex, physically constrained system is generic. Participation from numerous stakeholders with competing constraints and objectives requires a mix of economic and regulatory incentives to balance individual, organizational, and public goods under a wide range of environmental conditions. Although vocabulary does differ across verticals, the detailed examples used introduce many of the concepts required to support broader reading and study into specific kinds of energy systems or technologies.

The rest of this book looks at demand through a number of lenses:

- Part II examines historical demands to explain how the energy system evolved into what it is today. We use three frameworks to analyze energy consumption—energy balances, the energy supply chain, and the *wicked problem*. Wicked problems have no defined solution and typically involve multiple stakeholders with competing priorities or desired outcomes. We also define the six different silos introduced in Figure 1.1 and examine their intersections and how they influence each other.
- Part III explores critical demands facing energy infrastructure today. We apply the three previously introduced frameworks to decisions that managers face in five different energy infrastructures: electricity, transportation, waste, steam, and cyberspace. The cases and stories focus on how energy decisions can be catalysts, influencing and affecting stakeholders across large swaths of society.
- Part IV introduces changing demands to energy infrastructures as transactive systems and introduces complex adaptive system concepts as an important part of effectively managing future energy systems. We also investigate key attributes of energy risk management, infrastructure resilience to meet societal demands, and the movement towards energy-as-a-service. This section employs a mix of historic and recent case studies to illustrate key commonalities

with proven approaches managing critical energy delivery systems while simultaneously demonstrating that new requirements require procedural and technological innovation.

- Part V puts energy demand into a broader context via a targeted exploration of energy's role in the evolution of human civilization. An exploration of humankind's ability to harness ever more energy begins by tracing the evolution of energy and metallurgy through the Stone, Bronze, and Iron Ages. This exploration of energy's role in transforming our society underscores the importance of making educated energy decisions to meet our economic needs and societal values.

As we explore the energy issues, please keep in mind the entire supply chain from fuel through waste. Acknowledging the end-to-end requirements, costs, and constraints is often required to understand societal decisions about energy. Often, critical but sometimes non-obvious issues have significant influence over how we build the supply chain that ultimately delivers energy when, where, and in the form it is needed. The numerous interdependencies between these energy silos also means that seemingly simple changes can have ripple effects—both beneficial and detrimental—throughout the entire system. It is not uncommon for individual or groups of silos to advocate for opposing goals that sometimes impede the implementation of particular solutions.

1.5 Chapter 1 Summary

- This book explores the question "Why did our energy requirements result in the development of today's energy infrastructure systems?" Understanding the reasons behind the evolution of this large system is the first step in understanding potential ways forward. The efficacy of future changes depends on the understanding of end-use requirements and which products or services have direct substitutes or replacements.
- Infrastructure and supply-chain issues account for the physical nature of the energy system. Often, these components have engineer-optimized solutions that do not reflect the context of larger societal choices and values that become controlling.

- Societal issues arise between businesses, governments, and public advocacy groups as they interact with the infrastructure.
- A change made in any one energy silo will have ripple effects across the entire system. The effects—both good and bad—of changes across silos are not uniform.

PART II

Three frameworks

2 | *Energy: the product*

If you want to find the secrets of the universe, think in terms of energy, frequency and vibration.

Nikola Tesla

A fireplace releases heat into a cold room. An escalator brings a person up a flight of stairs. A glowing bulb provides light energy in a dark corner. Each of these examples describes a different form of energy: heat, motion, and light. They also subtly point to an important property of energy. Practical discussions about energy require context or a comparative. The cold room relates to an assumed context based on our human experience. Likewise, our experience allows us to visualize and relate to the energy-related actions of climbing a staircase or the illumination of a once-dark corner. The context is defining. Context informs us whether the energy is useful or not useful. A fire in a cold room sounds pleasing, while the same fireplace emitting heat in the middle of a Texan summer sounds miserable, if not hazardous. A fire by itself with no context of the energy environment is largely meaningless.

Practically speaking, *energy exists only in comparison with something else.* For more complex, real-world energy decisions, this same principle often applies. The context of energy consumption is critical to understanding whether or not to pursue a particular technology or process change. The same "solution" may be perfect in one situation and a disaster for another. In order to understand energy as a product, it is always important to begin by examining the ultimate purpose for an energy transformation and the environment in which it is being consumed.

2.1 Energy and power are not equivalent

Energy is frequently confused with power. Formally,

Energy = Capacity to do work

and

Power = Rate of work being done OR energy flow per unit of time

A deeper look is warranted.

Energy is used to perform work. It is most frequently measured in calories (cal), joules (J), British thermal units (BTUs), kilowatt-hours (kWh), and others. These units quantify the amount of energy *consumed* and are typically used to describe the energy content of a meal (calories) or the amount of heat content in fuel oils (BTUs).

Power describes how much work is being done during a specific time period. Units of power include horsepower (hp), watts (W), BTUs per hour (BTU/h), tons of refrigeration, etc. Specific communities often develop their own units to make relative comparisons within their field less time consuming and more intuitive. These units quantify the amount of energy *demanded*. A commonly misunderstood, but familiar, example is torque and horsepower.

Torque is used to describe how quickly a car can accelerate by applying force at some distance from the axle of the vehicle. Higher levels of torque mean greater acceleration and are achieved through gears in the transmission. No matter how high the torque, the maximum horsepower available always limits the total work done. To measure useful horsepower delivered to the wheels on a typical vehicle, torque and engine speed (in revolutions per minute) are measured on a dynamometer and fed into a formula. Sometimes we care more about the energy, and other times we care more about the power. We care about energy when storing electricity in a battery or heating an oven. We care about power when we are towing heavy objects or lifting boulders.

Much of the electric power industry has standardized around using kWh to describe energy and kW to describe power. These units frequently appear on utility bills as separate charges for energy and power. Their similarity and familiarity often lead to their being used interchangeably, despite meaning totally different things. Perhaps if

Table 2.1 *Comparing a sprinter and a marathon runner (Gallagher, 2013; SRM, 2013)*

	Usain Bolt, sprinter athlete, 100 meters	Alban Lakata, distance mountain bike athlete, 100,000 meters
Time	9.58 sec	3 hours, 37 min, 48 sec
Energy consumed	81.6 kJ	4,466 kJ
Max power	2,620 W	342 W

the industry had standardized on the watt and the joule, we would not confuse the two so easily.

Relationship between energy and power

In equation form, power equals energy divided by time.

$$\text{Power} = \frac{\text{Energy}}{\text{Time}}$$

One way to differentiate between energy and power is by analyzing a sprinter and a marathon runner. Usain Bolt of Jamaica set the world record for the 100 meter sprint at the Berlin World Championships in 2009. It took him a mere 9.58 seconds to travel 100 meters, equivalent to running over 27 miles per hour. He reached a maximum 2,620 watts of power and consumed a mere 81.6 kJ for his race of less then 10 seconds.

On the other hand, Alban Lakata won the European Marathon Championship in mountain biking in 2013. He took 3 hours and 37 minutes to travel 100 km. He consumed 4,466 kJ for his race, over 50 times the energy Bolt consumed. Yet his max power requirement was only 342 watts or less then one fifth of what Bolt needed. Lakata needed more energy than Bolt since he traveled further. Bolt needed more power to complete his feat. The difference between a sprinter and a marathon runner is in the balance between power and energy. Note that the sprinter, who trains for speed, needs more power. The marathon runner trains for distance and consumes more energy but at a slower rate.

2.2 The three laws of thermodynamics

The three laws of thermodynamics dictate the theoretical limit of what is possible based on physical laws. These laws are not only useful for science but have profound implications for understanding business and many other fields. In the simplest business cases, these laws give us an indication as to whether an innovation will require an increase in energy consumption or a decrease in energy consumption. The three laws of thermodynamics can be distilled, in their simplest form, to:

1. Energy cannot be created or destroyed.
2. One form of energy can be converted to any other form of energy.
3. Energy conversions always incur losses.

The first law: energy cannot be created or destroyed

The first law is based on the principles of the conservation of energy and the conservation of mass. It states that the total energy of a system remains unchanged for an isolated system. If energy seemingly appears or disappears, then the defined system is likely too small and needs to be expanded. A candle can be an example of a system. By burning the candle, we can cause it to disappear. This seems to contradict the first law. However, if we expand the system to include the room where the candle was burned, we will find that the room did not lose any mass. The energy contained by the candlewax and wick become light energy, heat energy and waste product–smoke and soot. The first law always holds true. The energy of an isolated system will always be in *balance*.

Since we know that the entire system will always be in balance, within the context of energy solutions, a change introduced in any one of the silos will propagate through all of the silos. For instance, when we shift from petroleum-fueled vehicles to electric vehicles, our actual energy needs remain the same. What differs is the form of energy consumed and the externalities (e.g. pollution) caused by the entire supply chain and ultimate resource consumption. A gasoline-powered vehicle consumes energy in the form of petroleum and emits fumes from the tailpipe. An electric vehicle consumes energy in the form of electricity and emits pollution indirectly from the power plant.

We have to be careful about how we define pollution during fuel shifts. What we consider pollution-free today may not be the case in the future. In the early 1900s, the horseless carriage was hailed as a pollution-free device for getting around town. After all, it had the potential to eliminate over two million pounds of horse manure that needed to be cleaned from the streets of New York City every single day. At the time, the odor of car exhaust was considered pleasant when compared to the smell of horse manure.

The second law: energy can be converted from one form to another

The second law describes how energy transformation occurs. We consume energy in many forms. These can be mechanical energy, light energy, chemical energy, electric energy, etc. Some of the terms used in energy describe the transformation taking place. For example, a *thermoelectric* power plant takes heat energy (thermo) and converts it to electric energy for transmission and eventual consumption. Table 2.2 depicts some common forms of energy and their naming conventions. Energy conversions frequently combine the two roots, such as a photo-thermal system converting sunlight into heat.

Most of today's consumption of energy depends on at least one conversion of energy into an alternate form. A sail on a boat converts wind into motion. A fire converts wood into a combination of heat and light. A steam engine has two primary energy conversions: chemical into heat, then heat into mechanical. Our bodies convert food calories into chemical and kinetic energy. Some geothermal systems convert heat from the earth into kinetic energy (moving water or steam) spinning a turbine to generate electricity.

Every energy conversion requires a dedicated piece of equipment to perform that task. Energy conversions tend to be narrowly tailored for their specific application based on the unique requirements for energy and power. During the Industrial Revolution, we built different steam engines to run factories and locomotives. More recently, we have built specially designed combined heat and power plants which use steam for heat energy and electricity generation in modern buildings. In 1859, John Inshaw even demonstrated a steam engine being used to run a clock (Upton, 2013). The same principle holds true for other energy transformation devices: an internal combustion engine designed to consume gasoline uses different technology than an engine running off natural gas or diesel.

Table 2.2 *Sample forms of energy and their roots*

Forms of energy	Roots
Electricity	Electro-
Heat	Thermal-
Light	Photo-
Motion	Kinetic-
Chemical	Chemo-
Earth	Geo-

Electricity freed us from the restriction and inefficiency of dedicated equipment by providing an efficient transport mechanism with an ability to be easily converted back into other forms of energy to do specific, useful work. While electricity is merely one form of energy, it is perhaps the most significant in terms of recent societal advances. Virtually every form of energy can be converted into electric energy. Examples include harvesting heat from a thermal power plant, mechanical motion from a wind farm, and light energy from a solar panel. Electricity can also be converted into any other form of energy. It is used for diverse purposes, ranging from heat for cooking to mechanical motion for transportation to light emission.

By using electricity, specific end-use devices essentially become interoperable or supportable by many diverse fuel sources. By generating electricity, organizations specializing in transforming energy can easily gain access to a large and readily available market without requiring specialized devices to be adopted by consumers. This makes electrically powered devices far more flexible than their fossil-fueled counterparts in most cases, as long as a reliable supply of electricity is available and conversion losses are tolerable.

The third law: energy conversions are always inefficient

The third law tells us that every conversion will incur a loss. In any conversion process, energy will be lost as heat. In many situations, this heat needs to be dissipated and released in order to keep the system running. Smoke stacks from a power plant release hot air into the atmosphere. A typical consumer's car takes in cooler air from the outside to help cool the engine and releases it back to the atmosphere.

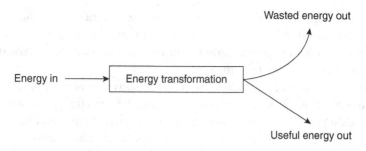

Figure 2.1 Diagram of the energy balance of a system

Figure 2.1 depicts the transformation of energy. The "Energy in" on the left side becomes "Useful energy out" on the right side. The third law concerns the value of the "Wasted energy out" and states that it will always be greater than zero. In effect, every energy conversion will incur loss.

All three laws are represented in Figure 2.1. The first law states that (Energy in) = (Useful energy out) + (Wasted energy out). The second law states that we can transform energy from one form to another. The third law states that any transformation will always result in wasted energy.

The measure of efficiency is a ratio of useful energy out to energy in for a system.

$$\frac{\text{Useful energy out}}{\text{Energy in}} = \text{Efficiency}$$

When we say that an internal combustion engine–powered vehicle is 20% efficient, it means that 20% of the energy content in the petroleum is converted into the desired form of useful energy–in this case kinetic (mechanical motion). The remaining 80% is lost as heat from the engine, friction from the road, and drag from the air. We can also calculate multiple transformation efficiencies. For example, we can use coal to create steam and steam to power a locomotive. The total efficiency of a steam-powered locomotive, which is normally non-condensing, is commonly around 15%. Alternatively, we can use any thermal energy source to create steam to generate electricity and then use electricity to create the kinetic energy required to run a train. In this case, we multiply the individual components. A typical efficiency of converting steam to electricity at a power plant is approximately 35%. The conversion of electricity to motion is approximately 60%. The total conversion for the system is 35% × 60% = 21%. Despite an additional transformation, the electrically powered engine is more efficient.

On the surface, it is counterintuitive why converting steam to motion is less efficient than converting steam to electricity to motion. Shouldn't a single conversion be more efficient than multiple conversions? To understand this, we need to understand the maximum theoretical limit of efficiency. This value represents the highest possible efficiency from a particular energy conversion. The larger the steam engine and the hotter the steam, the closer we can reach to the maximum theoretical limit. In looking at the two steam efficiency numbers above, 15% for a steam locomotive and 35% for a steam power plant, we can see that a power plant operates more efficiently. If the two steam engines were exactly the same, then it would be more efficient to convert the steam directly to motion. However, since the power plant is much larger than a steam locomotive, operates at a higher temperature, and typically uses condensing cycles to increase efficiency, we are able to achieve a net gain despite the extra step.

Not all energy conversions are created equal. Some conversions have a higher theoretical limit than others. In many cases, efficiency can be a red herring. For instance, a co-generation facility can operate up to 80% efficiency by making use of heat directly while also generating electricity. The differences in maximum efficiencies can be confusing. Photovoltaic cells are currently around 15% and internal combustion engines are about 20% efficient, making relative comparisons more difficult.

2.3 Thermal energy

The most common byproduct of an energy conversion is heat. When heat is generated, it needs to be consumed, released or eliminated. Sweat released during a workout is the body's way of dissipating waste heat. The fan on a computer dissipates heat from a CPU. The air intake and radiator of a car help cool the engine while it is running. While heat is often a waste product, it can also be the energy form we need or desire to harness and use. We use heat to barbeque food, to provide hot water or warm the air inside buildings, and to melt everything from mineral ore to metals to plastics.

Generating heat is easy; one can typically even harness the heat byproduct of an energy conversion. A fire converts chemical energy in wood into light and emits heat. Heat always flows from higher temperatures to lower temperatures. A bowl of hot soup will release heat and warm one's body. Insulation in the walls of a thermos bottle prevents heat from flowing out, keeping hot liquids hot. The same insulation keeps heat from flowing in, keeping cold drinks cold. "Cold" or "coolness" is used to describe the absence of heat, just like the word "dark" is used to describe the absence of light. Substantial amounts of meaning are derived from its context or an explicit comparison.

The laws of thermodynamics dictate that we cannot generate "cold" *per se*. Therefore, cooling an object requires some engineering effort and ingenuity. The challenge is to extract heat from objects that are already at a lower temperature. In a refrigerator, we are taking the heat from the inside and moving it to the outside. With a combination of equipment such as compressors, condensers and chillers, we trick the heat into flowing into a refrigerant that is at a lower temperature than the inside air. The heat from this refrigerant is then dissipated on the outside of the refrigerator. In older-style refrigerators, there was some type of grill in the back which was dissipating heat from the grill and was often dangerously hot to touch. The overall system still causes an overall increase in entropy, consistent with the laws of thermodynamics, but the cooled region benefits from the work done and reaches lower temperatures.

An even more counterintuitive case is using steam to run air conditioning units for cooling. Recall that heat is merely a form of energy and that we can convert any form of energy to any other form. A heat engine, such as a Stirling engine, converts heat energy into mechanical energy. The mechanical energy can turn engines and run the compressors, condensers, or chillers that are needed to cool a space. As strange as it sounds, a heat engine can easily be used for cooling!

2.4 Controlling factors are often unrelated to energy

The Columbia River Treaty between the US and Canada is a real-world example of how controlling factors of energy decisions are often unrelated to energy. These two governments signed the Columbia River Treaty in 1964, agreeing to share the responsibility and benefits of the Columbia River system. The first third of the 1,200-mile river is located in Canada, but the remainder of the system snakes through the state of Washington before forming the natural border between Washington and the state of Oregon.

The treaty was originally conceived for power and flood control purposes. The Canadian portion of the river supplies 38% of the water in the river and the Grand Coulee Dam in Washington depended on a steady supply of water for hydrostatic power production. The Canadians agreed not to build hydroelectric generators on their side of the river and to utilize their dams for flood control purposes instead. The compromise allowed the region to keep the Grand Coulee Dam at an optimal water level to maximize energy production. In return, the Canadians received 50% of all electricity generated by the dam.[1]

The treaty is scheduled for renewal in 2024 and in 2014, any of the parties could give 10-year notice of intent not to renew the treaty. To the US, the value of the Canadian Entitlement of electricity is approximately $214 million USD per year (Columbia River Treaty Review). To the Canadians, they would build their own electric facilities than be dependent on US resources. However, the Canadian Entitlement supports the general revenue account of British Columbia and without Canadian flood control, the US would be less likely to have a steady and reliable supply of water for the Grand Coulee Dam and reliable transport for goods along the Columbia itself.

Other factors come into play as well. The Native American tribes with management authorities and responsibilities affected by the Columbia River Treaty—many of whose ancestral fisheries (e.g. Kettle Falls) were also destroyed by the treaty—are currently demanding stronger fishing rights. Environmentalists, who have substantiated a decline in salmon population due to dams, want better protection to support the return of fish populations. Farmers in the Columbia Basin want the water in the river diverted for irrigation purposes. Industrial groups and internet giants with numerous regional data centers want to maintain their competitive advantage in the marketplace with their low energy overhead costs and liberal access to water.

---

[1] Canada originally resold its share of the power, known as the Canadian Entitlement, for a period of 30 years for $254 million but regained its entitlement in 2003 when the agreement expired (USACE, 2014).

The actual *energy* side of the Columbia River system is probably the simplest. Engineers only have to build and operate a dam, tracking and optimizing energy output based on available and forecasted water resources. Society's broader, and more contentious, challenge is balancing a wide range of equities that are affected. This context is often controlling.

2.5 Chapter 2 Summary

- Energy *context* is just as important, if not more so, than specific metrics, technologies, or issues about energy itself.
- Energy and power are different. Energy is the capacity to do work and power is the rate of work in a specific time period. A sprinter values power over energy, while a marathon runner values energy over power.
- Energy is always in balance throughout any isolated system. Changes made within any individual energy silo can often propagate through the entire system because of complex relationships and interdependencies.
- Every energy form can be transformed into another energy form. Facilitating this transformation generally takes dedicated and specific equipment.
- There is always waste in every energy conversion. The most common waste energy is heat energy.
- Electricity is a form of energy that is easily generated from other forms and can be readily converted into useful work. These characteristics have contributed to the societal demand for electricity which can be easily used in a broad range of applications and supports increased interoperability between individual energy systems.

3 | Energy: the delivery system

You and I come by road or rail, but economists travel on infrastructure.

Margaret Thatcher

Every year, Lawrence Livermore National Laboratory (LLNL) publishes a flow diagram of energy consumption for the entire United States.[1] A 2013 flow diagram is reprinted as Figure 3.1 (LLN, 2013). The arrows denote the direction of energy flows, while line width denotes the magnitude of the flow. This is perhaps the most singularly important graph to understand national energy demand, delivery, and supply. It is essentially a version of the energy balance diagram of Figure 3.1 for the US during the calendar year 2013.

Energy is balanced throughout the entire system. On the left-hand side, "energy in" is denoted by a plethora of different fuels, ranging from solar to petroleum. These fuels are also called *sources*. On the far right-hand side, "waste energy" is quantified as *Rejected Energy*, and useful energy is denoted by *Energy Services*. These correspond to "waste energy out" and "useful energy out" in our simple, original diagram (Figure 2.1). Rejected Energy includes generation losses, transmission losses, unused energy, and wasted energy. Energy

[1] More specifically, a Sankey diagram. Source: LLNL (2014). Data is based on figures from the Department of Energy (DOE) (2014). If this information or a reproduction of it is used, credit must be given to the LLNL and the DOE, under whose auspices the work was performed. Distributed electricity represents only retail electricity sales and does not include self-generation. EIA reports consumption of renewable resources (i.e. hydro, wind, geothermal, and solar) for electricity in BTU-equivalent values by assuming a typical fossil fuel plant "heat rate." The efficiency of electricity production is calculated as the total retail electricity delivered divided by the primary energy input into electricity generation. End-use efficiency is estimated as 65% for the residential and commercial sectors, 80% for the industrial sector, and 21% for the transportation sector. Totals may not equal the sum of components due to independent rounding. LLNL-MI-410527. For a full version of the chart, please visit Lawrence Livermore's website at https://flowcharts.llnl.gov/.

Services are the benefits of useful work, such as running appliances and lighting or powering vehicles. The US consumed 37 quads in energy services and generated 58.1 quads in rejected energy in 2013.[2] All told, 61% of the energy sourced during that year was waste energy.

The boxes in the center of the figure represent the different energy transformations performed throughout the energy delivery system. The four boxes denoted by *Residential*, *Commercial*, *Industrial* and *Transportation* are known as sinks or end uses. All fuels can be converted into *electricity*. For some fuels, such as nuclear and wind, electricity is the only practical conversion. This is apparent when following their output across the chart. Other fuels, such as natural gas, are very flexible and are consumed in a wide variety of end-use categories. Generally speaking, the source of energy is rarely located next to the sink of consumption. Transporting the energy from source to sink is big business.

There are several options to link energy sources and sinks. An increasingly dominant approach is to convert the energy to electricity, where it can be transmitted via the electric power grid. To consume the raw fuel, we need to transport it via containers across our roads, rails and ports to a specific device capable of transforming the source into useful work. This source-to-sink model represents a complete linear supply chain for energy.

This complete supply chain for energy, along with associated externalities, is the basis for metrics like "footprints" designed to help make energy intensity and environmental impact more understandable. When calculating energy footprint, it is important to differentiate between source energy and sink energy. *Source energy* includes the wasted energy lost at the generation facility as part of the footprint. *Sink energy* will only measure how much energy a facility consumes. Source energy gives a more complete environmental picture, since it takes into account generation waste and transmission loss.

Carbon footprint calculations are typically utility-wide averages because the actual number depends on a plethora of variables, including the efficiency of each power plant, how far a customer

---

[2] One "quad" is 1 quadrillion BTUs of energy. Each quad is approximately equal to 170 million barrels of oil. Generally, this unit is only used to describe energy consumption on a regional or national level.

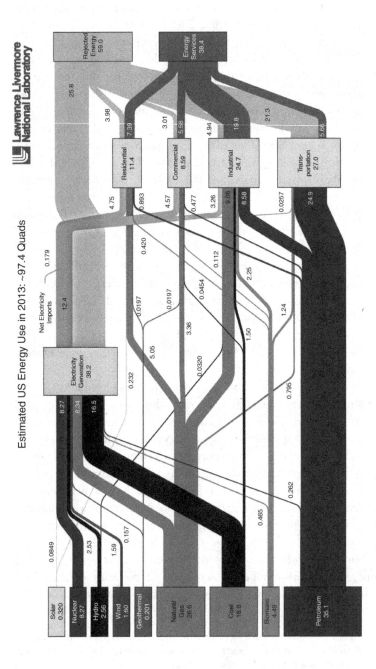

Estimated US Energy Use in 2013: ~97.4 Quads

Figure 3.1 An energy flow diagram for the US

is located from the transmission line, time of day, etc. When operating on common infrastructure, it becomes more challenging to link specific generators and consumers to one another in today's environment. On the other hand, sink energy is much easier to measure and track, since it is the energy consumed by end users. This value, commonly referred to as *behind-the-meter*, is essentially the value of the energy as listed on your utility bill. However helpful, calculating energy or environmental footprint from sink energy is inadequate. Sink energy–based accounting does not take into account the full energy or environmental impact of extraction, generation, transformation, and transport of consumed energy to the end user.

Conversion efficiency rates depend on the end use. The efficiency of electric generation is 32.5%, while the efficiency of the transportation sector is a measly 21.0%. If every gas-powered vehicle were replaced with an electrically powered vehicle there would be an immediate increase in the efficiency of the transportation sector by approximately 11.5%. This would decrease US national petroleum needs by 21.3%, a non-trivial amount. The number is less dramatic when the additional losses associated with mobile battery or supercapacitor storage are considered. Many issues would shift if such a plan were enacted. Energy consumed by transportation would suddenly need to be generated by the electric power system, nearly doubling its size. Depending on the shape of energy demand for electrically-powered transportation, many new generation, transmission and distribution assets would be required. Doubling the number of power plants and existing electric transmission infrastructure would be expensive, with an approximate cost of more than $10 trillion.[3]

Such a dramatic shift in fuel sources would have geopolitical and domestic political ramifications. As of 2010, the US imported 49% of liquid fuels and only imported 1.9% of coal thanks to rich domestic reserves (Department of Energy, 2012; Watson et al., 2011).

---

[3] This order of magnitude number is calculated assuming that we would need to double the number of power plants operating in the US. Currently, there are 6600 in operation. Assuming $1.5 billion to build a new power plant, the cost needed to build new plants would be $9900 billion. There are also approximately 450,000 miles of transmission wire in the US. Assuming $1 million per mile, an upgraded transmission infrastructure would cost approximately $450 billion. Thus, in total, doubling the size of the US electric infrastructure would run approximately $10,350 billion or over $10.35 trillion. For comparison, the US GDP in 2012 was $16.42 trillion.

Rapid shifts towards electrifying transportation are likely to increase US coal consumption for baseload generation but would reduce US dependence on foreign energy sources substantially. A major near-term shift to electric vehicles could eliminate our petroleum dependence and could double our coal consumption.

This energy flow diagram also helps to explain the intersections and overlapping responsibilities of many different regulatory bodies in the United States. For instance, in the physical infrastructure, the Nuclear Regulation Commission is the primary regulator for nuclear generation, while the electric transmission infrastructure is regulated by the Federal Energy Regulatory Commission (FERC), which approves standards for electric power system operation developed by the North American Electric Reliability Corporation (NERC). A publicly traded investor-owned utility would be responsible to the Securities Exchange Commission (SEC) for its financial activities, (FERC) for wholesale energy trading, and state public utility commissions (PUCs) for the setting of retail rates.

Even buildings are subject to substantial regulation. Buildings are the largest component of end-use consumption in the United States and must comply with extensive standards. As of August 2014, all but seven states have adopted energy codes for buildings. The DOE estimates that its Building Energy Codes program will save the US approximately 22.0 quads of site energy and 40.1 quads of primary energy between 2013 and 2040 (Livingston et al., 2014).

The rest of this chapter is dedicated to the energy infrastructure—namely, generation, transmission, and consumption. For most fuels, the supply chain is:

Fuel → Transportation → Consumption

For electricity, we need to also consider the conversion of the fuel to electricity and transmitting the electricity to end users:

Fuel → Transportation → Generation → Transmission/Distribution
   → Consumption

We will explore the infrastructure from the perspective of customer demand. We will begin by discussing consumption needs of energy. Then we will explore generation technologies that meet those energy needs. Finally, we will conclude with methods of transporting the energy from source to sink.

3.1 How we consume energy

Humans consume energy in dozens of forms, but in this section, we focus on three of the most significant forms of energy consumed in everyday living: thermal, mechanical and electrical. Heat energy warms our homes, cooks our food, and allows refrigeration. Industrially, we use heat to smelt iron ore and dissipate heat out of server farms and data centers. We consume mechanical energy—which includes both energy of motion and energy of position—when we use a car to move across the land or use an elevator to move vertically. We also consume large amounts of energy indirectly by moving objects to us, such as shipping products to our homes. We consume electric energy to run devices and appliances. A more detailed look at some specific end uses like metallurgy, data centers, buildings, and transportation helps provide context for deeper exploration later in the book.

Metallurgy

Metallurgy is the science of transforming ore and metals into useful alloys like steel, aluminum, and other metals. Metallurgical energy consumption is massive. Iron and steel refining comprise about 5% of all energy consumption in the US, while aluminum refining consumes another 1.2% (EIA, 2012a; DOE, 2007). Both metals are essential to modern societies and economies. Metals and the derivative products that require their use enable much of our energy extraction and transformation into work. Steel is a central ingredient in bridges, buildings, cars, and other forms of transportation. It is used for injection molding of plastics for cell phones and small electronic devices. Aluminum is used in everything from airplanes to soda cans. Despite becoming increasingly reliant on imported metal products, the US still produces more than 88 million metric tons of steel and 9 million metric tons of aluminum per year, equivalent to a per capita consumption of 558 kilograms of steel and 29 kilograms of aluminum.

Metallurgical processes take huge amounts of thermal energy. Extracting metal from raw ore requires temperatures of 1200–1600 °C (2192–2912°F). To reach those high temperatures, we primarily burn what is known as *metallurgical coal* in purposefully designed furnaces that can withstand the high heat. Metallurgical coal is a special form of coal which has a higher carbon and calorific value than the *thermal coal*, which is predominantly used for coal-fired power plants. Metallurgical coal is used in the production of the coke which is required to refine iron ore as part of steel-making. Making coke from metallurgical coal and then firing ore in blast furnaces using the coke is energy intensive and costly but necessary. Aluminum is no different. Bauxite, an alumina-rich rock is mined, ground, refined into alumina powder, and mixed with aluminum oxide while being passed through a strong electric field to create aluminum.

Creating metal from recycled products requires substantially less energy than producing metals out of raw ore. Aluminum recycling only consumes 5% of the energy and 3% of the water as compared to smelting raw bauxite ore (Robinson, 1986). Dramatic reductions are similarly achieved in steel, paper, and glass recycling. The lower energy requirements for recycling can be met by electricity that can be generated from a broader range of energy sources, including renewables. Recycling metals is not sufficient to meet our growing demand for these valuable materials, but it is an important part of long-term energy and resource strategy.

Data centers

Data centers now consume about 2% of the entire US energy generation, rivaling the consumption intensity of today's domestic metallurgical industry and representing 1.5% of global electricity production (Natural Resource Defense Council, 2014). Data centers are central to the current utility computing model, where central-ized operators provide large-scale computing as a service. These energy-intensive facilities are central to the operation of our economy and support telecommunications and information networks around the world. Without them, everything from cellular networks to the internet would cease to function. Data centers are growing at a rapid rate, with information traffic projected to grow at an annual rate of 31% through 2016 and with even more rapid growth likely thereafter, dramatically increasing long-term energy needs (Cisco, n.d.).

Data centers consume electricity, avoiding dependence on any specific fuel source. This differs from the steel industry which depends on both electricity and metallurgical coal for its heat content and role in the chemical reactions required for iron ore refinement. Data centers produce large amounts of unwanted heat as a byproduct, as electricity passes across the circuits and wires in servers and networking equipment. In fact, cooling costs for data centers are among the most significant operational costs. Without proper cooling, data centers would literally destroy hardware. Designing efficient ways to manage heat from data centers is big business, with a variety of techniques ranging from water cooling, to oil baths, to traditional air conditioning being employed by organizations of different sizes and sophistication. Some companies are experimenting with innovative designs to recycle the rejected heat and use it in other parts of the building, substantially reducing the amount of energy waste on site.

Buildings

A well constructed building provides the occupants with a comfortable environment and a well-lit workspace in order to maximize productivity and quality of life. In most developed countries, this requires a mix of heating, ventilation, and air conditioning (commonly referred to as HVAC); lighting; and power receptacles to allow for plug-in devices. As a rough order of magnitude, each of these five categories typically accounts for about 20% of a building's energy footprint. The specific percentages vary substantially based on the actual end use of the facility and local climate. A library in Boston will have a very different energy consumption profile compared to a grocery store in Phoenix, especially when comparing seasonal variations during summer or winter.

Large commercial buildings (and some medium ones) often have a central plant that is responsible for HVAC. A large mechanical room on the roof or in the basement provides significant amounts of heat or cooling capacity for the entire building. For example, a central boiler may heat water in the basement and use pumps to circulate it throughout the building via a system of pipes. After circulation, the hot water returns to the central boiler for additional heat in a continuous cycle. Many facilities also use a centralized system to circulate air. In this case, a desired ratio of fresh air is combined with internal air and heated or cooled to a comfortable temperature. Air is circulated through the building, and terminal units on each floor (e.g. variable air volume boxes) may perform some final temperature adjustments before the air is released into the environment. Returned air is then captured and sent back to the central system.

Heating and cooling are both required year round. Hot water is still needed during the summer, and many workplaces with heat-emitting devices like servers still need to be cooled in the winter. In fact, it is not uncommon to see steam rising out of the top of a tall building in the middle of winter as the building dissipates excess heat from computers and body heat from employees.

To increase efficiency, many centralized systems employ heating and cooling strategies that are materially affected by weather. For instance, one common strategy during a cold winter is to recycle as much of the inside air as possible, reducing the need to preheat the outside air to a comfortable temperature before circulating it to the rest of the building. When pursuing this strategy, it is also important to maintain a level of indoor air quality. Fresh air is important in maintaining a healthy work environment. Venting recycled air is critical in eliminating toxins that accrue inside buildings. Hence, most building codes require a minimum amount of outside air exchange, even though that is not the most energy-efficient strategy.

Lighting is another major component of consumption within the built environment. Different technologies of bulbs have their own efficiency for converting electricity into light. As shown in Table 3.1,

An incandescent bulb is only 2–13% efficient in converting electricity into light, while the rest of the energy is released as heat (Agrawal et al., 1996). This table compares typical consumer bulbs available from a national retailer, Home Depot (as of 2013), to enable an exemplary comparison of efficiency versus cost. Fluorescents and LEDs are more energy efficient and do not release as much heat. Sometimes, if a building is predominantly lit with incandescent bulbs, a measurable amount of the energy emitted from the bulb will actually be heating the building. After performing a lighting retrofit, the energy consumed for lighting may drop, but the energy to heat the building may rise. Although the building will achieve overall reductions in energy consumption, the rise in heating costs negates some of the benefit of the more efficient bulbs.[4]

---

[4] We are familiar with seeing lightbulbs rated by watts (W) but not by lumens (lm) or candala (cd). Lumens is a measure of the amount of light emitted by a source. Watts is a measure of how much power is needed to run the lightbulb.

Table 3.1 *A comparison of lighting technologies with similar lighting outputs*

	Incandescent	Compact fluorescent	LED
Power needed (W)	60	13	9
Efficacy (lm/W)	13.0	69.2	88.9
Approximate life (h)	2,000	10,000	25,000
Cost (per bulb)	$0.73	$1.38	$13.3

Source: Home Depot, 2013

Plug loads is a catch-all category for the various electronic gadgets and devices that consume energy but do not fall into the broad category of heating and cooling or lighting in typical structures. For most commercial and residential buildings, this includes computers, cell phones, toasters, refrigerators, TVs, and electronic appliances. Plug loads are always consumers of electric energy. A major energy consumption concern with plug loads is vampire loads or phantom power. Devices frequently continue to draw small amounts of electricity even when off. By doing so, the device can be in stand-by mode and turned on instantaneously rather than having to wait for the device to boot up. Common household items with vampire loads include computers, TVs, electric toothbrushes, microwaves, and baby monitors. Many manufacturers design virtually any device that is continually plugged into the wall to be instantly usable, creating numerous vampire loads. These loads are tiny individually but in aggregate can account for as much as 10% of a household's total energy footprint (Mohanty, 2001).

Transportation energy needs

Transportation systems throughout the world today primarily use chemical fuels. Vehicles use gasoline, diesel, and compressed natural gas. Planes use aviation gas or jet fuel, and the historic railways that opened vast tracts of land and resources up to society were fueled by coal. Some forms of transportation consume electricity, such as

Different technologies require different amounts of power to emit the same amount of light. An LED bulb consumes much less energy to produce the same brightness as a 100 W incandescent bulb.

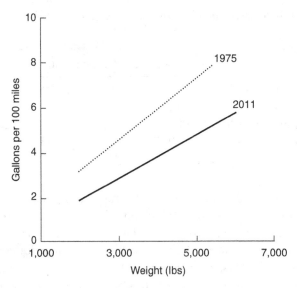

Figure 3.2 Fuel consumption versus weight of vehicle
Source: EPA, 2012

battery electric vehicles, subway systems, elevators, and escalators. Some vehicles combine chemical fuels and electric energy to create hybrid vehicles that can consume either. Even the energy used to walk, cycle or run is produced from chemical energy transformations we refer to as metabolism.

In the physical world, the mass of the object being moved is the largest factor in determining energy consumption. Heavy objects take significantly more energy to move than objects with less mass. Professional cyclists prefer super-light bikes and scrupulously monitor their own body weight. Airlines have implemented surcharges for overweight baggage to help recoup fuel costs caused by induced drag, which proportionally increases the lift required to fly. Heavier cars require more fuel to drive the same distance. A powerful graphic depiction of the correlation between fuel consumption and weight is shown in Figure 3.2. Although vehicles have become significantly more fuel efficient since 1975, this basic relationship between weight and efficiency still holds true (EPA, 2012). There is no escaping physics.

Since weight has such a large effect on fuel consumption, it is desirable to use the lightest possible fuel with the highest energy

Table 3.2 *Select energy densities for various fuels*

Fuel type	Energy density (MJ/kg)
Uranium-235	78,500,000
Gasoline	45.8
Coal (anthracite)	31.4
Lithium-ion battery	0.5
Lead-acid battery	0.18

Source: Ronneau, 2004; MIT, 2003; Linden and Reddy, 2002

content. If you have to carry all of your energy with you, your fuel source must be dense enough to move. The metric of energy density tells us how much energy content is stored in a kilogram of fuel. As Table 3.2 indicates, gasoline has a significantly higher energy density than a typical lithium-ion battery, which was the most common battery for electric vehicles in 2014 (Ronneau, 2004; MIT, 2003; Linden and Reddy, 2002). The higher the energy density, the less fuel we need to obtain the same amount of work. As a result, higher energy density fuels are more valuable.

Higher energy densities ensure that chemical fuels remain the fuel of choice for transportation. Facilitating a fuel shift in transportation requires a comparatively dense, stable, and mobile fuel source which is economically competitive. In cargo and passenger air travel, where lightness is critical, chemical fuels such as fossil fuels are currently the only viable option at any practical scale. Since chemical fuel markets can be highly volatile and the long-term supply uncertain, the aviation industry has spearheaded the efforts to create renewable biofuels and experiment with other byproducts from industrial and commercial processes that produce dense chemical fuels. These biofuels have similar energy densities to fossil fuels and can mitigate future uncertainty about access to reliable aviation and jet fuels.

3.2 Generation

Energy sources and fuels are chosen to meet our specific needs at the point of consumption. A plethora of chemical, mechanical, or solar fuels offers a vast array of choices for generating electrical energy.

Some energy conversions are better suited for specific applications, but no energy source is ideal for all purposes because of losses involved during the conversion process to other forms of energy or into useful work. Understanding where our energy sources come from is essential to grasping how these resources are transformed to meet our growing energy needs.

Some common conversion forms

Every form of energy can be converted into electricity, as shown in Figure 3.1, in a process known as electric generation. To say that this conversion is "generating energy" is really a misnomer. In reality, this is an energy transformation like any other with calculable losses, mostly given off in the form of waste heat.

The simplest way to generate electricity is to rotate a wire loop through a magnetic field while it is connected to a circuit. When the copper wire is in motion relative to the magnetic field, electrons flow in the wire. This flow of electrons is caused by the principles of electromagnetic induction and is known as a current, which seeks the path of least resistance through the wire loop and the rest of the circuit.[5] To induce a larger current, the strength of the magnet, the number of loops, or the speed of rotation can be increased. For a uniform magnetic field, this produces a simple, asynchronous alternating current generator.

Most electric generation technologies are variations on methods of turning the loop, which is known as a *rotor*. The permanent magnets—or electromagnetic windings for larger or higher-efficiency generators which generate the magnetic field—are, are collectively referred to as the *stator*. The rotor can be spun by any number of devices, but for large-scale generating facilities, turbine generators are most common. In a self-illuminating bicycle lamp, the wire loop makes contact with the bicycle wheel and provides enough electricity to illuminate a nighttime ride. Many newer technologies, such as concentrated solar power, co-generation, and some forms of geothermal generation, are also based on turbines to extract

---

[5] For more detailed information on how an electric generator works, please refer to the works of Michael Faraday on induction.

mechanical energy from heating a fluid. Photovoltaic (solar cell) technology is one of the few commercially available exceptions.

Using steam to turn a turbine generator

Thermoelectric generators, the most common form of power plants, use mechanical energy from moving steam to turn turbine-driven generators. For example, in a coal-fired power plant, the rotor is connected to a turbine that is rotated by the kinetic energy from hot gases moving from a high-pressure boiler to a lower-pressure condensor. As depicted in Figure 3.3, we first burn a fuel and use the heat to boil water. The boiled water creates high-pressure steam. The steam rises through a column which turns a turbine. The turbine spins a shaft connected to a generator. The end of the shaft is a rotor, which spins inside the magnetic field created by larger, energized coils of wire in the stator. This entire process generates electricity that is transmitted into the electric grid. After the steam passes through the turbine, it can be cooled and condensed back into water for another cycle or released into the atmosphere.

The essential ingredients for any thermal plant include a chemical fuel to burn, a fluid to boil and pass through the turbine, and a fluid to help carry away waste heat. The fuel can be any type of wood, coal, natural gas, petroleum, trash, or nuclear material.[6] The fluid for the actual turbine process can be self-contained to operate at higher efficiencies or at special pressures. For many systems, the fluid used in the turbine process is open and uses water. The cooling fluid needs to be abundant to help condense the steam after it passes through the turbine and to keep the rest of the plant safely cooled.

Carnot's Rule states that the greater the temperature difference between a hot and cold temperature reservoir, the more efficient the energy conversion. A thermal system has a maximum theoretical efficiency of approximately 70% (BCS, 2008). The larger the power plant, the better it can maintain enormous temperature differentials by heating fluid and cooling the condensing reservoir to operate closer to the theoretical maximum. In addition to the business and regulatory

---

[6] Nuclear material does not use a chemical reaction to generate heat, rather it uses the nuclear fission process to generate heat. The heat can also be used to create steam.

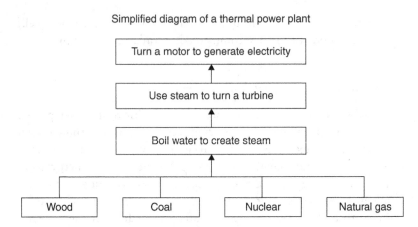

Figure 3.3 Schematic of thermal generation of electricity

reasons for consolidation within the electric power industry, physics has directly contributed to the development of a centralized electric infrastructure with few generation facilities. In practice, the efficiency of thermal plants in generating electricity is approximately 30% due to the compounding effects of many other losses throughout the production process.

The term *power plant* gives us an indication of what an electric generator provides on an instantaneous basis: power. Power plants are rated based on *nameplate capacity*, or the maximum amount of power capable of being generated at any given time. Nameplate capacity, measured in watts (W, kW, MW, GW, etc.), is defined based on normal operational conditions, and temporarily exceeding nameplate capacity for emergencies is acceptable.[7]

The turbines of modern peaking thermal plants, also known as *peaker plants*, can ramp production up or down rapidly. Some quick-start units—which usually only operate for a few hundred hours per year—can go from shutdown to full load in minutes, but baseload generation and intermediate demand-focused units take more time to

---

[7] While plants are rated according to nameplate capacity, their economic feasibility is closely tied to the *capacity factor*, which is defined as the ratio of its actual output over a period of time to its nameplate rating over the same period. For resources like wind, a minimum capacity factor of 30% is usually required to be economically feasible to develop.

adjust output. Because thermoelectric power plants are very controlled processes, they have a critical ability to provide *dispatchable power*. For instance, thermoelectric power plants can be operated based on the actual needs of the grid with great confidence in performance. Since the entire grid needs to be in balance at all times, this is critical to maintaining the stability of the system.

Dispatchable power plants increase generation to meet extra load or decrease generation when expected loads do not materialize. *Baseload* thermal plants have high capital costs but low variable costs, running for the majority of the year outside of planned outages for maintenance. *Intermediate* plants, like combined-cycle natural gas plants, often start and stop many times throughout the year and provide substantial amounts of the commercial-scale ramping for normal demand fluctuations throughout the day. Peaker plants are much less efficient but are designed to ensure that the grid can balance during peak periods; their primary purpose is rapid ramping, not efficiency.

Using mechanical forces to turn a generator

Generators can also be turned via turbines transforming the kinetic energy of wind or water into rotary motion. As wind blows over a ridgeline, across the plains, or off along the coast, we can harness this energy with large wind turbines. Similarly, we can build a dam across a river and use the velocity and gravitational fall of water to turn a turbine. Impulse turbines generally use the velocity of the water to spin a generator, while reaction turbines harness the pressure from water amassed behind the dam in addition to its motion.

In reality, both wind and hydro energy are ultimately dependent on the sun. Wind is created from the pressure gradient in the atmosphere caused by temperature differentials (between parts of the earth exposed to the sun and those shaded from its emissions) and the rotation of the earth. The sun's radiation heats the atmosphere unevenly, generating currents that stem from pockets of expanding and rising air and pockets of cooling and settling air. The most visible shifts occur at dawn and dusk, which is why winds (and windpower production) tend to be highest during late afternoon and pre-dawn hours. Water energy comes from a combination of solar power and gravity. The sun's radiation causes water evaporation and cloud formation, which eventually returns to earth via gravitational forces back on land. This precipitation, in the form of rain and snow, replenishes the water supply at high altitudes. Dams are used to control the discharge of these cyclical water resources as they make their way towards the oceans, attempting to sustain power production and navigation through drier periods by shaping and storing the flow in a series of controlled reservoirs.

The efficiency of these forms of energy conversion is mechanically limited by the flow and pressure of air and water. For the wind turbine, the theoretical performance limit is Betz's Law. It states that the maximum amount of wind energy that can be extracted is just above 59% (Betz, 1966). There are also some practical constraints in designing a wind turbine. They generally spin at approximately 10–20 rotations per minute and use a gearbox to step up the rotations for a small generator to spin at a higher rate. Although the number of rotations may seem small, the blades can be as long as 75 meters. The tip of these large turbine blades can move at nearly 200 mph, even at these low rotational speeds. The speed of rotation is an engineering optimization problem which attempts to determine the optimal tip speed ratio. Too fast and the blades would block the wind and divert it. Too slow and the wind would pass through without doing useful work. A realistic design goal for modern wind turbines is to capture 35% of the power in the wind stream after accounting for drag, friction, rotor tip losses, and small continuous changes to the direction of the wind.

The efficiency of hydroelectric facilities is limited based on the amount of pressure and motion which can be generated from the earth's gravitational pull on water amassed in reservoirs. The two most popular hydro turbines, the Francis reaction turbine and the Pelton wheel impulse turbine, can both achieve real-world efficiencies at or exceeding 90% when applied appropriately. Power output from dams is also constrained by water management requirements. Dams often have responsibilities for flood control and navigation in addition to power generation. On systems like the Columbia, during spring runoff when snowpack melt combines with rain, dams are often unable to store additional water safely. During these situations, dams allow excess water to flow freely over spillways without generating any electricity. During these periods, operaters must also balance water quality issues because of lower levels of dissolved oxygen in reservoir water as compared to a free-flowing river. If the oxygen level is low enough, spilling the water too fast can kill fish downstream and cause other ecological damage (Peterson et al., 2003).

One major difference between wind generation and hydroelectric generation is dispatchability. Wind energy is not considered a dispatchable resource because the generation of electricity from wind cannot be controlled: it is dependent on nature. Electricity is generated when the wind blows, which may not coincide with periods of high demand when the power is needed. As such, wind energy needs to be shaped and stored to provide steady power. Dams are natural partners, but other forms of shaping and storage, to include coordination with consumers, can also serve this function. Hydroelectric facilities are considered dispatchable because the turbines can ramp up or down. This is accomplished by controlling the rate of water allowed through the turbines. Hydrostatic resources are able to rapidly and cost-effectively adjust to changes in load or in non-dispatchable resources like wind or solar.

Photovoltaics

The last major form of electric generation is the conversion of light energy into electric energy. Light energy is also known as electromagnetic radiation and is the only form of energy that humans can actually see directly. The *photovoltaic* (PV) effect was first observed in 1839 by Alexandre-Edmond Becquerel, and the photoelectric effect was first observed by Heinrich Hertz in 1887. In 1905, Albert Einstein became the first person to publish a paper describing the photovoltaic effect using quantum physics, an accomplishment that would win him the Nobel Prize. Despite these much earlier developments, it wasn't until 1954 that Bell Labs produced the first modern silicon solar cell with 6% energy conversion efficiency (Bell, 1954; APS, 2009).

In laboratory settings, the efficiencies of solar technologies have approached 30% (Green et al., 2011). Commercially, solar cell efficiencies are more commonly in the 10–20% range. Solar efficiencies have evolved dramatically over the last several decades, and they generally follow two trends. First, there are classes of chemicals developed that are used as solar cells. Monocrystalline silicon cells are the most efficient but also the most expensive. Polycrystalline cells tend to be less expensive but also less efficient. There are also multijunction cells,

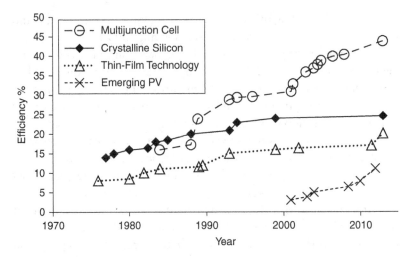

Figure 3.4 Evolution of efficiencies of select solar cell technologies
Source: NREL, 2013

thin-film technologies, and other emerging technologies which can be made more flexible or appropriate for more use cases.

Single-junction cells have a maximum theoretical efficiency of 34%, but multijunction cells could potentially reach much higher levels. The current world record for efficiency of a multijunction solar cell was set in 2013 at the Fraunhofer Institute for Solar Energy Systems at 44.7%. Each time a new approach is developed, the efficiency of the cell is usually between 10 and 15%. Efficiencies increase over time as scientific understanding improves. Figure 3.4 depicts efficiency trends for major solar technologies (NREL, 2013).

Although the efficiency of the solar cell has increased over the last few decades, the total cost of a solar installation is still closely linked to labor costs. The total cost of a residential PV system in the US costs approximately $5.21 per kW. The solar module hardware itself costs merely $0.75 per kW, or less than 15% of the total cost of the system. The balance of the system consists of finance charges, installation labor, legal costs, insurance, marketing efforts, and other hardware- and business-related expenses. While critical, the scientific work of improving solar cell efficiency only drives down the component cost of the solar module. Lowering the total costs of solar adoption depends heavily on innovations that lower the labor costs, simplify the financing mechanism, and reduce customer acquisition costs (RMI, 2013a).

Similar to wind, electricity generated from solar cells is not considered to be dispatchable. Electricity from the sun depends on many factors, including the cloud cover, time of day, and intensity of the sun (including the angle between the panels and the sun). These exogenous and external factors cannot be controlled like the fuel of a thermoelectric power plant or the flow rate of a hydroelectric dam.

Heat utilities

Heat utilities are one of the most commonly forgotten or misunderstood utilities. In the US this generally refers to steam utilities, while water utilities are more common in Europe. Some downtown regions, college campuses, and hospitals use a central steam plant that supplies heat to multiple facilities for a variety of purposes. Steam utilities generate and sell steam in a fashion similar to electric utilities. A steam generator essentially functions like a thermoelectric power plant, but without a turbine. The steam from the boiler system can be directly piped into a building's hot water or forced air heating system. When the heat from the steam is used up, water is typically released back into the sewerage system to be discharged normally. A central steam facility removes the need for a building to invest in and maintain its own boiler(s). This not only saves space which can be used for revenue-generating purposes but also allows the facility to benefit from higher efficiencies achievable only by large, centralized steam plants.

For many of the larger facilities, combining a steam plant and a thermoelectric plant in a combi ned heat and power (CHP) facility provides efficiency and some degree of resilience. Also called co-generation, these power plants use the waste heat from a thermal power plant like a steam utility instead of dissipating the energy as waste. CHP facilities first use the fuel to create steam and generate electricity via a turbine. The remaining heat then powers a specialized heat pump, chiller, or another thermal device. By using more of the energy byproducts from the boiler and turbine processes, CHP facilities can reach efficiencies of nearly 80%. This is substantially higher than the average 30–35% efficiency of an electric generator alone. Some countries have vigorously pursued CHP as an energy strategy, with Denmark achieving over 50% of its energy needs, drastically reducing dependence on imported fuels (IEA, 2011a) An in-depth discussion of steam utilities can be found in Chapter 8.

3.3 Transmission and transportation

Different forms of energy have different means to transport them. If the fuel is solid, such as coal, we typically use conveyors for short distances and specialized cargo systems built for trains, barges, and ships to travel farther. If the fuel is liquid or gaseous, we ship it through pipelines, trucks, or tanker ships. When possible, pipelines are the most economical and energy-efficient means of moving fuels out of these three options. Fuels can be moved many thousands of miles across continents and oceans based on the availability of the infrastructure of railways, pipelines, roads, and ports. In these situations, we are transporting *energy*.

Heat is more difficult to transport long distances. Steam plants are only able to economically serve customers within a 1- to 2-mile radius. Heat utilities that use water can efficiently transfer hot water for 20 to 30 miles. This is more common to Europe and countries of the former Soviet Union. The physical losses over longer distances do not make sense, explaining why typical steam customers are downtown cores and college and hospital campuses within a close geographic area. Steam is pressurized and sent through underground pipes to end users. We can increase the amount of heat transported by raising the temperature or by increasing the pressure of the steam. Although no longer common, power can also be transported via mechanical wires and cables, known as telodynamic systems. Examples of telodynamic systems include the cable car system of San Francisco or recreational ski lifts. Cable cars work by latching onto moving cables underneath the city's streets. As the cables move, the cars are able to climb very steep hills. It may seem arcane to us today to travel via cables, but many cities adopted cable cars in the late 1800s prior to the proliferation of electric trolleys. In these cases, we are transporting *power*.

Electric transmission of power

Electric power transmission and distribution systems transport both *energy* and *power* simultaneously. This property contributes to the confusion about differences between energy, electricity, and power. Early industrial processes relied on whatever energy transformations could occur on site. Initially, most of this power was obtained from waterwheels next to streams and rivers. Windmills allowed for slightly more flexbile siting, but desirable locations for harnessing mechanical forces remained scarce.

Engineers sought methods to improve power transmission over longer distances to power factories and processes located farther away from existing hubs. Numerous types of energy transformations were considered competitive until 1896, when the Niagara Falls Power Company successfully completed a 20-mile electric transmission line and cemented electricity as the preferred method of power transmission for the industrial era. The challenges faced by these engineers are detailed in Chapter 5.

Transmitting power based on air, steam, or cables is relatively easy to understand. We can increase the power of compressed air or steam by increasing the pressure. We can increase the power transmitted via cables by pulling harder or faster. In electric power transmission, we increase power by increasing the voltage or current of the system. Energy and power are related based on time. Electric energy can be measured simply by knowing how much power (kW) has been provided over a certain amount of time (kWh).

Electric energy and electric power are hard to visualize since they are intangible compared to obvious mechanical energy sources like a flowing river. One common analogy is that electricity transmission is akin to water flowing down a pipe. Energy describes how much water flows through the pipe over time, while power describes the size of the pipe. The pipe has a specific capacity that is easily understood by most people. In the case of water, this capacity is the volume of water passing through each second. A very small pipe has a small cross-sectional area and a small capacity. Forcing more water through the pipe is possible with pressure, but at the cost of serious drag, known as *head loss*.[8] A small wire is similar. The larger the electrical current, the more losses that are incurred from the resistance to the electrons flowing in the wire. A system that is limited in its ability to transmit electricity based on this principle is capacity constrained.

---

[8] Head loss is generally measured as the amount of energy contained in a static column of the (incompressible) fluid. See Bernoulis' principle.

Big wires that are good conductors have less resistance, meaning that they can more easily transport large amounts of electricity. If a system has big pipes which are not yet used to their maximum capacity—for a water system, this would mean that the pipes aren't full or are operating below their maximum pressure—then the system is energy constrained. Energy or capacity constraints can occur because of generation, transmission, or distribution infrastructure.

Capacity and congestion

Transmission systems are rated based on the capacity to transmit power. Transmission lines and assets use the watt (W, kW, MW, GW, etc.) as the unit of measure. The voltage at which transmission can be stepped up or down is via transformers in substations. Electricity naturally flows across circuits from higher voltage to lower voltage, just like water flows from a higher elevation to a lower elevation. Electricity flows inside the system follow the path of least resistance towards lower voltages.

The water pipe analogy really breaks down when we try to define the maximum capacity of an electric wire. The size of a water pipe does not change; a 1/2-inch pipe will always be a 1/2-inch pipe. However, electric transmission capacity can be stepped up or down based on changing the voltage on a particular conductor. Losses in transmission and distribution systems are linked to electric currents, making higher voltages (meaning lower currents) attractive for transmitting more power with fewer losses. Similar to nameplate capacity, transmission capacity ratings are determined based on normal operating parameters. Moderate and temporary exceedance of normal operating limits during emergencies is acceptable, but this generates heat. Exceeding the limit by too much or for a prolonged period of time results in excess heat from electrical resistance and can cause wires or transformers to melt.

Congestion occurs when higher cost generation assets must run in order to prevent a particular transmission line from exceeding its rated capacity. Under these conditions, the line is considered to be congested, as it is unable to transmit additional electricity without being exposed to the risk of physical damage. For instance, in the summer, the Pacific Northwest sends substantial amounts of electricity to California. The transmission lines connecting the regions often operate at or near maximum capacity. During these periods, any excess energy from wind farms or hydro facilities must be consumed or stored inside the region. This is an example of a capacity-constrained connection, which requires California to run more expensive peaker plants in another area. The capacity-constrained link between the Northwest and California is considered congested because its lower cost generation is unable to be utilized.

Losses

The main factor governing losses in transmission is the resistance of the transmission medium (e.g. the wire): the higher the resistance, the larger the losses. The resistance changes according to several factors. First, resistance of most metal conductors increases with temperature. The more electricity that passes through a wire, the hotter the wire gets. The hotter it gets, the more resistance increases along with additional waste heat. This can create a negative cycle if transmission or distribution lines are operated outside of their designed parameters. Second, longer wires have higher resistance. This was a primary obstacle in the early days of electric transmission when power could only be transmitted for 1–2 miles. Early power plants were located centrally within cities to serve the largest number of potential customers. Third, higher voltages reduce losses. Ohm's Law ($V = I \times R$) describes the relationship between voltage, V, current, I and resistance, R. Power loss is equal to the resistance times the square of current through the wire. Thus, by doubling voltage it is possible to deliver the same power with half the current. This means that power loss in line is inversely proportional to the square of the transmission voltage. As a result the longest transmission lines are also at the highest voltages. Long-distance transmission lines can be at 765 kV (765 000 V) as they travel across the country. One key task of substations is to step up and down voltages between consumers, distribution systems, and high-voltage transmission systems. High-voltage systems help minimize losses when transmitting large amounts of power over longer distances, while distribution systems balance efficiency concerns with safety and other practical considerations of serving end users.

AC versus DC in transmission

The benefits of alternating current (AC) and direct current (DC) have been debated for most of the existence of electric transmission, and both certainly have advantages for specific applications. Transformers to raise and lower AC voltage are common and easily built, giving the technology more flexibility in operations and use. Effective DC transformers can cost millions to construct and require much more maintenance. Like the original DC generators, called dynamos, DC transformers generally aren't worth the added cost and complexity. AC can be easily converted to DC at the point of use, satisfying the end-use needs of both AC and DC consumption with superior cost and reliability.

DC, on the other hand, is capable of reaching higher voltages than AC and transmitting the power over longer distances with fewer losses. Another advantage is that DC lines do not suffer from capacitance like AC lines, which limit most underground AC lines to a maximum distance of 40 miles (Faulkner, 2013). In this regard, high-voltage DC (HVDC) transmission lines are commonly used underground or underwater, including for major international connections such as Sweden to Germany and the United Kingdom to France. DC lines are also used as the backbone for very long-distance transmission where AC losses would be impractically high. As an example, the Xiangjiaba–Shanghai ultra high voltage DC line spans 1200 miles and can deliver up to 7200 MW of power (ABB, n.d.). Numerous HVDC projects are now ongoing, as new transmission technologies and the search for geographically disparate sources of renewable generation have made it economically sound to use commercially.[9]

Ancillary services

To maintain stability across the entire electric power system, transmission infrastructure operates within very tight specifications. The grid

---

[9] DC transmission requires additional infrastructure, such as converter stations to go from AC to DC and DC to AC. However, since HVDC systems require conversion and synchronization with each AC interconnect already, they are useful for connecting asynchronous power systems and can even connect systems operating at different frequencies such as 50 Hz and 60 Hz. Capacitance is the ability of any object or body to store an electrical charge, measured in farads after the English physicist Michael Faraday. For power lines, this value is a measure of how quickly the voltage at the far end of a line responds to voltage applied at the near end. Buried lines have much higher levels of capacitance than air insulated lines. Since AC lines reverse voltages 120 times per second at 60 Hz, the delay quickly becomes problematic if capacitance is too high. Since DC lines don't reverse voltages, capacitance has only a small transient effect.

is interconnected, and a fault in one area can rapidly impact the entire system. If managed improperly, major disruptions can propagate through entire regions in minutes, if not seconds.

To help maintain transmission stability, the not-for-profit international transmission regulatory authority, known as the North American Electric Reliability Corporation (NERC), requires all generating facilities to supply or purchase a group of products called *ancillary services*. There are over a dozen ancillary services that each utility serving as a balancing authority must ensure remain within specified bounds. An example is a *spinning reserve* service to provide additional emergency capacity within 10 minutes of notification. Typically, this is accomplished by keeping a generator running at less than full capacity. A *non-spinning reserve* is a generator that is capable of serving load but needs to be started up.

Two other important ancillary services of note are voltage and frequency support. In the US, voltage is maintained at 110 V with a frequency of 60 Hz. As load increases and decreases in real time, system-wide voltage and frequency will fluctuate. In order to maintain a constant voltage for transmission lines and steady frequency, a utility will employ *automatic generator control* to adjust the output of multiple generators across different power plants in response to changes in load. If system frequency is increasing, less output is needed; if system frequency is decreasing, then generation must increase.

Even though ancillary services are provided by generating facilities and owners of other specialized assets (e.g. shunt capacitor banks), they are not sold directly to end users. These critical energy, power, or reactive power services are used by power producers, distribution entities, and transmission system operators to keep the grid functioning.

Energy storage
Utility-scale electric energy storage is frequently touted as a holy grail of the electric energy industry (Lunde, 2012). The electric transmission infrastructure provides electricity *where* needed, but supply and demand must be coordinated in time. Energy storage technologies have the potential of providing electricity *when* needed. The current utility model predominantly varies supply to meet system load. With cost-effective energy storage, excess generation for capacity-constrained areas could absorb energy during times of

surplus and inject it back into the grid when there is a deficit. Effective storage technology would help shape and store many intermittent and distributed power resources, making them behave in a manner more similar to traditional dispatchable loads. Over time, this has the potential to help solve many constraints resulting from the real-time balancing needs of the grid.

Energy is easily stored in chemical forms, such as petroleum, or in mechanical forms like flywheels or compressed gas. Each form has its own energy density, which dictates how much work can be done. For example, from Table 3.2, gasoline has an energy density of 45.8 MJ/kg, and lithium-ion batteries have an energy density of 0.5 MJ/kg. This means that gasoline has 91.6 times more energy in the same weight. A 12-gallon fuel tank with a weight of approximately 33.4 kg contains 1530 MJ of energy. If we tried to achieve the same amount of energy in an electric battery, the battery would be massive, weighing in at roughly 3060 kg. Furthermore, the losses in the transformation process, the large capital costs of equipment, and the high ongoing operations and maintenance cost limit the scalability and adoption of electric battery-powered products.

Energy storage solutions are not one size fits all. Technologies for storage are often carefully configured for their particular application in the energy supply chain. Choosing an appropriate technology requires consideration of the capacity needed, restrictions on mobility, and the time periods over which energy discharge is required. Short, intense discharges of electrical energy are likely to employ tools like capacitors, while longer-duration discharges may employ batteries or pumped storage (hydro). Time periods may vary from several seconds for voltage and frequency response to many days for backup power. Other examples include the use of long-duration flywheels for small-capacity discharges over long periods of time, and superconducting magnetic storage can discharge large capacities for a very short duration at high intensities. A graphical depiction of these characteristics is shown in Figure 3.5 (EIA, 2011a).

Pumped storage hydro is the only proven technology for storing large amounts of electric power within the bulk power system. Conceptually, this is akin to shaping power output from dams. Operators pump water uphill when there is surplus energy—meaning when energy prices are low—and run the water back downhill through turbines when energy is scarce and more expensive. This solution

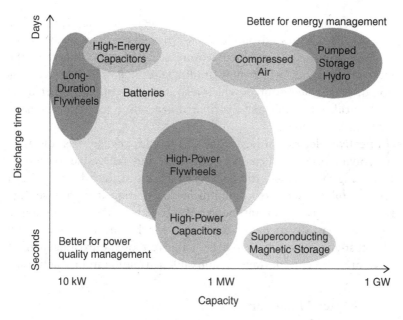

Figure 3.5 Energy storage technologies based on size and discharge time

has been adopted in areas where there is an abundance of water, because of significant evaporation, and plenty of elevation difference to exploit. Unfortunately, areas that tend to have both, such as the United States' Pacific Northwest or Switzerland, do not tend to be the areas with the most acute energy problems. Many of the areas most in need of electric storage are dry and flat.

Importance of dispatchability

Sources of generation are most valuable when they are dispatchable. A dispatchable source is one where the supply can be increased or decreased based on demand. They are used to provide *firm power*, where scheduled commitments must be physically satisfied to avoid financial penalty, a concept revisited many times throughout this book.

Thermal plants and hydro facilities are dispatchable resources that provide many of their own ancillary services to help maintain transmission grid stability. Wind and solar facilities, as non-dispatchable resources, have to secure additional ancillary services from other sources. This creates a dynamic where renewable resources buy energy services from fossil fuel-based resources in order to connect to the transmission grid, which only trades in firm power contracts. As non-dispatchable resources which also increase strain on other parts of the grid, renewable and distributed generation technologies make many utilities nervous.

Effective electric storage solutions could change this dynamic by solving the dispatchability problem for renewable resources. Increasing the reliability that a renewable or distributed resource can reliably fulfill its future commitments by pairing it with a storage technology has significant potential. Together, these technologies form a quasi-dispatchable resource, which can be more easily and cost-effectively integrated with the rest of the existing market structures and current transmission and distribution infrastructures.

3.4 Chapter 3 Summary

- Energy is always in balance. A change in any one of these silos will have ripple effects through the entire system. This is particularly true of electric power systems which balance (almost) instantaneously.
- Energy is produced and transported in order to meet consumer demand. Most of our end-use technologies consume thermal, mechanical, and electrical energy.
- Energy and power are two different commodities. Sometimes we value energy, and sometimes we value power.
- Electricity generation is rarely located next to energy demand. We transport power and energy differently depending on the physical characteristics, the entire supply chain, and end uses. Common transport mechanisms include wires, pipelines, trucks, trains, and ships.
- Energy storage addresses the issue of meeting variable energy demands by storing energy during times of surplus and discharging energy when energy is scarce.
- Pairing renewable and distributed generation sources could increase the economically sound adoption of new sources of generation.

4 | *Energy: the value proposition*

For every complex problem there is an answer that is clear, simple, and wrong.

H. L. Mencken

Imagine a future where an automobile can drive 100 miles per gallon (mpg). To an environmentalist this could be a dream come true: it would represent the quadrupling of fuel efficiency. Fuel consumption would drop, and we would become less dependent on foreign oil. Low-income families would not have to dedicate nearly as much of their income to commute to work. People would be able to travel more economically, and businesses could expand their products and service offerings to larger geographic areas with a lower fuel expense.

If such a future came to pass, these ostensible benefits would ripple across society with a number of foreseeable and emergent effects. Not all would be positive. Compare a normal car achieving 25 mpg to the fuel-efficient 100-mpg car of the future. Assuming standard driving patterns and $4.00 per gallon of gasoline, the normal car would consume nearly $2500 per year in gas. The fuel-efficient vehicle would only require $624 per year, representing a significant savings and benefit to the individual. In the US, the fuel tax is used to maintain the streets, highways, bridges, and tunnels. The fuel tax varies by state, and the national average tax is $0.48 per gallon. This means that an average driver would pay roughly $300 per year in fuel taxes. The fuel-efficient car would only pay $75 per year, not nearly enough to cover the upkeep of the roads. The government would be forced to make up the difference by raising registration fees, levying taxes elsewhere, or selling bonds to pay for infrastructure.

It is easy to see how a new technology applied to business-as-usual practices can lower fuel consumption. However, it is difficult to predict how a new technology will change future business practices. The Jevons paradox states that the more efficiently we consume a resource, the faster we deplete it. This was first observed in 1865, when William Stanley Jevons noted that the more efficient coal-fired steam engine introduced by James Watt caused the amount of coal consumed by the entire United Kingdom to soar as it was adopted (Jevons, 1865).

The autonomous vehicle is another technology being touted to reduce congestion and improve the driving experience, but it is also closely related to the economics of vehicle use and ownership. Setting aside the complicated questions about ethics for autonomous vehicles, a computerized driver could eliminate human error. The car would more reliably avoid accidents, have a more efficient driving pattern, deliver passengers directly to their destination, and park itself. Instead of spending time on a non-productive task, passengers could read, nap, surf the internet, or work while commuting. Such a car can achieve more economical results with constant route adjustments and optimal acceleration or deceleration. This kind of control is possible in the near future, with internet companies and auto manufacturers engaged in ongoing tests throughout Nevada, Florida, and California.

Combining these two technologies could have a number of unintended consequences. Setting aside the unresolved moral and ethical issues associated with practical adoption of computer-driven vehicles and assuming that a future autonomous vehicle allows people to take naps while driving, nothing prevents a passenger from spending the night inside of a moving vehicle. The cost of driving for eight hours at highway speeds is a mere $19.40 for fuel. This is significantly cheaper than renting a hotel for the evening at a respectable establishment anywhere in the US. As of 2013, the cost of renting a standard dorm room at Seattle's University of Washington is $32 per night. Any enterprising student on a tight budget (with a tolerance for moderate discomfort) will see that sleeping in an autonomous vehicle is a better deal. This student would then consume fuel at twice the rate of the national average.

It would indeed be tragic if the 100-mpg car became a meaningful contributor to increasing gas consumption. Yet technical solutions do not exist in a vacuum. Engineers optimize technical problems according to societal needs. Their solutions often touch all aspects of modern life and are used in ways that the original inventor could neither predict nor prevent. These sorts of issues have no defined solutions. They are often referred to as *wicked problems*, and this chapter is dedicated to their properties as applied to energy.

4.1 The properties of a wicked problem

Professors Horst Rittel and Melvin Webber first described the wicked problem in 1973 (Rittel and Webber, 1973). They were attempting to explain the complexities of social planning in cases where the problem extends beyond a society's historical experience. In the real world, information gathered is frequently incomplete and contradictory. The stakeholders involved can be difficult to recognize and organize. Seemingly inconsequential actions may have negative effects on vocal or highly underserved groups.

A wicked problem is termed to be the opposite of a *tame problem*. A tame problem has a clear ending condition, and reaching it makes the problem go away. For example, a checkmate ends the game of chess. A wicked problem does not have such a clear definition. In their paper, Rittel and Webber identified ten characteristics of a wicked problem. Table 4.1 applies these characteristics to the canonical "energy problem."

Energy is a classic wicked problem. Humankind will continue to consume increasing amounts of energy. Energy has numerous stakeholders, each possessing disparate goals and conflicting agendas. Complex relationships and competing interests hamper efforts to find solutions. We frequently find ourselves maintaining the status quo because it is simplest to do so. However, choosing to maintain the status quo is itself a decision with its own set of consequences and risks. For wicked problems, electing stasis is not without risk.

4.2 The clumsy solution

In theory there are three primary strategies to cope with and manage expectations within the wicked problem context; each is shown in Figure 4.1 (Verweij and Thompson, 2006). *Competitive* solutions identify the best solution by pitting one against the other. Businesses compete against each other to see whose solution generates the most profit or other desired metric for decision-making. *Hierarchical* solutions focus on processes to reach a solution. Governments enact

Table 4.1 *Characteristics of the wicked problem applied to energy*

Characteristic[1]	Applied to energy
1. There is no definitive formulation of a wicked problem.	The energy problem cannot be described succinctly. Frequently, people use carbon-free, fossil-fuel-free, net-zero, or energy security when describing certain attributes of the energy. None of these descriptions is complete.
2. Wicked problems have no stopping rule.	There is no point in time when society will cease needing to consume energy. Energy is intertwined with the advancement of society.
3. Solutions to wicked problems are not true or false, but better or worse.	Advantages and disadvantages to any energy solution are always stated relative to another solution. A wind farm emits fewer harmful pollutants and gases than a coal-fired power plant, but a wind farm is not dispatchable and is ultimately more expensive to build than a coal-fired thermoelectric generator.
4. There is neither an immediate nor ultimate test of a solution to a wicked problem.	We cannot isolate an energy solution to test it and fully understand its consequences. Since every solution causes ripple effects through society as a whole, solutions will always cause consequences outside of those foreseen by any preliminary experiment.
5. Every solution to a wicked problem is a "one-shot operation"; because there is no opportunity to learn	In a tame problem, we can test different solutions until one works and then scale appropriately. In a wicked problem, every solution changes the nature of the problem. With finite resources, we cannot simultaneously or iteratively test

Note: [1]From Rittel and Webber, 1973

Table 4.1 *(continued)*

Characteristic	Applied to energy
by trial and error, every attempt counts significantly.	all approaches. Replacing all coal-fired power plants with wind farms would commit resources that would preclude investments in competitive technologies or approaches for some time.
6. Wicked problems do not have an enumerable (or an exhaustively describable) set of potential solutions, nor is there a well-described set of permissible operations that may be incorporated into the plan.	There are an infinite number of potential solutions that could be applied to meeting energy needs. Two logical solutions in their own context could be at odds with each other when viewed in a larger context. For instance, raising retail electricity rates would make solar power more cost competitive, while lowering rates would help to alleviate energy poverty and enable more access to energy.
7. Every wicked problem is essentially unique.	There is no other problem exactly like the energy problem.
8. Every wicked problem can be considered a symptom of another problem.	Energy issues are part of other issues such as climate change, pollution, national energy security, nuclear proliferation, etc.
9. The existence of a discrepancy representing a wicked problem can be explained in numerous ways. The choice of explanation determines the nature of the problem's resolution.	Energy issues are often defined by their frame of reference, and solutions are valid only for that frame of reference. If one is concerned with climate change, one might advocate for nuclear power due to zero greenhouse gas emissions. However, if one is concerned with renewable energy, one might be an advocate for refining biofuels–even though its consumption will release greenhouse gases.

Table 4.1 *(continued)*

Characteristic	Applied to energy
10. The planner has no right to be wrong (i.e. planners are liable for the consequences of the actions they generate).	Every energy solution will always have negative consequences. Wind farms have a reputation of killing birds. Natural gas, which burns cleaner than coal, is condemned for other kinds of environmental damage. Dams are criticized for blocking salmon runs, impacting natural ecosystems, and affecting water quality.

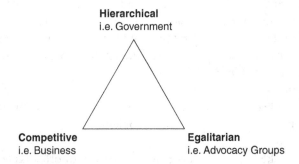

Figure 4.1 Diagram of the three forces in a clumsy solution (Rayner, 1995)

laws and enforce them through regulations. Rigid processes are created for granting approval. Such an approach does not place a value on any particular solution. *Egalitarian* solutions give all stakeholders an equal voice and promote the public good. Community organizations advocate for disenfranchised groups or environmental causes. Metrics for analyzing impact on the whole system and individual vested parties are critical in any case.

The clumsy solution recognizes the need to balance all three forces. If the solution becomes dominated by any one force, the corresponding negative consequences become more pronounced. Solutions that help bring the system back into balance will succeed, as much as might be possible. Under this framework, finding opportunities to make an impact become fairly straight forward. To maximize impact, one identifies the weakest force and creates a mechanism to strengthen it.

If society decides to prioritize solar power above all other options, businesses could design products and services that highlight the strengths of solar. To help, governments could create laws that incentivize solar panel production, installation, or investment. Policies allowing solar power to become more cost competitive to other resources might be enhanced. Processes used to schedule the electric power system could be modernized to minimize integration costs. Public interest groups could educate homeowners and the general public on the long-term benefits of increased solar power.

In practice, approaches are never so closely aligned. Successful competition from solar businesses could hamper the development of other important renewable technologies such as wind or hydro. Tax incentives skew investment towards economically inefficient outcomes. Installation incentives would drive up labor costs through monopolizing skilled workers. Government will need to manage diverse rules and system-level impacts for different types of installations. Industrial solar farms will need to be regulated differently than residential generators. Integration costs for small generators and risk of non-performance will create new challenges. Despite the obvious requirements, many existing government processes cannot tell the difference between very large residential solar projects or very small solar farms. Many regulators cling to historical roles and responsibilities for guiding utility incentives, despite clear differences between the problems faced when existing institutions were created and those faced today. Public interest groups would advocate for rate control mechanisms to prevent rate increases caused by increased solar contributions to the energy resource mix.

Wicked problems are not easily or quickly resolved. They require constant reassessment of priorities and usually involve haphazard movement towards a better future. The process, however awkward, is an approach to transparently expose and address the competitive objectives and interests of diverse groups.

A clumsy solution's three forces at work

A typical retail electric utility has all three clumsy solution forces in play. Retail utilities are in a heavily regulated business that guarantees equal access to all citizens. The government recognizes that electricity provides a common good for citizens and that the transmission and distribution networks have major capital costs requiring certain assurances to enable private investment. The "wires" side of the utility business is clearly a natural monopoly, and the government has embraced this structure. In return for its protected status, the utility must provide reliable and stable electricity for consumers within a geographic region. Any rate increase must be approved by a public utility commission (PUC) where citizens can voice their concerns. If the PUC determines that a utility does not meet reliability standards or that improvements to infrastructure fail to warrant rate recovery, they can impose fines or decline to offer financial incentives to the utility. If the utility is not able to recover their investment and cannot find new investors without additional assurances, generally via rate increases, the PUC can increase retail rates.

These forces result in restrictions of how electricity is bought and sold to most end users. First, because many utilities are local monopolies, they are the only entity allowed to buy and sell products based on dollars per kilowatt-hour ($/kWh). A homeowner with a solar panel cannot sell electricity directly to a neighbor. The homeowner must sell the electricity to the utility. The utility can then sell it to the neighbor based on the state-specific rules. Second, if the homeowner has faulty equipment or non-performing distributed generation and part of the grid fails, the utility gets fined for not providing reliable power. This is a disincentive to a utility for adopting decentralized generation. Third, the utility is mandated to provide electricity at the lowest possible cost to ratepayers under the "cost of service" theory of regulation. Recent efforts to decouple utility fixed cost recovery (return on capital investments) from the amount of electricity or natural gas sold are central to a more modern implementation of this concept. The goal is to fully incentivize capital investment in reliable infrastructure while ensuring that utilities do not suffer negative financial consequences from improvements to energy efficiency or distributed generation within their service areas.

4.3 Competitive approaches

Competitive forces identify solutions that achieve demonstrable financial success. According to economic theory, market forces are an efficient method to allocate resources. This is accomplished by allowing consumers to vote with their wallets. The solution that can best meet consumer needs while still generating a profit will be the most widely adopted and sustainable solution.

The electric infrastructure is very expensive to build and maintain. Under competitive forces alone, companies—especially in areas where they were the only viable service provider—might be incentivized to provide unequal service based on a customer's ability to pay or cut costs by not maintaining a reliable grid. Government regulators recognized this very early on and agreed to grant local monopoly rights to private corporations that wished to retail electricity. These utilities electrified the majority of the world and organized as vertically integrated entities, controlling virtually the entirety of electricity operations.

Monopoly status gave private investors a guarantee that they could attain a return on invested capital and generate an acceptable profit that is commensurate with an acceptable market rate for providing a needed societal benefit. In return, regulated corporate entities agreed to subject their electricity prices to government and public scrutiny. Under this structure, the corporation is required to invest in their infrastructure to serve all citizens with stable, reliable, and consistent electric power. If they fail in their societal mission, the government is able to impose fines against the company.

However appropriate at the time of adoption, the vertically integrated monopoly structure is undergoing major change. Many countries have recently, or are currently, restructuring electricity industries to enable competitive (and largely unregulated) generation services and competitive retail electricity sales, facilitated by regulated utilities operating the "wires" portions of the electric power system.

Financing generation and supply

A typical new combined-cycle natural gas plant costs hundreds of millions, and coal-fired baseload generators can cost billions of dollars (EIA, 2013d). Once built, the same power plant can often sell electricity for many decades with proper operation and maintenance. A private developer can finance the upfront cost of the project with debt, usually via a mix of bonds and equity investment. Municipalities and rural electric cooperatives can turn to bond markets. Developers attempt to guarantee a return on their investment by negotiating

long-term sales contracts for electricity early in the development
lifecycle.

A *power purchase agreement* (PPA) is a contract that guarantees
the purchase of electricity over a period of time. Some PPAs
can last more than 30 years. The price of a PPA is usually
determined by *where, when,* and *how much* power will be provided.
The contract will indicate the location for point of delivery, where
physical settlement is to occur. There will be a clear start and end
date for the contract. PPAs for utility-scale projects are denoted in
MWs and are referred to as *bulk purchases* or *wholesale electricity.*
Similar contracts are increasingly common for financing distributed
solar generation projects for retail customers with utilities or third
parties.

Price variations of wholesale electricity vary substantially based on
location. In 2012, the average wholesale electricity rates varied from
a low of $0.022 per kWh in the Pacific Northwest to a high of $0.046
per kWh in New York (EIA, 2012a). Wholesale prices also vary based
on fuel type and availability by region. The electricity from nuclear
resources has an average levelized cost of $0.096 per kWh. Coal elec-
tricity has an average levelized cost of $0.095 per kWh.[1] Renewable
resources are still comparatively expensive: solar can be as high as
$0.13 per kWh, and wind can be as high as $0.20 per kWh (EIA,
2014c). Because of the economics, fuel availability, and asset characteristics, coal overwhelmingly remains the primary source of baseload
energy throughout the world. The difference in cost between coal and
wind or solar is known as the funding gap for renewables. In essence, a
utility would have to pay a premium to procure needed electrons from
wind and solar as compared to a fossil fuel resource. This is especially
true when the fully weighted cost of ancillary services and integration
costs are considered at higher levels of renewable penetration.

A typical utility reselling power to retail customers has several
choices to purchase electricity. In vertically integrated markets, which
are disappearing around much of the world, it can build or buy a
power plant, own the asset, and operate the facility. In this model, the
cost of electricity would be based on the asset value over the useful life

---

[1] The levelized cost of electricity includes capital costs, fuel costs, fixed and
   variable *operations and maintenance* (O&M) costs, financing costs, and an
   assumed utilization rate for each plant type.

of the plant. Alternatively the utility can purchase power directly from another utility that has extra capacity. It can also purchase power from a private developer who is known as an *independent power producer* (IPP). IPPs are generally owned and operated by investors who make money from the electricity they generate and sell.

The vertically integrated utility model is evolving with the adoption of competitive power markets across the world. These are generally referred to as the bulk electricity markets. In competitive systems, regional balancing authorities are responsible for ensuring that the most economically advantageous power plants are selected to run. This prevents a utility from preferentially buying its own generators' electricity, despite it being more expensive to end users. In competitive power markets, some form of comparing the marginal cost of generation between available assets is used. In the US, this comes in the form of *security-constrained economic dispatch* (SCED). Other areas currently using the alternative power pool models, like the Northwest Power Pool, are considering implementing a full market operator-based SCED process in the near future.

The wholesale price of electricity is varied at each location based on the loads and generation assets available in each considered time period. Because electricity can be physically scarce due to physical limitations on generation or transmission, locational prices can vary substantially. To manage risks associated with capital investments, utilities leverage long-term agreements for providing or consuming power. These bilateral agreements are negotiated contracts which stipulate all terms and help ensure stable planning and operations for the grid.[2]

Metallurgical plants and other large industrial users may consume enough electricity to purchase power directly. Large purchases from consumers may arrange for power to be supplied at a particular generation facility and include long-term transmission rights for transporting the power to the end-use facility. They are able to buy electricity at wholesale prices for long durations by negotiating in advance, but depending on the nature of their purchases and regional bulk market structure, true costs generally are still impacted by the real locational marginal price at the sink and source locations at the

---

[2] More details on energy contracts and risk management are provided in Chapter 11.

time of fulfillment. Many industrial users will also be located near a power plant to provide for better power quality.

The real-time price of electricity can be very different from the long-term contract prices. During the summer of 2001, wholesale energy costs were sky high. This was due to many factors, including a hot summer that drove up demand, low snow pack that decreased supply, and market manipulations from power marketers. The aluminum smelters of the Pacific Northwest found it more profitable to sell their long-term electricity contracts to the California energy market than to operate their plants. As a result, nearly 43% of all US primary aluminum smelting capacity shut down (DOE, 2007).

Locational marginal price and transmission rights

In the 1990s, the Federal Energy Regulatory Commission (FERC), which oversees the wholesale electricity markets, decided to allow competitive generation and open access in the wholesale market. They required utilities to separate their transmission and generation businesses and allowed open access of transmission lines to any producer who could meet stability and reliability requirements. The FERC encouraged the formation of *independent system operators* (ISOs) and *regional transmission organizations* (RTOs). Although functionally similar, ISOs are generally within state boundaries, while RTOs generally expand across many state lines. As shown in Figure 4.2, not every state is part of an ISO/RTO (FERC, 2009). They manage the open-access same-time information systems (OASIS) for their geographic region.

To sell power, a utility needs to be able to secure adequate transport for generated electrons to reach the location of the load. The simplest solution, once employed by integrated utility companies, is to build a transmission line from the power plant to a load center (e.g. a city) at a current cost of approximately $1 million per mile (Brown, 2009). Transmission rights are a necessary part of enabling multiple entities to use common "wires" and avoid bearing such costs individually but, if improperly implemented, transmission rights can enhance buyer or seller market power in inefficient and non-competitive ways (Joskow and Tirole, 2000). The movement towards competitive electricity markets generated a great deal of controversy over the best market structures, especially how to manage the limited capacity of specific

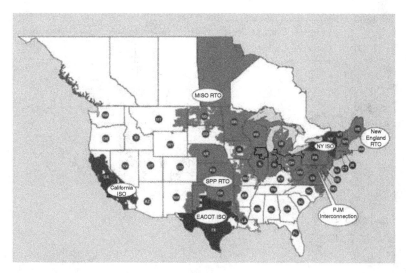

Figure 4.2 Map of North American RTOs/ISOs
Note: This map was created using Energy Velocity, August 2009.
Source: FERC, 2009

transmission lines needed by numerous competing providers to serve potential customers.

Physical transmission rights emerged from original scheduling procedures for transmission infrastructure. A transmission owner would sell the rights to move specific amounts of electricity across a physical asset, in a particular direction, during a specified time. Conceptually, this is not significantly different from securing a contract with a moving company or a shipping provider to haul personal effects from one place to another. In such an environment, an IPP located along the transmission line that wanted to sell power would have to enter into a contract for the right to transmit electricity over the wire. Existing rights holders or the owner could deny access, forcing the IPP to abandon its project or finance and build a separate line. As a result, this structure has largely disappeared.

These earlier approaches have effectively been replaced by market structures reliant on the absence of physical tranmission rights and the use of coordinated spot markets to maintain system stability and avoid overloading specific transmission links (Hogan, 2012). In a typical bid-based pool (e.g. PJM) with *financial transmission rights*, a holder

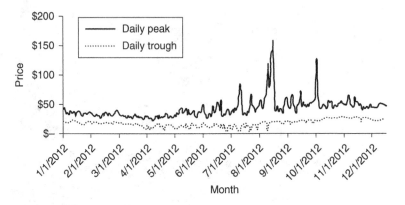

Figure 4.3 Exemplary peak and trough prices of electricity
Source: CAISO, 2013

is entitled to a stream of revenue (or charges) based on the hourly congestion price differences across a specific transmission path in the day-ahead market. In a well-run market, this ensures open access to all generators. The price of transmission across the infrastructure is the difference of the locational marginal price between nodes.

Each market region has many *nodes* representing connections between load centers, generators, or smaller transmission lines with the main transmission infrastructure. The specific set of available generators available at a particular time and the nature of the corresponding load determines how the transmission infrastructure is used. The general properties of *where*, *when*, and *how much* determine the price of electricity to balance the system via short-term markets. The *locational marginal price* is determined by the SCED process, using the cheapest generators to achieve energy balance across the system.

Figure 4.3 compares the highest cost of electricity per day to the lowest cost of electricity per day for node NP-15 in the day-ahead market in California (CAISO, 2013). The volatility based on system conditions is evident. The peaks and troughs correspond to the demand of energy at any specific time. Peak prices usually occur in the afternoon, while the trough prices are usually around 2 a.m. Seasonally, we can see that winter prices are fairly steady and that the price of electricity is approximately $40 per MW. However, on a hot August afternoon when air conditioning load is high, the demand of electricity can outstrip the ability for the grid to deliver supply. This causes a dramatic jump in the price of electricity to nearly $160 per MW. At night, when demand returns to normal, the trough prices of electricity return to normal.

Peak load management (PLM) attempts to alleviate these sudden spikes in prices by enabling demand-side participation. In some parts of the grid, peak prices can reach over $1,000 per MW during hot summer afternoons (Joskow, 2012). Some utilities incentivize consumers (mostly commercial and industrial) to shift usage to the evenings by offering lower rates and imposing retail tariff structures that penalize peak electricity use with higher rates. *Demand response* (DR) also addresses this issue by paying people to not use electricity when the system is capacity constrained. This makes financial sense to the utility only when the cost to pay people to defer their usage is lower than the cost to procure supply and maintain system balance.

Understanding utility rates

Retail utility rates (regulated or competitive) try to incorporate the highly variable factors of *where*, *when*, and *how much* electricity will cost in the bulk system into a static and predictable cost for the end consumer. Generally speaking, a utility is not allowed to price discriminate across a local geographic footprint. The cost of *where* electricity comes from is averaged out across all ratepayers. The cost of *when* electricity is consumed can be recovered by averaging different prices of electricity during different times of the day into the rate. In areas where congestion and peak pricing exist, the monetary differences can be substantial. The cost of *how much* electricity is consumed is determined by the rate schedule. If clients consume small amounts of electricity, they will be on a different rate than an industrial client that consumes large quantities.

A utility schedule is designed to reflect these factors. Charges for *energy* are based on kilowatt-hour (kWh) consumption. Charges for *power*, also known as capacity or demand charges, are based on the peak kilowatt (kW) consumed over the course of an entire month. There may be monthly or daily *flat rates* as a maintenance fee. Large commercial and industrial customers may see charges on their bill for *power factor* or *VAR*. These represent the costs for ancillary services–namely, reactive power consumption, which is measured in

VAR. Some rates are tiered based on when one uses electricity, and some are tiered based on how much energy one draws at any given time. There can also be additional surcharges for efficiency programs, low-income or credit-risk set-asides, rebates, and refunds.

Charges for *energy* are fairly easy to understand. The more kWhs consumed, the more paid. Charges for *power* are more complicated. Recall that the capacity of a wire is analogous to the size of a water pipe. Customer A concurrently runs the dishwasher, washing machine, refrigerator, and TV and charges an electric vehicle. He will have a very high instantaneous demand and will need a large pipe to deliver enough power. Customer B runs all those tasks separately. He will have a lower demand (i.e. peak power requirement) and a smaller connection, generally called a "service," will satisfy his need. Both customers consume the same amount of electric energy.

The utility is required to deliver enough power to satisfy the maximum demand of every customer. Unfortunately, the utility does not know *when* a customer will reach that maximum demand. As a result, the utility needs to maintain adequate capacity to satisfy both customer A and customer B's peak demand at any time. This is true even if customer A only concurrently runs all of his appliances once per month. The utility will charge the customer for the peak instantaneous customer demand, also known as a capacity charge.

Comparing three different rate schedules for three different utilities helps clarify. Assume we have a medium commercial client who has a maximum demand of 50 kW. A Seattle-based customer would be under Schedule MDC: Medium Standard General Service rate (Light, 2013). It is a very straightforward rate schedule. We immediately see the three main components of utility rates: a flat rate, a power charge, and an energy charge.

- Flat rate: Basic charge of $0.62 per meter per day
- Power charge: Demand charge of $2.13 per kW of monthly maximum demand
- Energy charge: Energy rate of $0.0566 per kWh

The same building in North Carolina which is served by Duke Energy would be governed by the Small General Service rate (Duke Energy Carolinas, 2013). Here, the basic charge and demand charge are both higher than in Seattle. The energy charge is tiered based on the capacity demanded per month.

- Flat rate: Basic charge of $18.13 per month
- Power charge: Demand charge of $3.7994 per kW above 30 kW
- Energy charge: Energy rate that is tiered based on total consumption per month

  - *Tier 1*

    o $0.116 per kWh for first 3000 kWh
    o $0.070 per kWh for next 6000 kWh
    o $0.072 per kWh for all over 9000 kWh

  - *Tier 2*

    o $0.067 per kWh for first 3000 kWh
    o $0.0596 per kWh for next 6000 kWh
    o $0.057 per kWh for all over 9000 kWh

  - *Tier 3*

    o $0.056 per kWh for all remaining kWh per month

The same customer in Southern California who is served by Southern California Edison would fall under Rate Schedule TOU-GS-2-A (Edison, 2013). This rate, effective June 1, 2013, is too complicated to reprint here. The prices for energy and capacity are different for every hour of the day, every day of the week, and every season of the year.

- Time of day (between 8 a.m. and noon, noon and 6 p.m., 6 p.m. and 11 p.m., and 11 p.m. and 8 a.m.);
- Weekday versus weekend;
- Summer season (June to October) and winter season (October to June).

Utilities with very simple rates, such as Seattle City Light, also tend to have very simple supply characteristics. Approximately 66% of their nameplate capacity in Seattle comes from their own hydro facilities that were built between 1917 and 1967 (Seattle City Light, 2012).[3] The Northwest Power Pool is a simpler environment with fewer competitive power producers, lots of hydro, and lots of public

---

[3] It is important to note once more that electricity is not routed per se, and that a Seattle customer will benefit from the utility's contribution to the regional system while consuming *actual* electrons from a more nearby generator when the system balances.

power organizations. On the other hand, Southern California Edison participates in the California ISO (CAISO) market and is subject to more intense seasonal price fluctuations, as shown in Figure 4.3. The ultimate rates passed on to customers are reflective of wholesale price patterns, resource mix and total costs.

Efficiency financing options

From a business owner's perspective, utility bills are complex, and rates are going up. Businesses need energy to run equipment, do work, and ultimately generate revenue. While many consumers don't feel particularly empowered to lower costs, there are actually a large number of options available to fund more efficient equipment in most areas. The greater challenge is in setting priorities and determining the right approach to meet the most pressing business need.

Utilities are incentivized to help customers buy energy efficiency equipment for a variety of reasons. Energy efficiency is touted as the least-cost resource for providing energy. A typical utility's responsibility is to procure supply to satisfy load within its service territory. Within a region, this can be accomplished by building new power plants when load grows or by incentivizing energy efficiency to reduce demand. Specific actions can include incentives for system-benefiting behavior or improvements to customer buildings to reduce or reshape load. The strategies are functionally equivalent – raising supply to meet demand or lowering demand to match supply. From an environmental and total cost standpoint, energy efficiency is preferred. It lowers total resource consumption and frees up operational capital for other purposes.

Depending on the utility incentive program, efficiency measures can range from replacing simple equipment, such as light fixtures and exit signs, to complex retrofit projects that replace boilers, chillers, and rooftop units. Some utilities pay up to 70% of worthwhile projects. The federal government, through US Department of Agriculture's (USDA) Rural Development Energy Program, also gives grants and loans to qualified individuals to pursue efficiency projects and develops renewable energy sources (USDA, 2013). The USDA also assists with completing some feasibility studies to determine the right strategy.

Community Development Financial Institutions (CDFIs) are enti-ties that provide credit and financial services to underserved markets.

Backed by the US Department of Treasury, CDFIs are mission-driven entities that focus on helping local communities. Many of them help finance low-income real estate development. Some are keenly aware of how energy efficiency can positively impact the ongoing costs of a housing project. Some have energy efficiency loan programs that can finance up to 100% of the loan with no money down and very low interest rates (Craft3, n.d.). Others, such as the New Market Tax Credit Program, have helped deploy solar panels in rural areas (NREL, 2012).

A consumer can also finance a project today based on projected future savings. An *energy services company* (ESCO) will pursue this strategy to help lower the upfront cost barriers. In this scenario, an ESCO will forecast how much they believe a structure can save after a retrofit. The ESCO will cover the upfront capital needed to perform the project. In return, they split the savings with the customer for a certain number of years. After the terms of the agreement, the customer enjoys the full savings for the rest of the equipment's life. This type of an arrangement is called an *energy savings performance contract* (ESPC). Many government organizations have adopted this as the preferred way of financing efficiency projects. The Federal Energy Management Program (FEMP) helps the federal government reduce its own energy footprint. They publish a guideline for how to structure an ESPC. Many states and other jurisdictions have adopted the FEMP guidelines for their own performance contracting.

Solutions such as on-bill financing or *property assessed clean energy* (PACE) try to alleviate some complexities by consolidating the repayment plan into an existing bill. Craft3 is a Seattle-based CDFI whose customers repay their loans as part of their utility bill, simplifying the need to write separate checks or manage another account. In the PACE program, the efficiency loan is repaid through property taxes collected by the county government for the same reasons. In both situations, if the building changes hands before the project is paid off, the new owners continue to repay the loan via the utility or property tax bills.

4.4 Hierarchical approaches

The electric utility industry is historically one of the most hierarchical industries in the nation. It faces a plethora of regulations, laws, policies, and taxes that influence individual and organizational behaviors. As a result of its close interaction with government and extensive oversight, many traditional utilities tend to be process oriented and usually have a very strict set of rules that guide interactions with internal business units and external entities. With well-defined process mechanisms inside utilities and oversight organizations, the policy-makers and utilities generally aim for conservative, incremental, and already-demonstrated change. Strong reliance on process can sometimes focus attention on matters of public interest but can also contribute to increased costs of doing business and the stifling of creative approaches to solving problems.

Governments use their full power to enact laws, set policy, and write regulations to set the conditions for providing economic and reliable access to energy. Laws are a system of rules and guidelines that govern the actions of individuals and organizations. Policies are principles or rules that guide decision-making to achieve stated outcomes. Regulations are the process of promulgating, monitoring, and enforcing the specific rules or exercise authorities established by legislation. In this section, we explore the regulatory structure, look at consequences of creating "green" laws, and explore some current energy policy initiatives.

Regulatory structure

As recognizable transmission infrastructure developed in the early twentieth century, electric generating companies quickly consolidated into localized monopolies that served entire cities and regions. State governments and public concern fueled anti-trust initiatives, following the massive consolidation of steel-makers, railroads, and oil under tycoons like Andrew Carnegie, Jay Gould, and John Rockefeller that helped lead to the passing of the Sherman Antitrust Act of 1890. A series of legislative, policy, and regulatory initiatives began extending the jurisdictions of commissions that were designed to regulate railroads into the electric industry. By 1914, 43 out of 48 states had oversight of electric utilities. The private companies were growing rapidly thanks to large infusions of cash from public securities markets and became known as *investor-owned utilities* (IOUs). By the 1930s, massive consolidation had led to the formation of vast public utility holding companies which were opaque and rife with accounting abuses and investor fraud. The rapid growth and large public participation in funding major infrastructure investments came to a head in 1932 when Samuel Insull's holding company, among the most prominent, was pushed into bankruptcy.

As transmission lines became longer, they increasingly linked together loads and generators across state borders. Growing utilities also led to local companies who were suddenly owned by entities located across state lines, where the financials of the parent company would not be subject to state review. Federal intervention was necessary, and in 1935, the Federal Power Commission (now the Federal Energy Regulatory Commission) was formed to regulate wholesale power sales and transmission infrastructure that crossed state lines (Brown, 2004). During the New Deal, the federal government also pursued hydroelectric projects in Tennessee and Washington state. The Tennessee Valley Authority (TVA) and the Bonneville Power Authority (BPA) report directly to the Department of Energy (DOE). The DOE also regulates all nuclear facilities.

There are several other common utility ownership structures in addition to the privately held IOUs. IOUs frequently neglected rural customers during their rapid expansion due to the cost of servicing them and less enticing prospects for growth when compared to more dense urban regions. *Public utility districts* (PUDs) were formed by counties and cities to serve these customers. Authorized by cities and counties, they generally report to a county commission or city council. A *cooperative* utility is similar, except all members of the cooperative have voting rights to elect the cooperative's board. In general, IOUs are large and can serve many millions of customers, while cooperatives are small and may serve only several thousand. PUDs usually reside somewhere in the middle, with some rivaling IOUs in size and others as small as co-ops. One notable exception is the New York Power Authority, which is the largest US-based public power organization, owning 17 generation facilities and more than 1400 miles of transmission lines.

Table 4.2 depicts the number of utilities in the US by ownership type. Most utilities are not the monoliths commonly ascribed to them. They each have their own hierarchy, reporting structure, culture, and needs. The shareholders of an IOU will have very different performance metrics and resources than a PUD's county-level commission. Successfully working with utilities requires a keen understanding of structure, objectives, and local constraints. The combination of

Table 4.2 *Makeup of utilities in the US (APPA, 2013)*

Participants	Number
Public utilities	2006
Cooperatives	873
Investor-owned utilities	193
Power marketers	181
Federal power agencies	9

regional and structural variations makes it particularly difficult to inspire change or drive technology adoption across 3200 unique US utilities (EIA, 2013b).

In essence, regulatory responsibilities can be summarized as follows:

- The federal government, through the FERC, is responsible for the wholesale electricity market.[4]
- The North American Electric Reliability Corporation (NERC) has jurisdiction over the standards for operating physical transmission infrastructure.
- The DOE is responsible for the hydro facilities in the TVA and BPA. It also has jurisdiction over all US nuclear facilities.
- State utility commissions have jurisdiction over retail rates for IOUs within their state boundaries.
- Public utility districts report to county commissions or city councils and generally set their own rates.
- Cooperatives are generally self-governing and are held responsible for rates and investment by their members.

This fragmented governance system creates a unique set of challenges for the utilities. Depending on the type of utility, executives report to different agencies and organizations who value different metrics. This also complicates the sales processes for vendors who seek to partner with utilities to reach end users. Simplistically, most utilities have two regulatory regimes—the federal government, who

---

[4] Except for the state of Texas, where ERCOT is exempted from FERC oversight by waiver.

regulates their wholesale purchases, and state or local governments, who regulate their retail prices.

Regulated cost recovery

Utility commissions are tasked with balancing two opposing forces when setting *prudent* retail rates. First, the rates need to be financially responsible to the utilities. They need to cover operational costs and satisfy debt burdens from past capital improvements. They also need to return a *reasonable profit* to shareholders. A reasonable profit attracts future investors who might otherwise invest in the stock market, mutual funds, etc. Without achieving an acceptable return, regulated utilities are unlikely to maintain access to sufficient capital resources to finance future upgrades and expansions. Second, the rates need to be fiscally responsible to ratepayers. The amount paid by end users must be reasonable enough for the entire population to benefit from the system. The public interest is usually represented by a state's Attorney General, citizen groups, and trade organizations.

Utility rates are generally based on the concept of *cost recovery* or *cost of service* regulation. The three primary costs considered are operational, capital, and a reasonable profit. *Operational expenses* such as fuels, worker salaries, etc. are recovered at or near their cost. *Capital expenditures*, such as transmission infrastructure and power plants, are recovered with a reasonable profit. Thus, for every $1 spent on new infrastructure, the utility can recover approximately $1.095 back. The $0.095 is considered the utility's reasonable profit that is approved by the utility commission and ultimately returned to shareholders.

A retail rate is determined based on all three of these costs. A commission adds the three values together, taking into account external factors like weather and historical performance, to determine the total amount of money a utility can recover during the coming year, commonly referred to as a *test year*. Simplistically, the utility rate is calculated as the sum of total allowed expenses and allowed return on investment divided by the amount of electricity forecast to be sold that year. At the end of the year, the commission can choose to adjust the rate for the next year by enacting a new rate case.

This system is intended to keep costs low. On average, utility costs today are only 4–5% of a household's income. Electricity is one of the few commodities that has gotten cheaper in real dollars over time. In 2012, the US average electric rate was $0.12 per kWh. In 1960, the inflation-adjusted rate would be $0.21 per kWh (EIA, 2011b). Access to affordable electricity has been a significant contributor to sustained, long-term economic growth and improvements in quality of life.

This structure is not universally positive. The incentive structure causes several odd side effects. For example, in order to act in the public interest, a utility commission needs solid evidence to justify increases or decreases in rates. Utilities and regulators are incentivized to be conservative stewards of the public good, contributing to reliance on traditional, proven technologies despite a number of advances which could improve operations and costs passed on to consumers if implemented properly. Other changes, like the development of concepts like the *smart grid*, are also challenging commissions to rethink their roles. Increasing prevalence of new technologies is forcing regulators and utilities alike to reimagine aspects of strategic planning and the rate-making process.

One area of regulatory growth is the ongoing work to incentivize utilities to reduce energy consumption via sponsored efficiency programs. The term "negawatt," coined by Amory Lovins, asserts that the cheapest kWh produced is the one not consumed. If a commission allows a utility to earn a profit from "buying" negawatts like other capital expenditures, then the utility can generate a reasonable profit. The commission needs assurance that the savings are actually achieved to validate that system-level benefits were accrued for the rate base. In the 1970s, some efforts to incentivize purchases of energy-efficient refrigerators backfired. Households bought and installed energy-efficient refrigerators and relocated their old refrigerators to the basement. This increased electric load rather than decreasing it, subsidizing a lifestyle change for participants but not helping to improve grid operations. Some commissions have very strict reporting standards for efficiency measures, while others do not allow any at all. Wide variations in incentives and local requirements has created a patchwork of efficiency incentives across the US.

Some utilities have successfully pursued conservation and energy efficiency measures in their territories. Sometimes they are so successful that they sell less electricity than their expected forecast. Since they did not reach their sales targets, they did not earn their expected reasonable profit. As a result, some utility commissions are faced with rate increases to recover lost margins as a consequence of overly successful conservation efforts. After all, a community has to pay for the transmission line, even if fewer electrons were sold from it.

Utilities generally do not pursue any significant research and development (R&D). Less than 1% of their revenues are allocated to R&D purposes and are commonly pooled into collective organizations like the Electric Power Research Institute. Money allocated to R&D increases the rate that utilities need to charge. Utilities are frequently unable to adopt new technologies unless they have already been fully vetted, tested, deployed, measured, and understood somewhere else. The incentives create a significant first-mover disadvantage. However, a utility is often able to pursue pilot projects in order to gather data needed to justify a rate case change. A pilot in the electric industry is also very different from that in other technology sectors. It is not uncommon to pilot a "new technology" that has been commercially available elsewhere for years.

The commission is focused on upholding its legislated duty to providing the lowest-cost electricity to citizens. This body is responsible for advising policy-makers while handling the rule-making which is responsible for the ongoing balancing of the fiscal viability of the utility with maintaining the lowest rates for ratepayers. Unless directed by legislation or policy, public utility commissions are not responsible for promoting renewable electricity, energy efficiency measures, or any specific form of modernization. When directed by law or policy, regulators can then consider the costs associated with goals other than end-user rates, including renewables and distributed generation, as part of rate recovery.

Laws

Laws are a system of rules which encourage, mandate, or require actions to take place. In general, good laws have strong and clear definitions and minimize vague or arcane language. Poorly defined terms or overly vague intent within the law can expose organizations and governments to litigation.

Most existing US federal and state laws governing renewable energy are fairly strong and precise. Many states have passed renewable portfolio standards (RPSs) requiring utilities to procure a certain amount of electricity from renewable resources. Utilities can usually choose to develop the assets themselves or buy renewable electricity on the open market. When these changes occur, regulators gain an additional constraint for the utility, whose cost of compliance must be worked into the rate case. In some states, such as California, the RPS can be as high as 33% of all resources by 2020 (DSIRE, 2013).

Because there is no discernable difference between an electron generated from a renewable resource and an electron generated from a fossil fuel resource, utilities and consumers use *renewable energy certificates* (RECs). An REC is a legal contract sold by renewable power plants that can be used to meet renewable portfolio standards or for the purposes of marketing environmental consciousness. The REC certifies the amount of electricity generated during some period as renewable energy. For every 1 MWh of electricity generated, a producer receives a 1 MWh REC. The electricity is sold into the wholesale market at wholesale rates. The REC is sold to any entity that wants the credit for producing these green electrons at the time and location they were produced. This certificate can be traded on the open market, and any entity can purchase it. The REC allows a renewable power producer to have two revenue streams: one for the sale of the electricity and one for the sale of the certificate. This additional revenue helps close the funding gap between renewable and fossil fuel resources.

Laws governing green construction issues can be vague. Some municipalities have adopted LEED® ratings as a standard for a green building. LEED®, which stands for Leadership in Energy & Environmental Design, has become the standard bearer for sustainable building design. It is a checklist system with four levels: Certified, Silver, Gold, and Platinum. The levels are determined by meeting specific criteria with established points. There are upwards of 100 points in the rating system. These levels can be useful to differentiate one building from another. But as a law, mandating LEED® participation or ratings can create problems.

One problem is that different cities have chosen different levels to mean a "green" building. This creates ambiguities in how to dispense consequences for failing to build a "green" building. A frequently cited case is Shaw Development v. Southern Builders. Southern Builders filed a lien claim against a condominium project for unpaid costs. The developer countersued for the contractor's failure to construct a LEED® Silver rating for the building, as stipulated in the contract (Vyas and Gentilcore, 2010). This case raises several important questions. LEED® ratings are awarded after the project is completed—usually after a contractor's work is done. Many of the points on the checklist are not within the contractor's usual responsibility. Contractors build to specifications given by an architect. Should the contractor be held liable for a process that includes other stakeholders? Even though the building did not attain a Silver rating, it did attain a LEED® Certified rating, one level lower. Is LEED® Certified not a green building? The answers are not clear, and sustainable law is one area of continuing evolution and development.

Policies

Energy policy can dramatically affect energy behavior. Germany has declared a national policy of shutting down all nuclear power plants by 2022. State governors promote renewable energy policies by supporting RPS legislation directly impacting new investments within their state (Baetz, 2011). At the same time, the US federal government has drawn criticism for continuing to lack a coherent national energy policy. The administration of George W. Bush touted clean coal and the administration of Barack Obama pumped billions of dollars into alternative energy research with mixed effects. Although federal policy is virtually nonexistent, the strategies pursued have been incredibly one-sided. Both administrations have focused their investments on developing supply-side technologies. Recent administrations have invested comparatively little in developing demand-side technologies or processes to improve energy efficiency or demand response. Appliance standards and efficiency targets were the primary demand-side efforts. The bias towards new generation is dramatic—for every $35 spent on developing new resources, the US spends only $1 on energy efficiency research (Wilson and Nemt, 2012).

Historically, US utilities were *vertically integrated* from source to sink. They owned and controlled their own generation, transmission, and load assets. In this mode, meeting energy, reliability, and stability requirements was relatively straightforward. The oversight of the system was relatively simple. Since the 1990s, the federal government has pursued a strategy of deregulation—or, more accurately, reregulation.

The new regulations created *horizontally integrated* utilities that could participate in any part of the supply chain. No longer does the utility control all assets from source to sink. New and unfamiliar players can enter the market. Tens to hundreds of thousands of homeowners have installed solar cells, wind turbines, electric vehicles, and other power electronic devices. Utilities are finding themselves with less and less control over where their electric supply comes from and where their electric load is going.

Decoupling is the policy of separating the amount of electricity sold from the reasonable profit attained. Under a coupled regulatory environment, a utility that invests in a new resource can recover its asset costs through the rate-making process. A utility that invests in energy efficiency has a harder time quantifying its effects and recovering the costs. Under legacy regulatory incentive structures, to grow the business and increase shareholder value, utilities needed to sell more electricity.

With a decoupled rate, the utility is allowed to recover the cost of invested capital independently of how much electricity it sells. Utilities governed by decoupled policies can now increase profits by lowering operational costs, passing savings on to ratepayers. In an ideal decoupled mode, the most effective way to maximize profits is to reduce energy consumption and minimize operating costs while remaining a reliable service provider.

Net-metering and *feed-in-tariffs* (FIT) are two common policies that allow residential consumers and private businesses to get paid for installing renewable resources. Under a net-metering policy, the utility meter spins forward when consuming energy and backward when generating energy. The electricity bill is calculated based on the net energy consumed that month. Under a feed-in-tariff policy, the customer signs a power purchasing agreement and sells electricity back to the utility. The monthly bill would include charges for consumption and a credit for electricity sold.

Superficially, a difference is that net-metering requires only one meter that runs forward and backward. The FIT requires two meters—a standard one measuring consumption and a separate meter measuring generation. More importantly, they have different financial structures. In net-metering, the consumer is being paid by the utility at the retail rate for power production. The monthly bill offsets consumption with every kWh generated. In an FIT, the cost of the electricity is commensurate with the cost of purchasing that power from other sources. Some jurisdictions allow the rate to fluctuate based on market forces, while others peg the value at specific rates. They can be as low as $0.02 per kWh, while peak prices can be as high as $0.39 per kWh. Contract lengths typically vary from 5 years to 20 years in duration (EIA, 2013a).

On the surface, the net-metering policies are simpler for a consumer and, in many cases, they may even receive more compensation. However, net-metering does not adequately account for the specific attributes of the power generated. Recall that retail rates are effectively averages of seasonal and intra-day variations. A solar plant producing power in downtown Los Angeles during the middle of summer may produce electricity that is much more valuable than the average rate awarded under net-metering. However, net-metering also creates other problems by displacing the normal electricity sales which include rate recovery for the large, fixed capital costs required to connect net-metered consumers to the grid infrastructure. The overall result is that a combination of utility rate increases or tax write-offs have been utilized to address utility incentive changes, with disproportionate negative effects on low-income households who are more sensitive to rate increases and least likely to participate in solar programs.

4.5 Egalitarian approaches

Regardless of who you are or where you live, all people have energy needs. Humans need heat for cooking, power to draw and move clean water, and energy to tend land and raise crops. In developed countries, these significant amounts of technology and infrastructure draw on electricity and petroleum to run stoves, pump water, move sewerage, and work land. In developing countries, these needs are still met primarily with human and animal energy. People, often assisted by animals, gather firewood for cooking, transport water with jugs, and painstakingly raise crops by hand. Reducing energy poverty is the single most important potential contribution to improving health, literacy, and a plethora of other issues faced by the poorest people on earth.

Some philanthropists focus extensively on addressing the needs of this disadvantaged segment of the population. Public interest groups can form as a way of addressing these differences that business or government solutions do not, cannot, or will not address. These groups are generally what we think of when we picture organizations focused on egalitarian solutions. Public interest groups are driven by values and missions to help others. According to the National Center for Charitable Statistics, there were over 10,486 registered non-profits whose main purpose or activity revolved around the environment in 2012. Of those non-profits, 355 classified themselves as Energy Resources Conservation & Development organizations (National Center for Charitable Statistics, 2013). These included ISOs which are concerned with efficient sales of electricity, winter sporting activists who are concerned with lower snow packs due to climate change, scientific policy advocate groups, efficiency incentive advocates, bioenergy industry groups, and many more. Each group is concerned with different facets of the proverbial energy issue.

There is no single solution to providing universal access to affordable energy. Fuel source and fuel consumption patterns are different between developed and developing countries. Supporting infrastructures, governments, and culture also have significant effects. Figure 4.4a shows that Nigeria consumes predominantly biomass, in the form of wood and dung, despite being Africa's largest petroleum exporter. Figure 4.4b shows that the US consumes predominantly fossil fuels, petroleum, natural gas, and coal (LLN, 2013). Figures 4.4c and 4.4d illustrate how the fuels are consumed on the African continent as a whole versus the US. In the US, transportation and industrial applications make up the majority of end uses. In Africa, household energy use is dominant. Energy solutions need to satisfy their basic demands before being diverted to higher-value functions. This is why developing surplus energy, at each stage of societal advancement, is so critical to accelerating the rate of economic growth and improvements to quality of life.

Energy poverty

In developing countries, energy or fuel poverty blatantly manifests itself as a lack of basic services. Energy or fuel poverty usually refers to the 1.3 billion people who live with no access to electricity (IEA, 2012). Of these, 590 million people without electricity live in sub-Saharan Africa, and 390 million live in India and Bangladesh. The lack of access to electricity has a profound impact on daily life, from

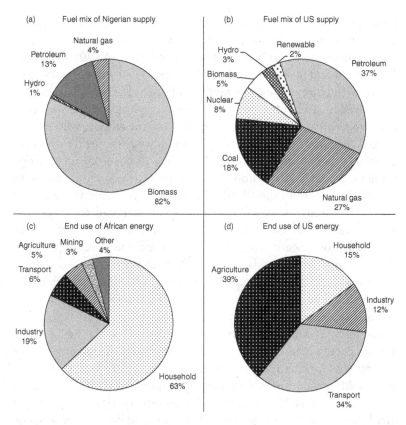

Figure 4.4 A comparison of fuel mix and end use between Africa and the US

cooking food and accessing clean water to health services and universal education. Energy poverty even contributes to systemic gender inequality, as women are disproportionately responsible for cooking and maintaining the health of the family. People in poorer countries also pay a *poverty penalty* where average costs of energy commonly exceed 10% or more of annual income (Bacon, 2010). Despite consuming much larger amounts of energy, people in developed countries typically only spend between 4 and 5% of annual income on energy (EIA, 2013a).

The total cost of universal electrification could be as high as $1 trillion. The International Energy Agency (IEA) estimated that extending access would involve investing nearly $49 billion per year until 2030 in order to give universal access to electricity. While significant,

these numbers seem surprisingly affordable when compared to the
$681 billion annual global cigarette sales (Marketline, 2014) or $1.8
trillion in annual worldwide defense spending (EIA, 2013b). As of
2013, the world invests only about $\frac{1}{3}$ of the needed amount to develop
the infrastructure required to support universal electricity access.

In developed countries, energy poverty typically affects the poor
who are already on government support and the elderly, who may be
on fixed incomes. In the US, energy prices have risen much faster than
inflation over the last few years. According to government statistics,
neither national GDP nor household incomes have grown as fast as the
rise in oil prices since the 1970s. This has a disproportionate effect on
low-income households who commit larger shares of their earnings
to basic requirements like getting to work. The poorest fifth of the
US population spends over 10% of their income on energy, while
the wealthiest fifth spends only 2.5%. Fuel price increases tend to
directly affect disposable budgets. In one estimate, if fuel prices this
past decade had stayed at 2001 levels, 2.6 million fewer Americans
would be in poverty (Houser and Mohan, 2011).

The US treats energy poverty as a poverty issue, while the UK treats
it as an energy issue. Fuel programs are based on the income level of
the household, akin to other social welfare services. Typically, qualified
households are ones whose income is no more than 150% of the US
poverty level (LIHEAP 2014). In the UK, a household is in fuel poverty if
they have an above average required fuel cost and if they were to spend
the amount, they would be left with a residual income below the official
poverty line. This enables the UK to use both energy prices and household
income as factors when determining fuel poverty assistance.

Although the differences are subtle, the different perspective has
manifested itself in different policy directions. Both countries have
identified energy efficiency as a primary need in reducing fuel poverty.
US fuel assistance programs grew out of the 1970s oil crises and
continued with their focus on the family, prioritizing households with
multiple children. The UK views fuel poverty as a social health and
climate change problem. Inadequate heat disproportionally affects the
elderly, who also often disproportionately live in poorly insulated
homes which require more fuel to maintain the same warmth.

Indoor air quality

Poor indoor air quality is one of the major symptoms of energy
poverty. Three billion people cook their meals with open fires burning

wood, dung, charcoal, and other biomass. Open pits and inefficient cookstoves emit carbon monoxide and greenhouse gases, constituting 25% of the worldwide black carbon emissions. The destruction of local forests for wood and charcoal contributes to the increase in mudslides during rainy seasons. Serious health issues are caused by perpetual exposure to smoke and carcinogens.

These issues disproportionally affect women and girls who are primarily responsible for cooking the food. While doing so, they have to breathe toxic smoke and fine particulates emitted from the rudimentary devices and are more frequent victims of fire-related accidents. Over 1.5 million premature deaths per year are attributed to poor indoor air quality, more than both malaria and tuberculosis (IEA, 2012). Women and girls are also burdened with gathering and transporting the majority of domestic fuel. The wood and vegetation closest to a village gets depleted first, requiring more and more effort to accomplish the same feat over time. It is not uncommon for a woman to walk more than 6 miles to gather fuel. These household chores prevent them from earning a living wage, going to school, or developing higher-level skills to advance their personal and family situations (IEA, 2011b). The United Nations estimates that the infrastructure required to shift fuels from biomass to alternative sources would cost $342 billion to build.

One immediately actionable approach is to switch to a cleaner-burning fuel. In Brazil, a concerted government effort enabled 98% of all households to access cheap liquefied petroleum gas (LPG), up from only 18% when the program began in the 1960s (Halff et al., 2014). Government-sponsored electrification reduced the need for wood-burning stoves. This lowered biomass pollution and reduced the risk of fires. Biomass consumption dropped, slowing the destruction of forests.

Another solution is to introduce new technologies, such as clean-burning cookstoves. A fire pit is hard to manage and may cook the food unevenly. A cookstove can extract more heat from less biomass, reducing the amount of fuel. It can burn the fuel more efficiently, reducing soot and pollution. It can also focus the heat, allowing the food to cook faster.

Before leaping to a particular technical solution, behaviors must be considered. New cookstoves need to be customized for particular regions based on prevalent cooking styles. Tortillas require flat cooking tops. Rice and curry need side-by-side arrangements. In some island nations, cookstoves have enabled fishermen to save time on cooking and make more money fishing. They can start cookstoves to make a pot of rice in the morning. When they return at noon, the fishermen can enjoy a pre-prepared hot lunch. On the other hand, if the stoves run for many more hours per day, more wood is burned to cook the same amount of food. Energy problems are ultimately human problems. At times, people respond to new technologies or energy resources in unknowable ways. Careful tailoring and experimentation are required to find commercially viable approaches to addressing practical issues around energy poverty.

Access to water

Extending access to clean water may be even more urgent than improving direct energy access. Contaminated water causes diseases like cholera and malaria. It is responsible for over 2 million deaths per year, and nearly 90% of victims are children. Women are frequently also the primary gatherers of water and in some areas walk 10 miles or more to reach a clean source.

The UN has eight development goals targeted for 2015, seven of which require access to clean water: eradicating extreme hunger, achieving universal primary education, promoting gender equality, reducing child mortality, improving maternal health, combating disease, and ensuring environmental sustainability. The UN estimates that a little over 5 gallons per day satisfies the basic need for clean water to meet all these goals. As such, the UN Development Programme (UNDP) has declared that access to 5.2 gallons (20 liters) of water per day is a universal right and is encouraging governments to achieve this goal (Watkins, 2006). The investment needed to supply clean water in developing nations is approximately $10 billion per year. Every $1 spent on water and sanitation creates an average of $8 of value for the economy and society (Watkins, 2006). Investments in infrastructure to deliver water could have a dramatic effect on poverty.

Access to energy helps address clean water issues. With enough access to economic sources of energy, countries can alleviate water issues. Drilling wells near villages and installing water pumps reduces the time and energy required to source potable water. Pipes and pumps can bring in fresh water and remove sewage. Separating refuse from the incoming water supply reduces disease. Water filters can remove contaminants before use. Unfortunately, water is heavy and is energy intensive to move. There are substantial global resources, but the high cost of processing and transport significantly impedes access. With growing populations and shifting climate and weather patterns, moving water to support populations will only become more important.

Subsidies

An energy subsidy tries to address imbalances or incentivize certain behaviors via financial rewards. Energy subsidies can be a direct cash outlay, a handout of fuel, or tax breaks. Formally, the IEA defines an energy subsidy as "any government action that concerns primarily the energy sector that lowers the cost of energy production, raises the price received by energy producers or lowers the price paid by energy consumers" (IEA, 1999). In essence, any money received from the government for activities related to energy qualifies as a subsidy.

To help address energy poverty in India, the government subsidizes fuel costs to make it more affordable. The price of kerosene is set below market value. The government then reimburses the providers for their losses. In 2011, the total petroleum subsidy in India was over $9.5 billion. India also has major subsidies for agricultural electricity used to support irrigation. Farmers pay only a fraction of the true cost of the electricity used, contributing to wasteful irrigation practices and the country's electricity challenges.

Artificial financial incentives caused by subsidies create distortions in consumption and investment behaviors. Cheap fuel is frequently diverted for unintended uses. For example, in India the purpose of providing kerosene is to displace biomass as a cooking fuel. Kerosene provides cleaner indoor air quality than biomass. However, only 1.3% of rural India uses kerosene for cooking. Since electrification is scarce, the fuel is instead used to provide light from kerosene lamps. The subsidized kerosene is also commonly mixed with diesel to be used as a transportation fuel (IISD, 2012). Kerosene subsidies were meant to help displace dirtier biomass fuel sources for cooking. Instead, kerosene has ended up displacing the cleaner-burning diesel that remains more appropriate for transportation purposes.

These problems are not unique to developing nations. In the US, the federal government administers the Low Income Home Energy Assistance Program (LIHEAP) through the Department of Health and Human Services. It also administers the Weatherization Assistance

Program (WAP) through the Department of Energy. The money is dispersed to states, with larger funds allocated to the colder states that have larger winter heating requirements. Since the system is run by states, the administration of the funds varies greatly. People who qualified for assistance in one state may not qualify in another. The budget allocated to LIHEAP also rises and falls based on the political climate. During the American Recovery and Reinvestment Act (ARRA) funding period, the budget reached over $5 billion. By 2012, the budget was returned to pre-stimulus funding levels of $2.57 billion, nearly 50% less than its peak.

On the production side in the US, subsidies have primarily focused on fossil fuel technologies, although the trend has been shifting. According to the Energy Information Administration, the US government spent over $37.1 billion on energy subsidies, of which $14.7 billion was part of the ARRA. The bulk of the subsidies was allocated to renewable technologies. This is a very recent trend, however, as fossil fuel technologies have received over 70% of federal energy subsidies between 1950 and 2010 (Management Information Services, 2011). Subsidies have kept fossil fuel prices low and renewable prices high. If all energy subsidies were eliminated, the cost of certain renewable electricity sources would be on par with the cost of fossil fuel electricity.

Much work has been done to reform, or phase out, energy subsidies in order to alleviate some of these distortions. The International Monetary Fund (IMF) identified several barriers to reform—namely, concerns for people in poverty, exposing the public to the volatility of energy prices, a lack of understanding of externalities of subsidies, and a lack of ability to administer a change in the program (IMF, 2013). Some suggestions were to depoliticize energy prices, create transparent plans and goals, and improve energy efficiency. Although subsidies were created to help alleviate energy inequality, they have also exacerbated waste and inefficiency, created large financial burdens on governments, and contributed to stifling commercial adoption of cleaner energy technologies (United Nations Environment Programme, 2008).

4.6 Chapter 4 Summary

- Energy is a wicked problem with many stakeholders and contradictory information and is beyond the experience of people.

- Clumsy solutions manage the interactions between business competition, hierarchical processes, and egalitarian access.
- Competitive forces are good at identifying winners but can create unequal distribution of wealth and resources if other interests have inadequate representation.
- Hierarchical processes seek to standardize decisions and approaches for a large population but can create a system of rules that are divorced from desired societal benefits.
- Egalitarian approaches seek equal access for the public good but can unintentionally create externalities and unexpected shifts in behavior.

PART III

Critical decisions

5 | The International Niagara Commission of 1891

Whether electricity is ultimately destined to supersede shafting, pulleys and other gear now commonly used for transmission of energy over short distances is a question which only enthusiasts or those imperfectly acquainted with the technical parts of the subject can be bold enough to answer.

Gisbert Kapp (1886)

In 1891, Edward Dean Adams, President of the Cataract Construction Company, convened a team of experts to pick the best way to harness the mechanical energy of the Niagara River for industrial use. Dubbed the International Niagara Commission, they were to announce the winner of an engineering competition to generate and transmit power from Niagara Falls to Buffalo, New York. The challenges were immense—no one had ever achieved commercial-scale power transmission over such a great distance. Three previous companies had already gone bankrupt attempting this feat. As a well-respected engineer, financier, and scientist, Adams was well suited to overcome these challenges.

Edward Dean Adams, at age 44, already had an illustrious career. Born in Boston in 1846, he graduated from Norwich University with a bachelor's of science. He became a student of the second incoming class at the Massachusetts Institute of Technology. While in school, he worked as a bookkeeper for T.J. Lee & Hill Stock Brokers and eventually became a partner at the exclusive Winslow, Lanier & Company in New York City (Jonnes, 2003). With a strong engineering and finance background, he made a name for himself in reorganizing bankrupted railroads (Carlson, 2006). He was a soft-spoken and well respected investment banker and a right-hand man of J. P. Morgan.

The US was at the tail end of the Gilded Age. The country was experiencing an enormous industrial and economic boom. New power-intensive industries, such as metallic ore refinement, were

93

just beginning to launch. Despite being violently cyclical, railroad development flourished and barons such as John D. Rockefeller, Andrew Carnegie, Jay Gould, and J. P. Morgan were approaching the zenith of power and wealth (Morris, 2005). The telephone had been in existence for nearly 20 years, and by 1889 AT&T became the overall holding company for all of the Bell companies. France had just given up on the Panama Canal after 11 years of hard labor. The United States was ready to jump in. The country was hungry for large projects and eager to prove itself as a world leader in technology and industry.

The rivalries between Thomas Edison, Nikola Tesla, and George Westinghouse are legendary among engineering circles. Edison was a major proponent for direct current (DC) technology, while Tesla and Westinghouse advocated for alternating current (AC) technology. Yet they only represent half the story. From a buyer's viewpoint, each technology had its own features and benefits.

At Niagara Falls, the key problem being addressed was the transmission of *power* over long distances, not the transmission of *energy*. Recall that power was historically measured in horsepower (hp) and is now (usually) measured in kilowatts (kW). Energy was historically measured in joules (J) and is now measured in kilowatt-hours (kWh). Power represents the rate at which work is done, while energy represents the ability to do work. The primary customers at Niagara Falls were factories that required a constant supply of power.

The experts were at odds over which technology was the most feasible, most profitable, and most reliable. Fourteen entries were received. Seven were for electric transmission of power, four were for pneumatics, and the others were for steam and telodynamics (a series of gears, pulleys, and levers). Perhaps unsurprisingly, the best minds of the day could not agree on the best solution.

5.1 The business of generating power

The first use of power at the Niagara River was around 1757–1758 by a Frenchman, Father Louis Hennepin. He diverted part of the river into a small canal that ran parallel to the river and fed back into it several hundred meters downstream. On the canal, he built a waterwheel and a factory that cut logs. At the time, most factories used waterwheels to harness positional and kinetic energy of water diverted from the river for mechanical power.

In 1814, the state of New York auctioned off 110 acres of land next to the river to develop water power for industrial use. Plans called for the construction of a canal that would travel from above the falls through the village of Niagara and discharge some distance below the falls. The first attempt to build this canal was in 1853; it was completed in 1861. However, the outbreak of the Civil War ruined its chances to generate a profit for the investors. By 1882, there were only seven factories consuming hydraulic power—two flour mills, three pulp mills, a silver plating company, and the village water works. All told, these factories consumed roughly 2,725 hp (Holder, 1882).

The manufacturing industry was keen to access the power along the Niagara River. The state of New York was also anxious to attract new power-intense metallurgical industries of the future. The state authorized the development of up to 200,000 effective horsepower from the Niagara River. This was to be a 100-fold increase in power generation along the river and one tenth of the power being generated in the entire US at the time.

Local civic leaders were wary of any development along the waterfront. Each factory would operate its own waterwheel, thereby monopolizing access to the river. This would impact the natural beauty of the region and impede access and use of the river by locals. These early efforts to industrialize also served as the catalyst for the formation of the the State Reservation at Niagara in 1885, which was established for the purpose of protecting the environment.

Thomas Evershed, an engineer with experience on the Erie Canal, tried to address both issues with a new plan. He proposed building 12 parallel underground canals deep inside the bedrock, each with their own turbines (Aiken and Aiken, 1962). Each turbine would generate power for the factories above. This plan had multiple benefits: it would preserve the above-ground scenery by placing the canals underground; it eliminated the need to cluster all of the factories along the waterfront; and it would also allow for nearly 300 factories to access the power generated by the turbines. Evershed's Scheme, as it was known, became the blueprint for the Niagara Falls Power Company.

Even early in the project, it was clear that the economics were simply not working out. Three different ventures went bankrupt trying to drill through the bedrock to build just one canal. Building 12 parallel canals became more and more economically infeasible. Further analysis showed that the revenue to be generated from the factories would not be enough to recoup the initial investments. It became imperative to find additional revenue sources and a new plan.

In order to reduce costs, the leaders of the venture compared the cost of drilling 12 canals to the cost of transmitting power. They concluded that it would be cheaper to dig just one tunnel and transmit power to the entire industrial park. The tunnel would divert water from the top of the river, and the water would be released at the bottom of the falls. All power generation would be consolidated into a central facility. A large, robust transmission system would distribute power to the factories on the surface. This would lower the cost of construction and still meet the cumulative needs of the same number of factories.

To find additional revenue, the Niagara Falls Power Company looked to nearby cities. If the power could reach other population centers, it could access more customers and generate more revenue. The village of Niagara itself was not a viable candidate. Niagara had only 5,000 inhabitants and could not consume the amount of power they expected to generate. On the other hand, the city of Buffalo, at a distance of 18 miles, had a population of 250,000 and was growing rapidly. The previous decade had realized a population increase of 66%, wealth of 89%, marriages of 78%, and births of 123% (Eng, 1890a). Buffalo was a city with flour mills, electric street lamps, and a network of street cars. Demand for power seemed insatiable. If power from the Niagara River could reach Buffalo, the financial viability of the project seemed assured.

This new plan created two problems. First, no one knew how to generate so much power in one facility. The largest power generator of the time could only generate 1,000 hp. Harnessing the full power available from the water diverted at Niagara would require 200 of these generators. Newer and larger generators were required. Second, no one knew how to transmit power over such a vast distance to *reach* Buffalo. Well-established technologies, such as steam or compressed air systems, could reliably transmit large amounts of power for one to two miles at most. Some demonstration projects in the era transmitted small amounts of power up to 30 miles, but never for more than a few days or months. No proven technology could reliably transmit large amounts of power for any considerable distance.

With these two unsolved questions William Birch Rankine went to New York City seeking investment to save the financially struggling Niagara Falls Power Company he had founded. The country was growing rapidly and ready for large and audacious infrastructure projects. The manufacturing industry was hungry for cheap power sources. Rankine met the capitalists of the era to raise the needed money. However, due to the previous failures, J. P. Morgan, along with John Astor, William Vanderbilt, and others, would only invest in the struggling company still requiring a technological breakthrough if Edward Dean Adams took over. Rankine agreed, and thus the Cataract Construction Company was born.

5.2 The International Niagara Commission

Within months of accepting his new position, Adams convened an international team of experts for advice. The team, known as the International Niagara Commission, was tasked with judging an engineering design competition that would determine how to best harness the power of the Niagara River. They were to travel the world and observe state-of-the-art techniques of generating and transmitting power. The commission, headed by Sir William Thompson (Lord Kelvin) of England, represented a broad swath of international interests and scientific expertise. Two members were from the United Kingdom, one from France, one from Switzerland, and one from the US. Their experiences ranged from electricity to compressed air to hydraulics. The commission's members consisted of:

- Sir William Thompson (Lord Kelvin), *President of the Commission, The University, Glasgow.* Sir Williams' work on the conservation of energy and thermodynamics would eventually elevate him to the House of Lords. He was an early proponent of electric power distribution and recommended that the Niagara project pursue direct current for primary transmission.
- Coleman Sellers, *Professor of Engineering Practice, Stevens Institute of Technology, Hoboken, New Jersey.* Professor Sellers was an expert at manufacturing. He had designed large turbines and generators as Chief Engineer for several manufacturing facilities. He advocated for alternating current electric power.
- Éleuthère Mascart, *Professor at the College of France and Director of Le Bureau Central Météorologique.* Professor Mascart was actively involved with the development of the compressed air system in the city of Paris. He was familiar with the challenges of large-scale city-wide power distribution systems.

Table 5.1 *Prizes for the Niagara Competition*

Prize	First place	Second place
Hydraulic projects to generate power by turbines or water motors	£200	£150
Projects for the transmission and distribution of power	£200	£150
Projects combining both development of power and transmission of power	£600	£500

- Theodore Turrettini, *President of the City of Geneva.* As a hydraulics engineer, Turrettini was best known for the development of water power of the Rhône River and the control of Lake Leman. He had the most direct experience in working with large-scale hydraulic projects.
- William Unwin, *Secretary of the Commission and Professor of Engineering at the Central Institute of the City Guilds of London.* Prof. Unwin was a leading expert in the testing of engines.

On June 25, 1890, the Commission sent invitations to leading engineering firms across the globe to submit plans for harnessing the power at Niagara Falls. In the invitation letter, contestants were invited to submit projects for the development, transmission and distribution of about 125,000 effective horse-power on the shafts of water motors at the Falls of Niagara. (Adams, 1890).

The Commission was authorized to award prizes in two parts: how to generate power from the waterway and how to transmit power over 18 miles. Both prizes were for £200 apiece. There was also a grand prize of £600 for the best proposal that combined both the generation and transmission components, as shown in Table 5.1. The contestants had to take into consideration the economics of the system, the scalability to build additional plants, the ongoing maintenance, and the amount of horsepower transmitted to the city of Buffalo.

In judging the entries, the Commission agreed on some points of primary importance. There was a general consensus that electricity would be the best technology for transmission. They were open to a hybrid electric/mechanical system, if one were to be proposed. Although not unanimous, they preferred DC over AC solutions

(Adams, 1927). The judges were further concerned with the safety of above-ground transmission lines. Electricity was new, and the safety of the technology was untested. It didn't help that both Edison and Westinghouse tried to demonstrate the danger of their rival's technology and the safety of their own. Edison, a proponent of DC power, went as far as to publicly electrocute an elephant using high-voltage AC power (Long, 2008). He tried to coin the word *westinghoused* as meaning *electrocuted*. Out of concern for public safety, the Commission advocated for underground electric transmission lines to Buffalo.

5.3 Generation and transmission technologies of 1891

Many technologies existed and flourished in developing and trans-mitting power within city limits. The Commission traveled to cities around the world, touring state-of-the-art power plants and transmission systems. This gave them a first-hand impression of the advantages and disadvantages associated with each technology.

Paris had just installed a city-wide compressed-air power system that replaced the need for centralized steam. Steam boilers were the primary power source for many factories of the era, but steam facilities generally needed a team of experts to operate. If run poorly, the boilers could explode, causing massive damage and killing or maiming workers. Compressed air could be directly substituted for steam power without any modifications to many types of machinery. If power from Niagara Falls arrived in Buffalo in the form of compressed air, no changes would be necessary to much of the existing infrastructure. Compressed air could easily be adapted to lift elevators in hotels, run street cars, and operate day-to-day factory equipment. The technology had its supporters, including George Westinghouse. In 1889, he recommended compressed air as the best mode of power transmission. After all, his first venture, the Westinghouse Air Brake Company, manufactured a brake for railcars that worked on pneumatics. However, long-distance compressed-air power transmission had yet to succeed. Most of the companies that tried failed to make any money (Adams, 1927).

Swiss engineers were pioneers of telodynamic technology, or the distribution of power by means of wire rope and pulleys. When they harnessed the power of the Rhine River, electricity was merely a theory at the time (Eng, 1890b). For Switzerland, telodynamic power was often the best technology available. The International Commission toured several cities that utilized this system. The primary objection to the technology was the inconsistency of power, especially during the cold winter months. However, in Belgrade, where the trans-mission distance was the greatest, there were no complaints about the technology (Eng, 1890c). The commission noted that the technology was practically limited to about 3 miles of use (Adams, 1927).

Electricity, the newest technology, held the record for being transmitted the farthest. Two years earlier, a 14-mile-long DC transmission line became operational in Oregon City Falls, Oregon. The Pearl Street Station, the world's first central power plant, built by Thomas Edison, had also been in existence for 9 years. It supplied power to downtown Manhattan to run lightbulbs and other small electric appliances. By 1891, there were approximately 1500 electric power plants in operation, 300 of which were using AC power. However, state-of-the-art electric generation technology could only create a miniscule amount of power. Edison's Pearl Street Station supplied 175 hp, a mere fraction of the required power at Niagara. Furthermore, reliable performance over large distances was still questionable. The Oregon City Falls line was destroyed during a winter snowstorm in 1890, after having operated for less than a year.

The chief advantage of AC technology was the lower cost to transmit power. With all of the technologies, the longer the distance, the larger the loss of power. High-voltage electric lines seemed the most promising but were also dangerous to workers.[1] In the ideal scenario, humans would interact with equipment operating at a safe, low voltage, while long-distance transmission wires would operate at efficient, high voltages. A voltage converter would be used to step up and down between voltages. Technologies existed to step AC voltage up and down as needed. Stepping DC voltage was complex and expensive. Existing DC systems only operated at low voltage, limiting their practical use to about 1 square mile around the power station.

The advantage of DC technology was the lower cost to generate power. Coal-fired DC power plants only required 10 lbs of fuel to generate 1 kWh, while AC power plants required 20–25 lbs of coal to

---

[1] Although Edison had claimed that AC power killed the elephant, it was actually high voltage that electrocuted the elephant.

generate the same amount of energy (Anon., 1890). DC power could be easily stored in batteries, and generators could be ramped up or down based on demand. AC machinery could never be shut down during periods of light load, no matter how small the load (Parsons, 1939). At the time, AC generator technology was only piloted at any scale with experimental Westinghouse generators in 1890. Proving that AC power could be used to generate steady power and would not damage end-user equipment was critical to adoption.

Frank J. Sprague, developer of the electric street car and inventor of the DC motor, summarized the sentiment of the day, stating that:

The whole question seems to me to be solved by a comparison, where long distances are used, between the two systems [namely, DC and AC], and in this case the alternating current distribution unquestionably has the advantage.

However, he was not optimistic about the Niagara Falls Power Company's prospects:

I do not think the problem to transmit power by electricity from Niagara Falls to several points at varied distances up to 20 miles, a sound one, commercially. (Foran, n.d.)

5.4 Received proposals

As the commission began reviewing proposals, only one thing was certain—no one agreed on how best to generate 125,000 hp and transmit it a distance of 18 miles. The commission received 17 proposals from companies all over the world. Of these, three were deemed incomplete and not judged. Of the remaining 14, seven were for electricity; four were for compressed air; one was for hydraulic transmission; one was for telodynamic power; and one used a combination of electricity, air, and water. Of the electric proposals, five were for DC power and two were for AC power. A summary of the proposals is given in Table 5.2.

5.5 No award

It was unsurprising that the commission declined to award the grand prize, declaring that none of the plans successfully solved the problem. Second place went to Cuenod Sautter & Co. of Switzerland for their

Table 5.2 *Entrants in the Niagara competition*

Entrant	Country	Plans
Generation and transmission		
Cuenod Sautter & Co.	Geneva, Switzerland	Complete plans for generation and electrical (DC) distribution to Buffalo.
Prof. Vigreux and M. Leon Levy	Paris, France	Complete plans for generation and aerial electric conductors (DC).
M. Hillairet and M. Bouvier	Paris, France	Complete plans for generation and electrical (DC) distribution.
Mr. G. F. Deacon and Messrs. Siemens Brothers	London, UK	Complete project for generation and electrical (DC) distribution.
Messrs. Ganz & Co.	Budapest, Hungary	Project for generation and electrical (AC) distribution.
Prof. Riedler and M. Victor Popp	Berlin, Germany, and Paris, France	Plans for generator machines and transmission via compressed air.
Mr. H. D. Pearstall	Orpington, UK	Utilization of power and distribution by compressed air.
Prof. Lupton and Mr. Sturgeon	Leeds and Chester, England	Complete arrangement for motors and compressed-air transmission.
Messrs. J. J. Rieter & Co.	Winterthur, Switzerland	Pressure turbines and telodynamic or wire rope transmission.

Table 5.2 *(continued)*

Entrant	Country	Plans
Generation only		
Prof. Vigreux and M. Leon Feray	Paris, France	Design for a group of turbines for hydraulic distribution of power.
Messrs. Escher, Wyss & Co.	Zurich, Switzerland	Hydraulic part only of a project with electrical distribution (electrical portion not submitted in time for judging).
Transmission only		
The Pelton Water Wheel Company	San Francisco, CA, USA	Design for a power distributed by pumps, hydraulics, and air compressors.
Prof. G. Forbes	London, UK	Electrical part only for a system of electrical (AC) distribution.
Norwalk Iron Works Company	South Norwalk, CT, USA	Project for distributing power by compressed air.

DC distribution system design. No AC proposals won any second- or third-place awards.

Interestingly, two obvious applicants were absent—Westinghouse and Edison themselves. When asked about it, Westinghouse responded "They [Niagara Falls Power Company] were trying to obtain $100,000 of engineering work for a $3,000 prize" (Foran, n.d.). He went on to say that he would be ready to submit a proposal when the serious work began. Sure enough, when the Catalyst Construction Company sent out the first half of a Request for Bids later that year, they only received three proposals. Two proposals were for AC power from Westinghouse and the Thomas-Houston Electric

Company. One proposal was for DC power from General Edison. No bids were received from any of the original 17 competitors.

By the time the second half of the Request for Bids was due in 1893, only two proposals were submitted, and both were for AC power. The DC proposal had disappeared. In the intervening years, several changes occurred. First was the Columbia Exhibition in Chicago. Westinghouse underbid all other vendors to build the US exhibition hall, where he illuminated the entire building with AC power. The brightness of the lights impressed all who entered the hall, cementing AC's reputation as a viable option. Second, an AC electric line in Germany between Lauffen and Frankfurt successfully transmitted electricity for over 100 miles. This demonstrated the great distances that AC power transmission could attain and reduced the risk of adoption by Catalyst Construction Company.

Most importantly, the AC market was rapidly growing, while the DC market remained largely stagnant. As the DC market share shrank, Edison lost control of his company. Since he was a huge proponent of DC technology, his company did not hold any patents in the AC industry. To enter the AC market, General Edison sought to acquire a company that could compete against Westinghouse. Thomson-Houston, a leading supplier of electric arc lamps, had a large portfolio of AC patents. The resulting merger formed General Electric, and the new company submitted an AC proposal.

When the Niagara Falls Power Company adopted AC power on May 6, 1893, it was not without protest. Sir William, who was now Lord Kelvin, sent a telegram in March prior to the decision: "I do not believe that alternating current will be found to be the right solution of electric transmission to a distance" and "It would be an awful mistake to choose the alternating current system for Niagara." After the opening of the Niagara Falls station in 1896, he embraced the superiority of alternating current, writing "Very much pleased to see the great success here achieved as a result of courageous undertaking and originality of invention and skillful design and construction" (John Fritz Medal Board, 1926).

5.6 Modern-day lessons

The "War of Current" between AC and DC power is an early example of a standards war. We frequently forget that other options existed as well. The best solution was driven by demand based on how many customers it could serve, when it could serve them, and how far they could be from a generator not by other ancillary technology features. The business need was to transmit power to reach a customer base in Buffalo 18 miles away. At the end of the day, only AC power transmission was able to span the distance in a cost-effective manner.

The story also highlights real-world differences between power and energy needs and delivery solutions. Power came from spinning water-wheels and was used to drive factories. Energy came from fireplaces and provided warmth. The two products were different. At Niagara, the engineers were concerned with the generation and transmission of power. The initial design was to transmit the power with water, building twelve parallel canals so that factories could each build their own waterwheel. When they decided to consolidate the waterways into one underground tunnel, they had to find alternative means to transmit the power. Fortunately, electricity is good at transmitting both.

Adams hoped that convening a team of worldwide experts would alleviate some of the technology risks. Yet the commission didn't help clear the muddled picture. Their fears of AC were unfounded, and so was their recommendation to bury transmission lines underground. This highlights the risky nature of making long-term infrastructure decisions with minimal information. It took several years of demonstration pilots to prove that AC was the proper solution before Niagara Falls adopted it. Even then, the company still hedged its bets. Adams only approved AC power for the transmission portion of the project connecting the generators to Buffalo. They planned to build a compressed-air system for the village of Niagara Falls, as well as a DC plant. However, after the successful completion of the Buffalo transmission line, the other plans were abandoned in favor of AC power transmission. Making the commitment to build the infrastructure required significant capital; and selecting the wrong technology would almost certainly have bankrupted the company.

Utilities are currently faced with managing long-term risk with the evolution of customer demands and technologies. On the demand side, today's utilities face a deluge of smart cars, electronic devices, and information technologies full of uncertain promise. On the generation side, they face intermittent renewable resources and growth in distributed generation. The electric grid was not designed for either of these challenges. The utilities know that they need to evolve, yet are hesitant to make changes for the same reasons that Adams was circumspect in adopting AC power. Any decision they make has the potential to last many decades. Committing capital to infrastructure is significant, they cannot afford to make the wrong choice, and their relationships with regulators and the public interest make experimentation and innovation difficult at best.

Utilities prepare for a "long tomorrow" where their decisions can impact future generations. Although Niagara set the standard for AC power transmission, there were still over 1,500 DC power plants in the country. Certain electronic devices, such as elevators, plugs, etc. were designed to run on DC power. This required DC power plants to continue operations until the devices were phased out. It was not until 2007 that Consolidated Edison (ConEd) shut down the last DC power plant in the US. Decisions made by utilities can have long tails, affecting operations far into the future.

There are still debates over AC and DC, but the context is now different. Over a century ago, the Niagara Falls system was designed to operate at 2,000 V—high voltage for its day. Long distance was considered over 15 miles. Under these constraints, AC has technical advantages over DC. Today, the bulk power system operates transmission lines at over 600,000 V and regularly transmits power over 1,000 miles. Under these design constraints, DC has technical advantages over AC. As system demands change over time, DC and AC technologies will play different roles where they enjoy comparative advantages.

This is also a cautionary tale about generation-only solutions. Upon completion, the Niagara Falls Power Plant generated 10% of all power in the United States. Prior to the construction of the transmission system, experts predicted that the project could solve the country's energy problem. Leading experts claimed that Niagara Falls could send power to New York and Chicago and light up the entire Eastern Seaboard. Instead, the access to affordable power propelled the nascent electrochemical industry to relocate to the vicinity. Within several decades, steel, aluminum, chemical, graphite, and other factories moved to the Niagara Falls area. Very little power actually left the region. This goes to show that the canonical *energy problem* cannot be solved with exclusively generation- or supply-side approaches. As energy becomes more abundant, society will find additional ways to consume the resource. Real energy "solutions" require balanced approaches that consider both demand- and supply-side issues.

As the grid evolved, centralized power met its goal of making energy affordable to all who demand it. Power has only become cheaper over time. Niagara Falls provided power to Buffalo at approximately $12.00 per hp. Adjusting for inflation and converted to the more familiar kW, this is equivalent to paying nearly $320 per kW for capacity, an astronomical number. As a comparison, the demand charge for Puget Sound Energy is approximately $10 per kW or 32 times cheaper than it was 100 years ago. As the grid developed, providing access to affordable power was a chief concern. This has influenced utility business models, government regulatory structures, and many capital investment decisions in the past century. The electric infrastructure was designed around service affordability for a high-capital-cost, low-marginal-cost industry and successfully reached its engineering goal.

Edward Dean Adams was well recognized for his achievements. He was awarded the John Fritz Medal in 1926, the highest honor among the four senior US national engineering societies. In the invitation letter, the medal cited "achievements as Engineer, Financier, Scientist, whose vision, courage and industry made possible the birth at Niagara Falls of hydro-electric power" (John Fritz Medal Board, 1926). He joined the ranks of previous winners, including Lord Kelvin, George Westinghouse, Alexander Graham Bell, Thomas Edison, Alfred Nobel, and Orville Wright. The preface to Adams' book, *The History of the Niagara Falls Power Company*, even states (Adams, 1927):

This history of the Niagara Falls Power Company is the story of the development of the pioneer hydro-electric system, forerunner of modern utility power service. It records the great step in the transition from the century of mechanical power to the century of electrical power. [...] But we can scarcely appreciate the meagerness of the experience upon which so many important decisions were based and the seriousness of the problems which were yet unsolved.

5.7 Chapter 5 Summary

- In the 1890s, people were primarily trying to determine how best to generate and transmit power, and less concerned with large overall transfers of energy.

- Electric transmission of power was considered a new and unproven technology. Proven and functional technologies included compressed air (Paris) and telodynamics (Zurich).
- Capital investments in infrastructure have long tails, sometimes affecting operations for multiple generations. The gravity of technology decisions in infrastructure, the lack of incentives for adequate experimentation, and the relationships with regulators all contribute to cautious decision-making.
- Energy problems cannot be solved exclusively with supply-side approaches. Energy systems require balanced approaches to address demand- and supply-side requirements.
- AC and DC power both have important roles in modern electric power systems due to different performance characteristics that yield comparative advantages for particular applications.

6 | *Trash as treasure*

American cities are like badger holes, ringed with trash—all of them—surrounded by piles of wrecked and rusting automobiles, and almost smothered in rubbish. Everything we use comes in boxes, cartons, bins, the so-called packaging we love so much. The mountain of things we throw away is much greater than the things we use.

John Steinbeck, *Travels with Charley: In Search of America*, 1962

The Italian city of Naples is well known for its waste management crisis that began in the mid-1990s and continues to re-erupt every few years. During the initial crisis, most landfills were already over capacity, and municipal workers refused to collect trash. The garbage piled up on the streets and caused rampant discontent among the citizens. Even the Italian prime minister, Silvio Berlusconi, became embroiled in this municipal issue. Disruptions to waste management processes can occur because of human action or because of dependence on energy infrastructure. In a number of places, waste management processes also feed back into energy infrastructure via various types of electric generation. Waste management and energy infrastructures, especially electricity, are examples of commonly co-dependent systems.

Most of what we buy—from pens to computers to dishwashers—is eventually thrown away. This trash gathers in large landfills where the embodied energy of the goods is not put to use. Proper waste management is a critical component of any society's well-being. A well-run waste management system keeps a city clean—crucial to dense environments remaining both pleasant and hygienic. Waste management is also an important part of an end-to-end energy strategy. Yet, removing trash from a city merely shifts the problem to someplace else. Storing trash in landfills outside of a city removes the problem from dense population centers but leaves toxic chemicals behind that can still leach into potable water supplies if a liner fails.

109

As cities and suburbia expand, areas that were previously landfills may become future building sites.

On human timescales, landfills last indefinitely. Modern archaeologists frequently use rubbish pits from ancient civilizations to understand everything from diet to social habits to economic structure. Burying our trash today often only postpones disposal problems for future generations. Landfill sites also require maintenance, even when a landfill is closed or not accepting new trash.

A better trash management system fully utilizes the energy and material content of the waste. From an energy perspective, significant embodied energy goes into manufacturing products. In the US, we rely on large numbers of trucks to transport products to stores, our homes and businesses, and ultimately to landfills. The embodied energy required to create the product does not disappear simply because it is out of sight. From a materials perspective, waste is merely another resource to manage. The lifecycle of waste is just like a familiar product supply chain with manufacturing, transportation, and consumption. Waste goes through a process of creation, collection, and treatment. If managed properly, the energy and materials of waste can be further utilized to the benefit of society.

Germany has successfully created a waste management system that has successfully diverted 95% of waste away from landfills. By adopting a comprehensive strategy of avoidance, minimization, reuse, recycling, power generation from incineration and, finally, landfilling, the country significantly increased net societal benefits. The Germans simultaneously increased recycling, developed a low-cost source of electricity, and created many synergistic benefits to better manage their resources. This chapter demonstrates how management through dedicated priorities can help create systemic benefits, even in complex systems.

6.1 The creation of municipal solid waste

Most products are thrown away at the end of their useful lifecycle. Studies have indicated that a whopping 90% of everything purchased becomes trash at some point in the future (Wall, 2007). In 2008 389.5 million tons of waste were generated in the US—or nearly 1 ton per person per year. The US waste industry, valued at approximately $55 billion is responsible for managing this resource (Waste Market

Overview and Outlook, 2012). As the population increases, the rate and volume of generated waste will only continue to grow.

The US federal government classifies waste as hazardous or non-hazardous. Hazardous waste disposal is strictly regulated by the US Environmental Protection Agency (EPA) to mitigate environmental impacts. Non-hazardous waste can be disposed of by any available means, with fewer restrictions. Non-hazardous waste is classified as industrial waste or municipal solid waste (MSW). Industrial waste tends to be large quantities of scrap materials that are relatively uniform in material content. For instance, a furniture shop may throw away many tons of wood chips per day. These can be repurposed as mulch, wood pellets, etc. MSW is what we commonly refer to as trash. It is a mixture of metals, plastics, paper, food scraps, and a variety of other materials. MSW requires sorting before treatment. Selecting appropriate management techniques for waste depends on both the quantity and quality of the waste generated.

The total quantity of waste is strongly correlated with income level. Countries with the highest incomes generate nearly half of the global waste. These countries have consumer-driven economies where people can afford to buy goods and products. As a result, wealthier individuals also generate more waste per person on average. As world wealth is projected to grow, waste generation is projected to grow accordingly. Current estimates forecast that waste generation will grow by 42% by 2025 (Hoornweg and Bhada-Tata, 2012).

The quality of waste depends on the economic value of the material being thrown away. Scrap steel is more valuable than a banana peel. The material composition of waste is also dependent on the income level of a country. Materials which are particularly scarce, involve energy- or process-intensive treatments to produce (e.g. metals), or are easily recycled into useful materials are valued at a premium.

The composition of waste is different between high-income and low-income countries, as shown in Table 6.1. Organic food waste averages 64% of the trash in countries with low GDP. Manufactured goods—paper, plastics, glass, and metals—average a mere 19% of the total trash composition. On the other hand, high-GDP countries' trash consists of 55% manufactured goods and only 28% organics on average. Although the percentage of organic waste is lower, people in high-income countries still generate three times more organic waste than people in low-income countries, since they throw away far more

Table 6.1 *Typical waste composition based on GDP*

Waste	Low-income country	High-income country
Organic	64%	28%
Paper	5%	31%
Plastic	8%	11%
Glass	3%	7%
Metal	3%	6%
Other	1%	17%

Source: Hoornweg et al., 2012

stuff. This means that developed countries throw away 28 times the paper goods, 6 times the plastics, 11 times the glass, and 9 times the metal of a developing country (Hoornweg and Bhada-Tata, 2012).

The most effective way to avoid landfilling is to throw away fewer goods. Disposable products or single-use products pile up quickly. Products that are more durable can be reused many times before ending up as trash. Another strategy is to purchase second-hand products, such as used textbooks, equipment, or vehicles. This delays the product's entrance into the waste stream. A third is to reduce the packaging content that is thrown away immediately after purchase. Companies like Amazon have made significant strides towards minimalist packaging.

Regional variations in materials, products, and end uses of goods require that strategies for managing waste be tailored to the local conditions. They must account for both the quantity and the type of waste generated. Strategies for managing manufactured goods, such as recycling, will be different than strategies for managing food waste. The amount of waste feedstock generated will dictate the economics of any solution.

6.2 The collection of rubbish

Garbage trucks are the predominant means of transporting trash in the United States. There are over 179,000 garbage trucks, each traveling over 25,000 miles per year. On average, these trucks obtain less than three miles per gallon. Most trucks consume diesel which generates particulate matter and also contributes to traffic congestion (Cannon, 2006). The process of rubbish collection alone is energy intensive.

Some waste transport systems are ideal for the homeowner, and others are ideal for treatment facilities. *Single-stream* collection requires one bin and one truck. This system is optimal for the consumer. Homeowners throw trash into a bin, and a truck transports it to a separation center. The sorted trash is sent for treatment—recyclables are sent to a recycling facility, food is sent to a composting facility, etc. However, recycling and composting facilities prefer their trash to arrive uncontaminated by other trash. *Multi-stream* collection requires the homeowner to sort their trash into multiple bins for multiple trucks. This increases the hassle for the homeowner but results in higher-valued products after treatment. At the end of the day, the transportation mechanism is merely a middleman. It is either optimized for the homeowner or optimized for the treatment facility. The appropriateness of single- or multi-stream collection is often linked to the transportation costs, which correlate with population density.

Minimizing the environmental impact of garbage trucks revolves around two factors: the distance traveled and the fuel consumed. One popular solution is the municipal garbage can with a built-in compactor. In some areas, these are commonly powered by solar energy. A compactor minimizes the volume of refuse by compressing the trash within the receptacle. A single truck can fit more trash per trip, reducing the number of miles driven and improving overall efficiency. These compactors can also send a signal to the garbage collector when they are full, thus dispatching a truck only when the compactor needs to be emptied. An alternative approach is to optimize garbage truck routes with scheduling and mapping software. Optimization tools have reduced the size of fleets needed to serve an area. Other approaches to mitigating environmental impact include efforts to reduce emissions and improve truck efficiency by converting collection vehicles to natural gas from diesel.

6.3 The treatment of trash

In general, there are four approaches to the treatment of waste. We can *reuse* the material through recycling. We can *transform* the material into another product, such as through composting. We can

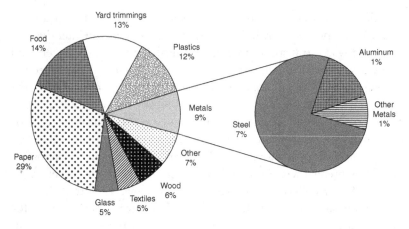

Figure 6.1 Waste composition in the United States
Source: EPA, 2011

extract the embodied energy content to power the electric grid through waste-to-energy techniques. Finally, we can *remove and forget* the material in an incinerator or landfill.

Each solution depends on the source material and chemical makeup of the resource. As waste is composed of many different materials, no single, monolithic solution will meet the needs of the entire waste stream. Figure 6.1 depicts the composition of the US waste stream (EPA, 2011). Recycling is a good strategy for paper (29% of MSW), plastics (12%), metals (9%), and glass (5%). Composting and fermentation are ideal for food and yard trimmings (27%). Waste-to-energy is ideal for materials that, for whatever reason, cannot be recycled.

Recycling

Recycling is perhaps the most well-known alternative to landfilling. It breaks down the specific material and reuses it for future generations of product. Materials that can be recycled include metals, glass, paper, and plastics. Since the chemistry of the materials is different, the methods of processing and the number of times they can be recycled vary significantly.

Virtually all metals can be recycled an infinite number of times. Recycling of copper, aluminum, and steel is very common and creates a valuable end product. Scrap metals are harvested, melted into a liquid state, and recrystallized into new metal parts. Most scrap metals originate from construction and industrial waste. Recycling of aluminum cans is probably the most well-known recycled metallic consumer good.

The prices of common metals have been rising recently, contributing to an increase in their scrap value and an increase in recycling. According to the American Metal Market, the value of scrap steel quadrupled from $100 per ton to $400 per ton between the years 2000 and 2010. Other metals saw similar increases in prices. This has created an underground market for metal theft. It is not unheard of for thieves to rip up train tracks, bridges, copper plumbing, or electric wires to sell the metal for scrap. This has delayed traffic, caused power outages, and damaged numerous facilities, resulting in many millions of dollars in damages. Some thefts have gained media attention for their notorious and audacious nature. In 2008, in Philadelphia, over 600 iron manhole covers were stolen, resulting in several injuries when people fell into the hole (*Newsweek*, 2008). In 2011, thieves stole a bronze sculpture worth £500,000 by Barbara Hepworth from a park in the United Kingdom (Andrews, 2011). The scrap value of the bronze was merely £750—nothing compared to the item's actual value, but valuable enough to the thieves. Even cemeteries have not been spared. Over 1000 bronze plaques were stolen from a concentration camp memorial just outside of Prague, Czech Republic (Spiegel Online, 2008).

Glass is also infinitely recyclable as long as it is sorted. Removal of non-glass, such as metals, and non-recyclable glass, such as Pyrex and some tempered window panes, is essential to the process. The recyclable glass is then crushed into small pellets, referred to as cullet. The cullet needs to be separated based on color prior to processing. Otherwise, colored glass can contaminate clear glass, much like mixing colors of paint.

Paper can be recycled approximately five to seven times before it must be discarded. The quality of paper depends on the length of wood fiber in the starting material. The highest-quality paper is made from dense hardwoods that have longer fiber lengths. Short fibers, such as bagasse (sugarcane agri-waste), are useful to make toilet paper and other non-durable soft papers. The lengths of the fibers become shorter and shorter with each generation of recycling. This process reduces the grade of the paper since it involves shredding, pounding, and soaking to create the pulp. As the fibers become shorter, they become suitable for uses as cardboard or paperboard. Most recycled paper is a mixture of new and recycled fibers in order to reinforce the paper's strength.

Plastics can typically be recycled once. Plastics originate as long-chained polymers made from petroleum byproducts. The recycling process degrades the long chains into short ones through a chemical process and heat. Short-chained polymers cannot be made long again. Polymers with chains too short to be recycled are discarded. For instance, a common long-chained polymer product is a plastic bottle. It can be recycled into a short-chained polymer product, such as a fleece jacket. The fleece jacket cannot be re-recycled back into a plastic bottle.

Even when specific recycled goods are possible to make in a cost-effective manner, recycling is sometimes secondary to other concerns. Products that come into contact with food have very stringent health-related constraints. US Food and Drug Administration (FDA) approved, food-grade packaging needs to be free of biological and toxic contaminants. Recycled metals and glass do not pose a problem. They are processed at high heat, killing any biological contaminants. Recycled paper and plastics, on the other hand, have the potential to be contaminated. They are generally processed at lower temperatures and use toxic chemicals to break down the fibers. The FDA requires testing of recycled paper and plastics to ensure that they are safe for consumers (FDA, 2006).

A robust recycling program is an essential component of a sustainable energy system. By reducing the energy required to produce goods, recycling can be a powerful contributor to reducing overall energy consumption. The benefit of recycling extends beyond the material transformation. Recycling metal is more energy efficient than extracting the metal from mines. As shown in Table 6.2, it requires 95% less energy and 97% less water to recycle aluminum than to refine it from raw ore (Robinson, 1986). Perhaps more significantly, the energy fuel source can be different. When making virgin steel, one needs to extract iron from ore. The process requires specific chemical reactions and temperatures so high that metallurgical coal, transformed into coke, is used. Recycled steel can be produced by melting scrap metal using electric arc furnaces that use electricity as their energy source.

Table 6.2 *Reduction in energy consumption from recycling for select materials*

Reduction	Aluminum	Steel	Paper	Glass
Energy use	95%	60%	50%	20%
Air pollution	95%	85%	74%	20%
Water pollution	97%	76%	35%	
Water use		49%	58%	50%

Source: Robinson, 1986

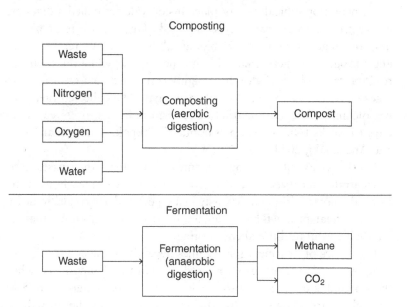

Figure 6.2 Process flow of composting and fermentation

Treatment of organic waste

Organic waste, such as food scraps and yard trimmings, can be composted or fermented. These are biological processes that create valuable byproducts that can be sold into other markets. Technically speaking, composting is an aerobic process, while fermentation is an anaerobic process. In essence, composting requires plentiful oxygen and water to create compost. Fermentation requires the absence of oxygen to create methane. Figure 6.2 is a schematic that shows the difference.

Composting involves combining organic waste with biological material. Food scraps and yard waste are combined with microbes, fungi, enzymes, and even algae. With a little heat to start the process, the biologics react with water, oxygen, and nitrogen to speed up the decomposition of the organic material. If done properly, the heat generated by a compost heap can be just enough to sustain a continuous reaction. Too little heat will not create compost, while too much heat may cause the heap to spontaneously combust. The ideal compost heap will reach temperatures of 40–50°C (104–122 °F) during the reaction and take approximately six months (Trautmann, 1996).

Fermentation typically takes place inside a device called a digester, where gases can be carefully controlled. Organic waste is placed into the digester along with biological material. Water, oxygen, and nitrogen are then removed from the digester while the chemical reaction takes place. Under these highly controlled environments, the digester creates methane, alcohol, or biogas. The collected fuels can be sold into the energy market. The fermentation process can take days to weeks (Nipon Pisutpaisal and Sirisukpoka, 2014; Gang Luo and Angelidaki, 2011).

Another way of utilizing organics is to turn the waste into concentrated fertilizer instead of compost. Aerobic microbes can be carefully mixed with organic waste in a digester. Early results indicate that the final product has a higher nutrient concentration than is attainable from traditional composting.

The choice of treatment for organics depends on the end product—compost, methane, or fertilizer—and the end use. Compost enriches the soil and is used as a topsoil. Prices run approximately $6 to $60 per cubic yard (McEntee, 2005). Fertilizer enhances the growth of plants primarily by providing nutrients—nitrates, phosophates, and potassium. Pricing of fertilizer is based on composition of nutrients and is traded on a commodity market. For instance, phosphate prices varied from $353 to $505 per metric ton between August, 2013 and 2014 (Ind, 2014). The market size is drastically different between fertilizer and compost. In 2011, US farmers purchased over 60 million short tons of fertilizer and a mere 94,000 short tons of compost (USDA, 2013).

Waste-to-energy

Waste-to-energy is the controlled burning of trash to harvest the embodied chemical energy. A waste-to-energy plant is similar to most

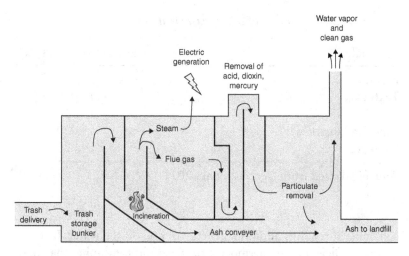

Figure 6.3 Anatomy of a waste-to-energy plant

thermal electric generators (adapted from Ecomaine, 2011). As shown in Figure 6.3, the plant takes the trash from the receiving bunker and transports that trash to the incinerator. After the burning process, the trash is separated into three components—namely, the ash, flue gas, and steam. The steam is utilized to generate electricity and can also supply heat if it is a co-generation facility. The flue gas is captured and processed several times to remove heavy metals, such as mercury. It is also treated to remove acid gas and other particulates before being released into the atmosphere. The last element, bottom ash, is chemically inert and is sorted for metals that do not burn. This metal, which makes up 6% to 10% of the bottom ash, is then sent to recycling facilities. The leftover residual ash is collected and used for bulk material, commonly finding its way into concrete, asphalt, or as part of bottom and top caps of landfills (CEW, 2011).

In a state-of-the-art facility, the feedstock and residuals are carefully controlled. To generate electricity effectively, the incineration temperature needs to be relatively high. Fortunately, the high temperatures break down most toxins and dioxins. Different kinds of feedstock also have different energy content, making some more suitable for combustion, as seen in Table 6.3. Plastics are petroleum-based products and are ideal for combustion at the end of their recyclable life. Textiles, wood, and paper are the next best feedstock for combustion. Organic biomass and food scraps have high water

Table 6.3 *Energy content of select waste materials*

Material	Energy content (kJ/kg)
Plastics	30,481
Textiles	15,020
Wood	13,250
Paper and board	13,020
Biologics	4,620

Source: adapted from Dehoust et al., 2011

content that requires additional energy to boil off the water. It results in lower efficiency for thermoelectric generation, meaning that these materials should generally be removed from the process (Dehoust et al., 2011).

Waste-to-energy conversion has its share of detractors. The burning of plastic and synthetic rubber can emit carbon dioxide, nitrous oxides, mercury, dioxins, and other particulates. Without proper controls, these may be released into the atmosphere. Large amounts of water are needed to run the factories, and the discharged water can pollute nearby water systems. The bottom ash and other byproducts require proper disposal, and the plant may have odor problems if organics are not handled well. These problems are typically more prevalent in older plants.

Waste-to-energy proponents argue that extracting energy from waste material is preferable to burying the embodied energy in a landfill. The burning process reduces the volume of trash by 90%, minimizing the space needed for storage (EPA, 2014). The volume of metals extracted from the bottom ash is a valuable commodity. In 2010, 12 European Union (EU) countries produced approximately 12.2 million metric tonnes of bottom ash through waste-to-energy. This included approximately 1.2 million metric tonnes of metals that were returned to recycling facilities. The amount of metals recovered was equivalent to over 16 Golden Gate Bridges![1]

[1] Metal weight of the Golden Gate Bridge: 75,000 metric tonnes (GGBHTD, 2014).

The electricity generated by waste-to-energy plants is considered baseload, as it is dispatchable and coupled to a reliable source of raw fuel. Recall that baseload electricity is typically supplied by coal or nuclear fission plants. Substituting waste for fossil fuels helps reduce the dependence on foreign energy sources. Across the EU, trash has replaced the need to burn 7–40 million barrels of fossil fuels per year (CEW, 2014). In Germany, waste-to-energy plants supply enough energy to meet the needs of over 2 million residences, or approximately 2.4% of the population (Dehoust et al., 2011).

Governments are recognizing the importance of waste-to-energy as a component of both waste and energy policy. In the Energy Policy Act of 2005, the US federal government classified waste-to-energy as a renewable energy source, enabling it to qualify for renewable energy credits (RECs). The EU considers waste-to-energy as an important component of their carbon reduction goals.

Incineration

At first glance, incinerating trash without energy capture looks comparable to waste-to-energy. At the end of the process, incineration also reduces the volume of trash by 90%. The similarities end there. In incineration, to save on fuel and costs, the trash is burned at low temperatures. Dioxins and toxic chemicals are not reacted completely at low temperatures and are found in the flue gas. These known carcinogens are then released into the atmosphere. The system never reaches a critical temperature where the fire is self-sustaining. Incineration plants must continuously consume additional fuel to keep the system going (Hoornweg and Bhada-Tata, 2012). The feedstock is not sorted for recyclables or organics that could be better used in other processes. The end product is a very low-quality ash that can still have harmful and toxic residuals and needs to be disposed of in a landfill. In essence, incineration without energy recovery addresses the volume problem but creates many other issues.

Landfill

Landfilling is the default US solution to waste management, primarily because land is cheap and plentiful. It essentially involves placing waste products inside a large hole in the ground designed to manage three harmful byproducts—noxious odors, methane, and acidic leachate. Landfills are different from dumps, which are holes in the ground with no management of odors, methane, or acidic leachate. These byproducts are formed by different materials in the waste stream. Landfills need continuous maintenance to ensure proper operations (BMU, 2006).

Odors are caused by the decomposition of organic waste. Food scraps and yard waste decompose when exposed to air, water, and microbes. Covering the landfill with a plastic tarp helps keep the air and water out. Every day, a new tarp is laid down over the fresh trash that arrives.

However, when air and water are removed from organics, the trapped organic waste can still undergo fermentation and produce methane. The methane is now trapped under the lining and can build up. If left unmanaged, the methane can ignite and explode. Some landfills see this as an opportunity to produce methane and syngas. They have built methane recovery systems to sell the product as fuel.

Acid leachate is formed when rainwater permeates the landfill body and reacts with plastics, glass, and metals. The resulting leachate slowly drips and collects at the bottom of the landfill. Modern landfills have a collection layer to monitor the leachate concentration and treat it before it can leak into the ground. If it does leak, it has the potential to contaminate the groundwater supply.

6.4 Prioritization and economics

As with other large and complex systems, the diversity of approaches and tradeoffs make it difficult to determine an ideal solution. Comparisons are often challenging, at best. Some people advocate for improving or adopting a single element of waste management, such as plastics recycling. These point solutions are critical but do not describe the whole picture. Others promote an "all of the above" strategy that tries to capture the advantages of all of the solutions. Although this approach seems comprehensive, it often results in insufficient resource allocation to make meaningful progress on specific initiatives and neglects identifying what to pursue first. Practical progress requires clear identification of what actions to take or what investments to make. In fact, advocating for an "all of the above" approach, without prioritization, can be just as meaningless as not doing anything at all.

There are two primary approaches to waste management. In the *consumer pays* model, the economics are driven by consumers who manage the removal of garbage. In the *polluter pays* model, the economics are driven by manufacturers who are responsible for the garbage they generate. Their results differ drastically.

Consumer pays

The US has generally embraced a consumer pays system. The economics of waste are driven by fees set by landfills and charged to garbage trucks. These fees vary based on land availability, labor costs, and environmental regulations in each region. For instance, in Idaho, where land is plentiful, the tipping fee can be as low as $10 per ton. In densely populated Massachusetts, the tipping fee averages $105 per ton (Wright, 2012). Garbage collection companies pass these costs along to consumers in the form of trash collection fees.

When a landfill is full and no longer accepting new trash, it still requires money to maintain. Landfills still need to monitor the groundwater, pay for cleanup, manage methane gas issues, etc. These costs last for many decades, if not centuries, and are rarely included in the tipping fees while the landfill is active. Instead, landfill closure costs are paid by current customers. For instance, in Washington state, although land is fairly plentiful, the tipping fee is $75 per ton, higher than the regional average. This is because over the next 30 years, Washington state is expected to spend approximately $2 million per year to maintain nine closed landfills (Wright, 2012).

The consumer pays strategy has seen uneven improvements to waste handling. Consolidating landfill operations helped keep marginal costs down and tipping fees low. Between 1988 and 2005, the number of landfills dropped from 7,924 to 1,654, while the capacity increased by nearly 2 billion tons (EPA, 2006). This occurred primarily by closing smaller and more inefficient municipal dumps which were consolidated into larger regional landfills. With fewer landfills, garbage trucks have to travel farther for disposal. This increases transportation costs and emissions. The city of Seattle, for instance, ships all of its garbage via train to a landfill in eastern Oregon at a distance of 250 miles (City of Seattle, 2000).

The consumer pays strategy has helped with a decrease in landfilling. Between 1990 and 2010 US landfilling decreased from 70%

PRIORITIES

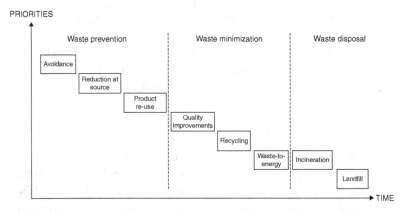

Figure 6.4 Prioritized steps of waste management
Source: Hottenroth et al., 2004

to 54% of MSW. Recycling rates nearly doubled to 26%, and composting quadrupled from 2% to 8% of MSW. The final 12% of MSW was used to generate enough electricity to run approximately 2 million homes. However, during that time, the US population grew by 18%, and the volume of waste grew by 17%. This has still resulted in an increase in volume being sent to landfills

Polluter pays

In 1996, Germany adopted the Closed Substance Cycle and Waste Management Act, which formalized a polluter pays principle of waste management. Under it, waste is considered the responsibility of the industry that created it. Germany's waste strategies are prioritized under three broad stages—*prevention, minimization,* and *disposal.* As shown in Figure 6.4, the order of the stages is designed to maximize the value of a given material before the material reaches the landfill.

Waste prevention, the first strategy, was already practiced. Germany adopted a polluter pays policy in 1991 which required producers to be responsible for the disposal of their packaging material. The EU adopted the strategy in 1994.[2] Waste prevention is managed through a Green Dot (Grüner Punkt) system. Manufacturers

---

[2] European Parliament and Council Directive 94/62/EC on packaging and packaging waste OJ L365/10.

pay into the system based on the weight of their waste. All items marked with the Green Dot logo are recycled and managed separately from the MSW stream. Manufacturers are therefore incentivized to design products that produce less waste in order to lower their payments (Der Grüne Punkt, 2014).

Waste minimization, the second strategy, depends on the material content of the waste. As we have seen, different materials require different treatment techniques. The organic waste, which causes odors in landfills and contamination in recycling, is removed first. It is upcycled—i.e. turned into a more valuable commodity—through composting or fermentation. Next, recyclable material is removed, sorted, and processed. Finally, when a material is at the end of its recyclable life, the energy is extracted via waste-to-energy to generate electricity.

After all prevention and minimization strategies have been pursued, the material is slated for disposal. By the time the material reaches this stage, all usefulness has been extracted. All waste that normally causes problems in landfills has been removed. The organics that cause odors have been composted. The plastics that cause acidic leachate have been recycled or combusted. The only material shipped to a landfill is the chemically inert bottom ash. This material has no long-term maintenance requirements for future generations.

The prioritization of prevention/minimization/disposal methods helped create an order of which technologies to pursue first. It acknowledges that no technology or method solves the problem completely. It also acknowledges that every technology has its place in the ecosystem. The methods are put into an order that minimizes cost while maximizing utilization of the resource. Waste-to-energy as a technology has its place in the value chain as well. It causes more pollution than recycling and reusing but causes less pollution than landfilling. These characteristics make waste-to-energy an appropriate technology to pursue after exhausting recycling efforts and before burying waste in the ground (Hottenroth et al., 2004).

The polluter pays strategy has been wildly successful, as seen from the EPA and Eurostat data provided in Table 6.4. Although the population stayed nearly the same between the mid-1990s and 2010, waste generation decreased by 7%. Initially, critics predicted that waste-to-energy would divert waste away from the recovery techniques of composting and recycling. This ended up being

Table 6.4 *Comparison between polluter pays (Germany) and consumer pays (US) waste management policies*

Germany	1995	2010	% change
MSW generation	50.9 m tons	49.2 m tons	−3%
Composting	13%	17%	22%
Recycling	26%	46%	42%
Waste-to-energy (inc. incineration)	18%	37%	52%
Landfill	43%	0.4%	−10 154%
US	1990	2010	% change
MSW generation	208 m tons	250 m tons	17%
Composting	2%	8.1%	75%
Recycling	14%	26%	46%
Waste-to-energy	14%	12%	−17%
Landfill	70%	54%	−30%

Source: Eurostat, 2014; EPA, 2011

unfounded. Recycling quantities doubled over the same period, while composting saw a healthy increase of 35%. Since waste-to-energy displaced fossil fuel electric generation, the waste sector helped reduce Germany's carbon emissions by 50 million tons of CO_2 (Jaron and Flaschentreher, 2012). Germany now landfills less than 0.5% of its waste, while the US landfills 54%.

The strategy has also been an economic boon. In 2006, it accounted for almost 200,000 jobs with annual revenues of over €50 billion or 2% of GDP (Jaron et al., 2006). It spawned a new sector of materials recovery experts that further reduced Germany's dependence on international raw resources. In addition, the Germans are continuing to push the boundaries of waste management. In 2012, the latest amendment to the Closed Substance Cycle and Waste Management Act came into effect. It now considers waste management as *resource management* with a focus on treating waste as valuable material to be managed (Friederich et al., 2011).

6.5 Chapter 6 Summary

- Focusing on priorities for waste management helps improve comparisons between technologies that all have distinct performance characteristics and externalities.
- The first step of waste management is waste avoidance, or preventing the generation of unnecessary waste.
- The second step is to minimize waste by repurposing or transforming the material into other uses. Processes include recycling and composting.
- The third step is to release the embodied energy of the material into electricity via waste-to-energy plants.
- Finally, the residual waste is responsibly disposed of via landfills.
- The German systematic waste management system has successfully diverted 99% of all trash away from landfills. Germany's efforts have substantially reduced the country's CO_2 emissions.

7 | *Paying for pavement*

Finland has produced so many brilliant distance runners because back home it costs $2.50 a gallon for gas.[1]

Esa Tikkanen, NHL Hockey Player, 1979

The Highway Trust Fund that built the US highway system generates revenue from a federal fuel tax, but is running out of money. According to the Congressional Budget Office, outlays have exceeded revenues by more than $52 billion between 2004 and 2014 (CBO, 2014). The fund is responsible for the construction and maintenance of the highway system in the US. Although the system makes up only 1.1% of the nation's roads, it carries over 24% of the nation's highway traffic and is an important backbone of the country's economy. At the same time transportation makes up nearly 28% of US energy consumption. Infrastructure decisions and land use policy can have a large impact on the energy required to live in a society or a community. This chapter focuses on Washington State, which needs to address many of the same problems faced by the federal government, as a more manageable case study for exploration. In May 2013, the Skagit River Bridge in Washington state was struck by an over-height truck, causing a section of the truss to collapse. Although no one died, it cost over $6 million to repair and caused many more millions of dollars of lost economic output stemming from disruptions to traffic. Bridge collapses are dramatic but far from the only form of roadway failure. Roadway disruptions are common, and much of the existing transportation infrastructure in the US is in dire need of

---

[1] At the time, the average gas price in the US was $0.80 per gallon.

substantial investment. The American Association of Civil Engineers 2013 Report Card for America's Infrastructure gives the nation a D$^+$ and recommends $3.6 trillion dollars in capital investment by 2020.[2] Roads were given a lackluster D rating by the professional society.

Increases in road maintenance costs have plagued Washington and other states for many years. In 2012, Washington faced $10–$50 billion in backlog maintenance and repairs. The state primarily funds road repairs with the gasoline tax. Yet gas tax revenues have steadily declined as vehicles have become more efficient due to stricter fuel efficiency standards. Hybrids and electric vehicles, which are steadily gaining in popularity and consume little to no fuel, further complicate revenue collection for roadway maintenance. The confluence of factors shows how the engineering problem of achieving greater fuel efficiency and the political subject of taxation can be intertwined. This chapter explores the situation faced by Washington state to elucidate some of the more subtle relationships between energy products and systems and other infrastructure.

7.1 Transportation basics

Roads and rail are critically important parts of the economy, with large ongoing maintenance needs and limited options for generating revenue. Priorities around transportation investments are made in our political institutions, but the decisions made there can have just as large an impact on energy consumption as engineering solutions that impact available technologies and actual implementation. If done well, they can encourage good behavior. If implemented poorly, they can reinforce bad habits. As with any system, the ideal solution is not clear in such a rapidly changing environment. The roads of Washington state are under the jurisdiction of the state, counties, and cities. The responsibilities to develop and maintain roads are different based on their form and function. Some roads are under the jurisdiction of the state, others the county or the city. As Table 7.1 shows, different government entities have very different portfolios based on both mileage and usage (Washington State Department of Transportation,

---

[2] Considered categories include aviation, bridges, dams, drinking water, energy, hazardous waste, inland waterways, levees, ports, public parks, rail, roads, schools, solid waste, transit, and wastewater.

Table 7.1 *Washington state roadway jurisdiction and mileage*

Jurisdiction	Centerline miles		Daily vehicle miles (thousands)	
State (total)	7,061.66	8.38%	86,179	43.84%
Interstate	764.27	0.91%	41,937	21.33%
County	39,868.65	47.31%	24,928	12.68%
City	17,696.69	21.00%	41,522	21.12%
Other	18,877.85	22.40%	2,030	1.03%
Total	84,269.12	100%	196,596	100.00%

2009). All told, there are over 84,000 miles of roads in Washington state that see over 196 million vehicle-miles driven daily.

The state maintains major highways, thoroughfares, and the interstate system. The system's importance to state commerce and economy cannot be understated. These roadways are critical to the movement of freight and mobility of personnel. Washington companies rely on their efficient operation to do business and attract cargo into the Seattle–Tacoma port system. Although they are only 8.3% of the roads in the state, state-maintained roads make up over 43% of the daily use.

County roads in Washington have the lowest number of daily vehicle miles traveled but represent the largest system of roads by length. County road networks reach many rural areas. In many cases, these roads are the "last mile" connecting individual communities to the rest of the state. Counties also own and maintain the largest portfolio of bridges, more than 3300. One fifth of them are already structurally deficient or functionally obsolete. Safety on these large, geographically disparate, and lightly traveled roadways is also a problem. Over 40% of all speeding fatalities take place on county roads even though they account for only 13% of vehicle miles traveled.

Approximately 63% of Washington's population resides in cities which generate nearly 90% of the economic output. City roads are more than just the asphalt. Roads also include the pavement for pedestrian traffic, bicycle lanes, and landscaping. Proper storm drains help maintain waterways, prevent flooding, and maintain the environment. Streetlights, where economically justified, offer security and safety after dark. Below ground, roads often have co-located or crossing utility wires, water mains, sewerage, and natural gas pipes, taking advantage of dedicated rights-of-way to reach consumers. The total cost for a complete city street can easily exceed $15 million per mile (WTF, 2011b).

7.2 Roads: a means to an end

Robust transportation networks like roadways are unsung heroes of thriving economies. As of 2012, Washington state has a population of 6.9 million residents, and virtually every resident utilizes the transportation system in some fashion for work or pleasure (U.S. Census Bureau, 2012). Washington state law RCW 47.04.280 defines the transportation system as supporting six policy goals, including economic vitality (Washington State Legislature, n.d.):

1. **Economic vitality**: to promote and develop transportation systems that stimulate, support, and enhance the movement of people and goods to ensure a prosperous economy.
2. **Preservation**: to maintain or preserve and extend the life and utility of prior investments in transportation systems and services.
3. **Safety**: to provide for and improve the safety and security of transportation customers and the transportation system.
4. **Mobility**: to improve the predictable movement of goods and people throughout Washington state.
5. **Environment**: to enhance Washington's quality of life through transportation investments that promote energy conservation, enhance healthy communities, and protect the environment.
6. **Stewardship**: to continuously improve the quality, effectiveness, and efficiency of the transportation system.

The economic impact of roads is described by the multiplier effect. This is an effect when an investment cascades into broader increased economic activity. There are several impacts that can be measured:

- **Direct impact**: investments in the project itself. Examples are construction jobs or transportation operations.
- **Indirect impact**: investment and growth by suppliers. Simple examples include food services, tool suppliers, and clothing shops that sell to construction workers.
- **Induced impact**: income spent broadly by the economy. Examples include increased sales from businesses due to the road.

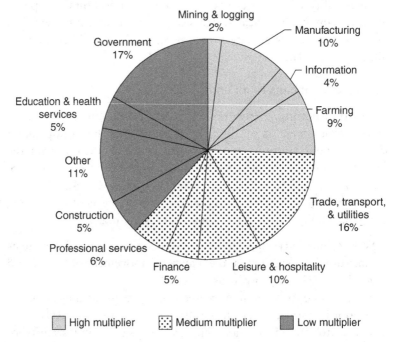

Figure 7.1 The economic multipliers of the transportation system (Melford, 2012)

- **Economic stimulation:** growth in the local economy. Examples include a bustling economy due to a robust transportation system.

One such study estimated (using 1992 dollars) that $10 million invested in transportation results in 570 short-term jobs in transit operations and $30 million gains in sales by local businesses (Cambridge Systematics, Inc., 1999). Some experts caution against treating this number too rigorously. Quantitative estimates for actual multipliers are extremely difficult to calculate in practice due to the complexities of measuring indirect and induced economic activity (Wachs, 2011). Qualitatively, one can describe which industries may be more dependent on the transportation infrastructure than others. Washington state has depicted the dependence on transportation infrastructure by industry, as shown in Figure 7.1.

About 25% of Washington jobs require a well-maintained transportation infrastructure for their primary economic driver. Washington's booming manufacturing industry depends on well-maintained roads to deliver goods quickly and on time. The agricultural industry requires the roads to transport the harvest to markets on demand. Washington state is also a major trading hub for trans-Pacific trade. The Puget Sound ports are located near the Pacific Ocean and at a similar latitude to major Asian trading partners (e.g. Japan, China, and Korea). Over $70 billion of international goods flow through the air- and seaports each year (Johnson and Paulson, 2011). These ports make Washington a non-trivial player in the $4.6 trillion global maritime transportation market (as of 2007) (OEC, 2011).

Another 37% of businesses are able to thrive due to a robust transportation system. Vacationers who visit Washington arrive and depart through the airports, seaports, or border stations. They travel many miles of back roads to reach tucked-away communities, giving a much-needed boost to the local economies. Other services, such as finance and professional services, utilize roads to travel to customers, meet clients, and perform business. Washington is also home to several large, international corporations whose executives need to travel globally to conduct business. The investment in transportation infrastructure has a medium multiplier effect on other industries.

The remaining jobs created depend on the mobility enabled by a highly functional road system. Private sector and government employees need roads to commute to work. Safe transport for students is important to meeting state educational goals for future voters. Emergency and health services need roads to respond to incidents and to transport patients to receive care. In these industries, the primary economic driver is not the roads, but increased efficiencies contribute to profitability and marginal productivity improvements.

7.3 The pressures on spending priorities

According to the following projections from a recent transportation task force, the demands on Washington's transportation system are expected to rise in the next 10 years (Hammond, 2011). Between 2010 and 2020, the population of Washington state is projected to grow by 28.6%, mostly in urban areas. Annual vehicle miles traveled is also projected to grow from 56 to 60 billion miles. Ferry ridership is

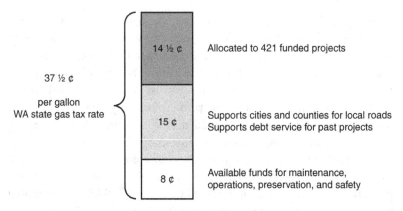

Figure 7.2 Allocation of the gas tax in Washington state (as of July 1, 2008)
Source: WTF, 2011b

expected to increase by 14.5%, and transit ridership across the entire system is expected to double by 2040. All of these are going to create immense pressure to build new roads and interconnects. These create different challenges depending on the type of road.

For the state, most of the funds are already committed. They are funding specific state-wide improvement projects, passed through to cities and counties and paying off debt from past projects. Figure 7.2 shows the allocation of the gas tax to various projects and committed funds. Only 21% of the gas tax, approximately $52 million, is available for regular maintenance and operations. The backlog of maintenance projects compete for this revenue. At a minimum, the state needs to collect approximately $10 billion per year to maintain all roads at 90% of fair condition. This would address a backlog of bridge maintenance, replace all ferry vessels to maintain current level of service, continue to provide subsidies to rail systems, and comply with current stormwater environmental permit requirements.

Cities face a challenge of allocating funds between new construc-tion, ongoing maintenance, and providing other public services like police and fire departments. A growing population raises the pressure for new roads, and most competitive grants from the state are for new construction. This results in a preference for construction projects over maintenance and operations. This can be a problem for smaller cities where roads are used as corridors for commuting workers who do not pay local taxes. In these cases, most of the wear and tear is not caused by the local community, even though it may be responsible for paying for it. If all city and county improvement projects were included in the maintenance requirements, the 10-year cumulative budget would reach $50 billion (WTF, 2011a).

7.4 The erosion of revenue

The state, county, and city tax schemes to collect and raise funds for roads all differ. The state's primarily revenue source is the gas tax, which generates a nominal amount from registration and licensing fees. Most counties in Washington depend on property taxes to pay for roads. Cities in the region use a combination of sales and other local taxes. Both counties and cities depend on the state for a share of the gas tax revenue. Nearly three quarters of the funding sources for public transportation comes from sales and local taxes. Figure 7.3 depicts the different funding sources for roads (WTF, 2011b). Note that the gas tax ultimately makes up only a small portion of roadway funding for counties, cities, and public transportation.

The gas tax in Washington, like in most other states, is a flat tax levied against gallons sold. The US corporate average fuel economy (CAFE) standards dictate the fuel efficiency of vehicles. In the 1980s, cars were achieving 25 miles per gallon (mpg). By 2012, new vehicles were attaining averages exceeding 30 mpg. The latest standard is scheduled to increase to more than 45 mpg by 2030. As the CAFE standards become stricter, fewer gallons of gas are needed to travel the same distance. The revenue decrease due to higher standards has been mitigated by the fact that people are driving farther as cars become more efficient. Figure 7.4 depicts the two competing factors—the rise in fuel efficiency standards and the rise in average vehicle miles traveled. More recently, the vehicle miles traveled per licensed driver have flattened and are forecast to remain relatively static, while CAFE standards are expected to increase.

Perhaps more importantly, the tax is a flat tax and is generally not adjusted for inflation. This means that while revenue has remained largely unchanged, the purchasing power of the revenue has decreased substantially over the years. In 2011, Washington state collected nearly $250 million in base gas tax. When inflation adjusted to 2001 dollars, this was equivalent to the purchasing power of merely $150 million (WTF, 2012). The gas tax is a statutory tax that requires a vote

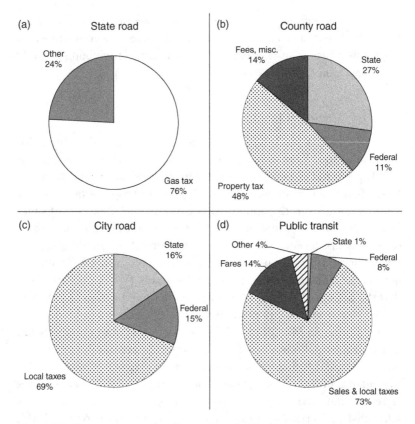

Figure 7.3 Transportation funding sources in Washington state

of the legislature to modify or raise it. With the current political and public pressure, the legislature has been reluctant to hike the rate.[3]

Washington, like most other states, has seen several economic factors add additional stress to an already fragile funding system. First, the 2008 recession and the housing bubble drastically reduced the revenue from property and sales taxes on which counties, cities, and public transportation systems depended. As the cost of fuel rose, people drove less and purchased less gas, lowering overall gas tax revenue. Furthermore, as the cost of driving increased, more and

---

[3] Although the legislature enacted a $0.03 hike in the gas tax in 2005, the funds were already committed to 261 projects already, so none of the funds were available for operations or maintenance costs.

(a)

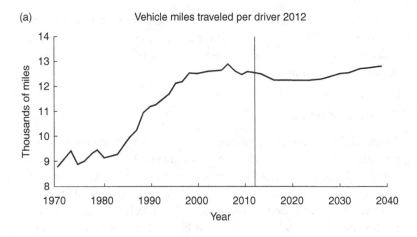

(b)

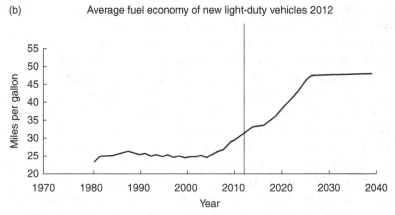

Figure 7.4 Average vehicle miles driven per licensed driver and average fuel economy of new light-duty vehicles

Source: EIA, 2014f

more people began utilizing the public transportation system, placing an additional burden on a system that also faced a combination of decreasing revenue and increasing ridership.

These factors have contributed to significant revenue deficits when compared to the 2007 projected revenues for the transportation budget. This deficit is forecasted to be as much as $5.04 billion dollars by 2023 (WTF, 2012). As a result, Pierce County Transit reduced services by 35% due to budget constraints in 2011. This was especially impactful on low-income families. Not only were they hurt by a poor economy, but they were also affected by transit cuts that increased costs for commuting and decreased the viability of working without a personal automobile.

7.5 Role of taxes

Taxes and fees are the primary means for the government to raise revenues. There are several categories of taxes. In general, economists agree that the most effective tax is a *consumption tax* or a *user fee*. The taxed amount is based on a consumer's use of the system. People pay into the fund based on the benefit they receive or damage caused to the system. In such a system, the higher the benefit or the higher the damage, the higher the tax (Williams, 2007). Sales tax and the value-added tax (VAT) are examples of user fees.

Tax rates can be a *flat* or *proportional* percentage, as applied to the economic activity. A sales tax is a flat tax. It is accessed at the same rate regardless of income. A *progressive tax* is a tax rate that increases as the taxable base amount increases. An example is the income tax, where the tax rate increases as a person's income increases. A *regressive tax* is a tax where the rate decreases as the amount of tax paid increases. Regressive taxes place a heavier burden on low-income individuals who have to pay a higher proportional rate for the same public goods. An example of this is the US Social Security tax, which is capped at 6.2% for gross compensation over $102,000. This structure means that while a comparatively wealthy person may pay significantly more in real taxes, a low-income family will experience a higher burden or impact to their ultimate discretionary income. A proportional tax is a mixture of the two and avoids any penalties or benefits based on income level.

A *Pigovian tax* is a tax applied to a negative externality. First proposed by Arthur Pigou in 1920, it attempts to address societal costs due to damaging behavior. For instance, alcohol sales may require a larger police force, additional medical support, and additional court costs within a metropolitan area. These additional costs can be a drain on a community's resources if paid for by all residents, many of whom may not partake. A higher tax on alcohol can raise revenue by increasing the cost and funding the community's burden associated with these externalities. Environmental pollution is frequently cited as a negative externality that can be corrected by a Pigovian tax. To control for acid rain, the US has implemented a cap-and-trade system for sulphur oxide (SOx) and nitrous oxide (NOx) emissions from power plants. This resulted in a more than 50% reduction in those emissions between 1980 and 2008 (EPA, 2009a).

There are several attributes of taxes to consider, including how much revenue is expected to be generated, the ease of administering collection and compliance, and expected behavioral changes. Taxes that affect larger sectors of the population will generate more revenue, but taxes only applicable to a small group are simpler to collect, audit, and enforce.

Taxes are well known to influence behavior. Changing behavior in order to avoid paying taxes can create unexpected consequences. Between 1696 and 1851, the United Kingdom levied a window tax. The reasoning was that wealthier individuals would likely possess larger homes with more windows. Tax collectors could simply count the number of windows to assess the taxes due. An unanticipated consequence was that many buildings were constructed with brick-filled cutouts in place of windows. The tax changed incentives in a negative way for the public and increased health-related problems due to lack of light and fresh air (Tiley, 2011).

Taxes can also send price signals that create positive changes. Cigarette taxes artificially increase the cost of smoking. This price increase has multiple positive effects from a societal perspective. Creating a revenue stream for the government, decreasing the number of smokers, lowering smoking-related medical costs, increasing longevity, and lifetime tax receipts from a productive workforce all benefit the broader community and are well-documented effects. Cigarette tax increases in France and South Africa resulted in halving cigarette consumption over a 20-year period (Jha and Peth, 2014).

Taxes that support the transportation sector can have many different goals. They sometimes generate revenue to cover for infrastructure investments. Rates can be segmented so that heavy vehicles that cause more damage pay higher taxes. Vehicles that consume specific fuel sources can also be taxed differently in order to influence emissions. Taxes can be used to manage congestion and shorten commute times. Tax can also be used to improve mobility and transportation options. Different tax schemes can achieve one or several of these goals.

Taxes by the gallon

The gas tax is a prime example of a user fee. The gas tax is statutorily set aside exclusively for improvements to the roadway system. Ever since Oregon first introduced a US gas tax in 1919, it has been one of the most popular taxes with legislators and the general public. Since gas tax funds are set aside for capital improvements, citizens can easily associate the fees and benefits.

The gas tax also had other advantages. It is simple to administer, and tax collection is effective and efficient. People paid at the pump, and the government collected the revenue from vendors. The tax is also directly correlated to emissions. People who drove more emitted more carbon and other pollutants. Thus, the tax serves as a simple proxy for externalities that are hard to measure and quantify.

As the use of gas tax funds became more muddled, there has been an erosion of support for increases in the gas tax. In the early days, when the gas tax was used to build new infrastructure, citizens could see the improvements in new roads and infrastructure being built. As the country became saturated with roads, further increases in the gas tax were allocated for maintenance purposes. Operations and maintenance costs are significant, and the results of these expenditures are often unnoticed by the general public. Several studies have pointed to the regressive nature of the tax. Lower-income groups can pay up to 15% of their pre-tax income on the fuel tax as compared to the 2.8% of the upper tax groups (Poterba, 1989).

Broad-based taxes

Sales and property taxes are examples of broad-based taxes that are used to raise funds for roadways. These taxes apply to a large segment of the cities and counties, and public transportation systems depend on broad-based taxes as their primary funding source. The infrastructure, collection, and administration of the tax are already in place, thus reducing the need for a new collection methodology or auditing. Since broad-based taxes apply to a large swath of the population, they also generate the most revenue.

Broad-based taxes as a whole are progressive in nature, even though specific taxes within the system can be regressive or progressive individually. For instance, estate, individual, and corporate income tax are progressive in nature. On the other hand, Social Security and Medicare payroll taxes are regressive. When all taxes are combined together, the lowest quintile has an average effective tax rate of 1.1% of their income as compared to a 26.2% rate for the top quintile (TPC, 2012).

Sales and property taxes are usually collected in a general fund, making it difficult to commit the revenue specifically to transportation needs. When taxes are amassed in general funds, transportation systems must compete with other priorities. City council, county commissioners, or state legislatures can reallocate money at any time, increasing risk to ongoing projects. Tax revenues can also be volatile, with revenues correlating with economic conditions. Ongoing operations and maintenance costs must be supported regardless of the relative strength of the economy.

Taxes by weight

Although trucks account for less than 1% of vehicles, they are responsible for nearly 30% of the costs for the transportation system (FHA, 2000). Heavier vehicles cause more damage to roads than lighter vehicles. Vehicles with fewer axles cause more damage than vehicles with many axles. Heavy vehicles with few axles are the worst combination of the two.

The costs of heavy vehicles are not limited to the direct damage they cause to the system during normal operation. Bridges are designed to support specific weights. Heavier bridges cost more to design, build, and maintain. As trucks get heavier, older bridges often require retrofits or upgrades. Traffic disruptions due to larger vehicles also tend to have more direct and indirect costs. Truck accidents simply cost more to clean up than passenger vehicle accidents. They also tend to result in larger and longer disruptions on roadways, stalling the flow of commerce and other economic activity.

Taxes by weight are based on values taken from certified weight stations, commonly visible adjacent to major highways. A truck driver or owner pays tariffs based on the maximum combined weight of vehicle and cargo. Some states have a weight-distance tax that is a combination of the actual weight being carried over the total miles traveled. However, administration of weight-based programs is often linked to rule enforcement for proper licensure and safety but can still be difficult to implement and enforce.

To simplify the administration of the tax on large vehicles, some states have implemented a truck tire tax. They reason that larger and heavier trucks have more tires, providing a more efficient way of gaining similar data. However, this can have unintended effects on vehicle purchases at break points between vehicle classes. Such an approach incentivizes truck owners to buy vehicles with fewer tires when possible and may increase the amount of load carried by individual tires to cause more road damage.

Regardless of the specific approach, any tax on trucks is part of the total cost of delivery. These taxes and tariffs are ultimately passed on to consumers through higher prices of retail goods and services. Since trucks are widely used to transport the majority of common goods in the US to retail centers, this ultimately becomes a regressive tax.

Taxes by mile and by time of day

Taxation by mile traveled correlates driving behavior directly to system deterioration. The most common mileage tax is a toll. A driver pays a specified fee to travel a certain distance along a thoroughfare. Common transportation assets with tolling infrastructure include bridges, tunnels, and highways. Once labor intensive to collect, new technologies like RFID tags can identify vehicles and trigger automatic payments. For major arteries, there are significant advantages to adopting increased automation. Pre-paid accounts can ensure toll payment, and traffic flow disruptions from toll collection are greatly reduced.

Some cities have applied tolls to entire downtown cores. Cordon pricing, or congestion charges, was pioneered by Singapore in 1975. Vehicles are charged a fixed fee once per day to traverse through the center of town. This has since been implemented in London, Oslo, and Stockholm. London experienced immediate benefits from levying a congestion charge, reducing roadway congestion by 20% and eliminating 10–15 minutes per round trip from a typical intra-city commute (TRB, 2006). The money raised from the fee was dedicated to improving and expanding the public transportation system. Reduced congestion also improves air quality, which is of great interest in many large cities. For example, Paris announced a 2014–2020 effort to reduce small-particle air pollution linked to the high concentration of diesel vehicles in France.

Some jurisdictions have entered into public–private partnerships to maintain and develop roadways. Governments sometimes work with private companies to finance and construct a highway, bridge, or other large infrastructure project. In these private development agreements, the company usually gets to collect a toll from all infrastructure users in accordance with a regulated rate structure. Private roads and bridges are not a new concept. In the mid-1800s, there were over 10,000 miles of private roadways in the US (Fisher and Babbar, 2007). In 2006, the state of Indiana approved a deal where a private company would operate the Indiana Toll Road for 75 years. In exchange, the private company paid the state $3.8 billion upfront that was used to invest in other transportation projects (FHA, 2010). Such long-term investments are not without risk. On September 22, 2014, the private consortium that runs the highway filed for bankruptcy protection, having accumulated $6 billion in debt during the first eight years of private operation. The company cited lower than forecasted usage following the 2008 recession as a major factor in its demise (Randazzo and Fitzgerald, 2014).

Once a tolling infrastructure is in place, rates can be changed based on the time of day. By charging more during peak travel times, the toll can reduce traffic congestion, one of the main hindrances to economic development. Dynamic tolling rates on the 520 Bridge in Seattle rise and fall based on peak transit times. This has reduced the number of cars on the bridge, shortening the time it takes to cross the bridge. Congestion charges to drive through central London and a number of other cities also vary based on the weekday hour. Similar approaches have, or are, being considered in other major cities such as New York and Paris.

Both tolls and congestion charges suffer from the same problem. The revenue generated is usually applied to specific sections of roads, bridges, or tunnels being taxed. The transportation system as a whole rarely benefits. Rural networks, which have relatively lower use, are not able to take advantage of tolling strategies. Tolling requires dense, capacity-constrained transport corridors to avoid just incentivizing drivers to bypass an area via alternate routes.

Vehicle-mile traveled tax

Recently, there have been efforts to combine these taxation methods into a comprehensive pricing scheme that takes into account the type of car, weight, and vehicle miles driven. Called a vehicle-mile traveled

(VMT) tax, this charge varies based on the time of use as well as the type of vehicle being driven on a particular roadway. Pioneered in Oregon state, the program demonstrated that it was viable, and 91% of the volunteers who participated said that they would agree to pay the mileage fee instead of a fuel tax if the program were available.

The program administration and tax collection use the existing infrastructure. Motorists self-report odometer readings when they pay at the pump. The tax is calculated based on miles driven and gasoline consumed. This kind of tax would be incorporated into the cost of gasoline and paid at the pump. The rate of the gasoline would be set to account for road maintenance, environmental externalities, and other concerns. A more rigorous system could be implemented that used GPS in vehicles and charged the car based on miles driven and time of use on specific congested roads. This would create a system of direct user fees with pricings that are better aligned with the cost of travel.

Although the VMT combines the best attributes of all of the above taxes, adoption, or toleration, is by no means guaranteed. While such approaches are ideal from a centrally managed engineering perspective commonly adopted by hierarchical and process-driven governments, there are significant detractors. Major concerns about privacy have already been raised about existing tolling and license plate capture systems. Commercial discounts for metered auto insurance enabled by similar data collection mechanisms have already been demonstrated to have privacy risks and drawn the ire of the Amercian Civil Liberties Union and the Electronic Frontier Foundation (Allyn-Feuer, 2009).

Annual fees

In addition to usage, governments can generate revenue from registration and licensing fees. Cars need to pay an annual registration fee, and drivers need to renew their licenses every few years. Generally, these rates are set based on the type or value of the vehicle. Consumers who own large, heavy trucks may be charged more for annual registration than owners of passenger vehicles and may pay 'tonnage' registration surcharges. In some states, electric vehicles have a higher annual registration rate than gasoline-powered cars. These fees are simple to administer and easy to audit, but this approach is incapable of separating out different driving behaviors. Consumers who travel many miles pay the same as consumers who travel on occasion. This tax also has a regressive effect, as the fee needs to be paid regardless of income.

Carbon taxes

Carbon taxes have a large effect on the transportation sector, as most vehicles require fossil fuels. Where enacted, this tax tries to capture the negative externality of CO_2 emissions by artificially raising the price of the fuels based on estimated chemical emissions. By paying more to consume the fuel, people would be incentivized to find alternative means of transportation with lower overall costs. Revenue generated from carbon taxes could be used to invest back in roads and/or alternative transportation modes.

Administering and collecting the tax directly is rare, because direct measurements are difficult and expensive. Instead, the tax is levied on the carbon content of the fuel and paid by the producer. Carbon taxes and policies have not been uniform, to say the least. Prices on carbon range from just a few cents to over $100 per ton of CO_2 (Sumner and Smith, 2009).

7.6 Energy effects of spending priorities and taxes

Spending priorities and land use policies have a direct effect on energy consumption. There is a distinct difference between cities that consider transportation to be a problem with road infrastructure and other cities that consider transportation as providing mobility. Priorities could include new roads to reach new destinations, creating alternative routes between already-served destinations, or improving the quality or size of existing roads.

Cities in the US predominantly invest in roads to solve transportation issues, as shown in Figure 7.5(a). Tyson's Corner in Northern Virginia is an example of the post-war suburban boom in the United States. Transformed from a rural agricultural community in the 1960s, Tyson's became an 'edge city' and grew to support more than 120,000 commuting employees and a resident population of more than 19,000 by 2014. Tyson's, in a potential model for other suburban communities (especially in the US), is now attempting an ambitious redevelopment effort to become a livable

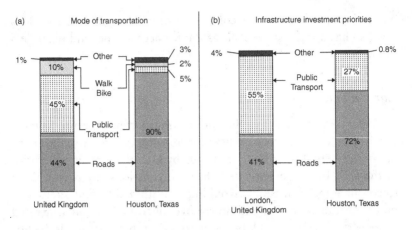

Figure 7.5 A comparison of infrastructure investment decisions and modes of transportation between London and Houston (apadted from Houston-Galveston Area Council, 2007; UKDOT, 2011, Piggott, 2007; US Census Bureau, 2012)

community by 2050. European cities tend to invest more in public transportation in the form of rail, buses, etc. as compared to their US counterparts. For instance, while the city of Houston spends 72% of its transportation budget on roads, the United Kingdom spends 55% on public transportation in the form of railways, subways, and buses (Houston-Galveston Area Council, 2007; UKDOT, 2011). The investment results in different commuting behavior, and these differences are also partially due to higher population densities in most European cities. Differences in commuting behavior are evident in Figure 7.5(b), where only 44% of Londoners drive to work, but 90% of Houston residents depend on the road network in their daily commute (Piggott, 2007; U.S. Census Bureau, 2012).

Commuting behavior directly affects the amount of energy needed. Both Houston's and London's transportation infrastructure offers comparable average commute times for their citizens, and both move a similar number of people per day: 25 minutes with 0.9 million people and 32 minutes with 1.1 million people, respectively. Yet Houston requires *three times more energy* than London to satisfy its daily commuting requirements. As a result, the transportation carbon footprint is drastically larger. The average Houston household is responsible for 15 tons of CO_2 emissions per year to satisfy transportation needs. That same household in London would consume a mere 1 ton of CO_2 (Greater London Authority, 2012).

Taxes can have a great influence on energy consumption without mandating behavior changes. The Canadian province of British Columbia implemented a carbon tax in 2008. The price on carbon had a direct impact on transportation energy footprints. Within 5 years, the per capita consumption of petroleum dropped by 17% (Elgie and McClay, 2013). By comparison, it took the US over 20 years to raise the fuel efficiency standards by 17%.

Like most public policy issues facing Washington state, the best approach to reliably funding transportation infrastructure is unclear; the stakeholders have different priorities, and the choices are contradictory. Deciding how to best generate revenue to serve the community is never an easy task. Tax policies can be aligned to maximize revenue, minimize externalities, or mitigate negative behaviors. Determining what to tax and the appropriate level of taxation has vexed governments for many centuries. Likewise, deciding on what to purchase with societal dollars is just as difficult. Within the transportation sector, the state can choose to build new roads, widen existing thoroughfares, pay for public transportation, or invest in other modes of mobility. Each investment choice brings its own energy characteristics and externalities. Only one thing is clear: the transportation sector needs to change as the needs and wants of the community evolve.

7.7 Chapter 7 Summary

- We frequently talk about the features of energy infrastructure, such as energy consumed per gallon or carbon released per mile. When investing in infrastructure, we ultimately decide based on social values and needs, not the underlying quantitative analysis of actual energy decisions.
- Governments can raise revenue for roads and other transportation infrastructure via a variety of mechanisms, including fees, taxes, and tolls.
- Taxes and fees are a powerful mechanism to change consumer behavior to maximize net benefits to society without mandating specific behavior *per se*.

- Investment decisions based on improving roads result in different consumption behaviors than investment decisions based on increasing mobility.
- Interdependencies between infrastructures, such as waste management dependence on reliable roadways and rail systems, complicate investment decisions. Some investments, or lack thereof, can cause significant downstream costs in related energy infrastructure silos.

8 | *Heat without the hot air*

Thermodynamics is a funny subject. The first time you go through it, you don't understand it at all. The second time you go through it, you think you understand it, except for one or two small points. The third time you go through it, you know you don't understand it, but by that time you are so used to it, it doesn't bother you anymore.

Arnold Sommerfeld, Physicist

As a result of our familiarity with electricity and gasoline in our daily lives, many discussions about energy are really discussions about these specific commodities. However, there are many forms of energy that can satisfy common energy consumption needs. Heat is a historically important example of a useful form of energy. Thermal energy, in the form of hot water or steam, can be used to complement electric energy in ways that can improve efficiency, lower costs, and offer better services in certain situations.

In Copenhagen, heat is treated as a valuable resource along with electricity and petroleum. All factories are required to feed their excess heat into a central water loop that circulates heat through the entire city. The steam runs Sterling engines for mechanical motion, heats water for domestic use, cooks food, and sterilizes hospital equipment. Using heat provides desired end-use energy benefits at a much lower cost than using electricity alone. This chapter explores how heat and electricity can be used together to great effect.

Heat energy is probably the most elusive form of energy. We constantly experience thermal energy providing the experience of comfort in our buildings and vehicles. We tend to notice heat energy when it is absent or pervasive: when we are uncomfortably cold or hot. Other forms of energy used to produce light or motion are easier to see or experience visually.

Even though we cannot see it, a great deal of energy is required to manipulate heat energy. Air conditioners, heaters, and other thermal devices maintain thermal comfort inside buildings. In aggregate, these devices consume approximately 61% of our commercial energy needs and 70% of our home energy requirements (EIA, 2003, 2009).

Heat is also the largest waste product of our energy sector. Every energy conversion creates heat, and waste heat removal is a necessity for all devices. A power plant vents and removes over 60% of its energy in the form of heat. Industrial processes lose anywhere from 20–50% via exhaust gases and cooling waters (BCS, 2008). Cars allow air to enter the front engine compartment in order to remove waste heat. Computers have fans atop the CPUs to cool the chip, preventing overheating and eventual failure. In aggregate, the energy lost to waste heat in the US is enormous. By one estimate, the industrial sector alone generates 16×10^{15} BTU/year of waste heat (BCS, 2008). This is over twice the entire annual energy consumption of the state of California (EIA, 2012c). Clearly, there are large opportunities to better utilize the resource.

8.1 Properties of heat

Heat energy is perhaps the most difficult form of energy to manage. First, we will discuss the quality of heat based on generation technologies. Next, we will look at the fluids used to transmit heat. Finally, we will examine the value proposition of how a customer benefits from consuming heat energy. Before we begin, two fundamental concepts must be understood: how to measure heat and how heat travels.

Measuring heat is complicated. We cannot measure it directly. Its value depends on a complex calculation based on the temperature, pressure, and volume of a fluid. Temperature has become the *de facto* shorthand to describe heat. Although this is sufficient enough for everyday uses (and for this chapter), keep in mind that this is an incomplete picture. Heat travels by moving down temperature gradients, from higher temperatures to lower temperatures. If there is no temperature gradient, then there is no movement of heat energy. It is rather cumbersome to always indicate two temperatures to demonstrate heat flow. For simplicity, engineers frequently assume that one of the temperatures is room temperature (70 °F or 21 °C). This gives a tacit reference point.

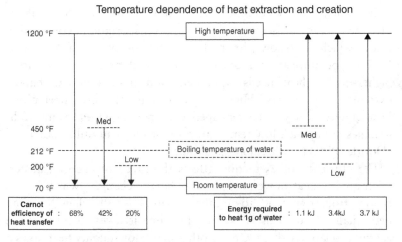

Figure 8.1 Thermal extraction efficiency and energy storage load for selected temperature gradients (adapted from BCS, 2008)

The quality of heat

The quality of heat depends on the size of the temperature gradient. The quality is determined by how *much* heat energy is available, how *efficiently* it can be extracted, and how *quickly* it can be transferred. The combination of these factors helps determine the quality of heat.

1. Temperatures above 1200 °F (649 °C) are considered to be high-quality heat. Common applications include metallurgical and other industrial factories.
2. Temperatures between 450 and 1200 °F (232 and 649 °C) are considered to be medium-quality heat. Common applications include industrial processes and preheating.
3. Temperatures below 450 °F (232 °C) are considered to be low-quality heat. By one analysis, over 60% of all waste heat is low-quality heat, making it pervasive (BCS, 2008).

The higher the quality of heat, the more energy it can store. This is known as thermal load. Larger thermal loads mean that more energy can be extracted. It also means that the efficiency of extraction can be very high. As shown in Figure 8.1, high-quality heat has an extraction efficiency of approximately 65–80%. Low-quality heat, on the other hand, can have an extraction efficiency as low as 10%.

Since high- and medium-quality heat is above the boiling point of water, steam is used as the fluid for transportation. Low-quality heat, which is below the boiling point of water, uses water. High-temperature steam is associated with high pressures. Therefore, higher-quality heat needs equipment and pipes that can handle extreme environments. High-quality heat usually uses steel pipes. Low-quality heat can be managed with regular pipes. Some technologies are at such low temperatures that they can utilize polyvinyl chloride (PVC) or other non-metal pipes.

The efficiency of extraction affects the speed of energy transfer. High-quality heat moves very fast. A heat exchanger, which moves energy from one fluid to another, can be small when handling high-quality heat. This is because the time it takes to transfer the energy is relatively short. On the other hand, low-quality heat takes a long time to transfer. As a result, heat exchangers for low-quality heat are typically larger and slower.

Heat energy extraction occurs by cooling the steam or water, while heat energy is stored by heating the fluid. The same principles of temperature gradients govern heating. Large quantities of heat can be extracted from large temperature gradients. Conversely, a large amount of energy is needed to create a large temperature gradient. This principle demonstrates the importance of preheating. As shown in the example of Figure 8.1, it takes 3,772 joules of energy to raise room-temperature water to 1,200 °F (649 °C). It would take only 876 joules of energy to raise medium-quality heat to the same temperature of 1,200 °F (649 °C). Preheating can be accomplished by recovering waste energy, significantly reducing the need to generate new heat energy.

High-quality heat is more efficient at generating electricity. Fossil fuel power plants use high-quality heat and generally require a capital cost of $1,100–1,500 per kW. Power plants from medium- or low-quality heat are more expensive and less efficient. Capital costs for electric power generated from medium- or low-quality heat can be as high as $3,500 per kW (BCS, 2008).

Thus, high-quality heat moves faster, stores more energy, is more efficient to extract, and costs less when generating electricity versus low-quality heat. High-quality heat can be used for both high- and low-temperature applications. Low-quality heat is more accessible but can only be used for low-temperature applications. This makes

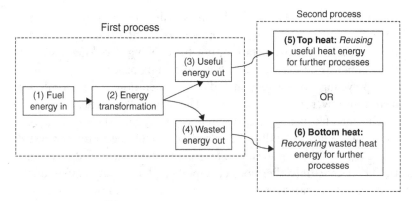

Figure 8.2 Different pathways to reusing and recovering heat

high-quality heat more flexible within dense areas. Low-quality heat is less flexible but more abundant and generally safer to use because of its lower temperature range. As a result low-quality heat is common to large parts of Europe and the states of the former Soviet Union.

Creating heat: top and bottom cycling

Heat is generated by virtually every piece of equipment that consumes energy, as per the energy balance replicated in Figure 8.2. All energy conversions require (1) fuel energy and are transformed by (2) a piece of equipment. The output is (3) useful energy and (4) waste energy. Heat energy can be useful energy or waste energy. We consume heat as useful energy when cooking food on a stove or heating a skyscraper during winter. The primary purpose of the stove or boiler is to provide heat. These processes consume useful heat energy and are called (5) *top cycling*. On the other hand, heat is generated as a byproduct when driving a car, operating a computer, or producing steel. There are two ways to manage the byproduct: the waste heat can either be disposed of or recovered for future applications. Processes that recover waste heat are known as (6) *bottom cycling*.

Let us look at some cooking examples to illustrate top and bottom cycling. Roasting a turkey is a top-cycling process. The primary purpose of the oven is to create heat and roast the bird. On the other hand, motorcycle enthusiasts may be familiar with cooking a steak while on a bike ride. By wrapping the steak in many layers of foil, it can be tied to the engine. The steak will slow-cook from the heat generated during the course of the journey. The primary purpose of the engine is for transportation, not for cooking. Cooking a steak in this manner is an example of a bottom-cycling process.

Top-cycling processes include co-generation and combined-heat-and-power (CHP) facilities. The names "co-generation" and "CHP" are frequently used interchangeably, and their difference is small. We will use CHP to refer to both technologies for the rest of this chapter. These power plants generate both heat and electricity. Unlike electric-only thermoelectric power plants that remove excess heat, CHP captures the heat and uses it for additional purposes. This results in a more complete use of the fuel. Efficiencies of CHP plants can reach 80% energy efficiency, as shown in Figure 8.3. These technologies can lower costs by requiring less fuel to provide the same amount of energy as an electric-only power plant.

Top-cycling systems generally use high- or medium-quality heat. They need fuel to run the process. The choice of fuels can range from fossil fuels (coal or natural gas), biomass (waste wood), or waste-to-energy. A common choice is to install equipment that can utilize multiple fuels, such as both biomass and natural gas. This gives the operator more options to purchase the cheaper fuel at the time.

Bottom cycling can use low, medium, or high qualities of heat. Metallurgical processes that reach very high temperatures provide high-quality heat. Cement kilns and drying furnaces provide medium-quality heat. Low-quality waste heat comes from drying ovens and exhaust gases. In a commercial building, it can be found in coolant from HVAC systems and leftover heat from the processing of hot water or air (Sundberg and Nyman, 1979).

8.2 Transmission constraints

Heating systems are designed based on the type of heat needed and the distance it has to travel. Some transmission fluids are ideal for short distances, while others are ideal for long distances. Generally speaking, shorter distances have more flexibility in the choice of heat quality, generation methods, and transmission mediums.

District heating refers to transmitting the heat across an entire neighborhood or city sector. Some utilities, such as Seattle Steam, are steam-based and serve a downtown area or other defined region of

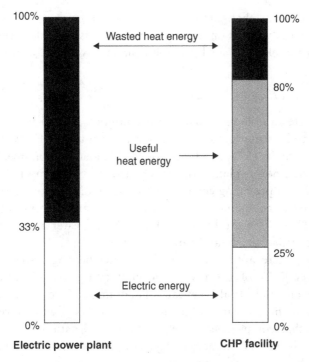

Figure 8.3 Comparison of efficiency between a typical thermoelectric power plant and a co-generation power plant
Source: Sunberg and Nyman, 1979

a city. Others, such as the city of Copenhagen, have implemented water-based systems that serve large areas of the city and suburbs. As one can imagine, the size and scope of systems offer their own benefits and challenges.

Local heating and backup power

Transmission of heat is a low concern when the distance traveled is small. Many hospitals and colleges with large campuses have a central plant that provides steam and/or hot water. Under regular operating conditions, the central plant supplements the amount of electricity purchased from the grid. Under emergency situations, it provides backup power. Many of these systems are designed to be *islanded* from the municipal grid. During a city-wide power outage, the campus may choose to separate itself from the rest of the grid and operate strictly based on its own power. This is possible because the systems are relatively small and self-contained.

District heating with steam

District steam utilities predate electric utilities. Birdsill Holly is considered to be the inventor of the district steam utility. Holly founded his company in 1877, five years prior to Thomas Edison opening the first electric power plant. For many years, the two competed against each other in providing energy to customers in New York City. Over time, the two companies reorganized, merged, and eventually became part of Consolidated Edison (ConEd), serving greater New York City with electricity, steam, and natural gas (CEC, 1911).

Today, there are more than 800 district heating systems in the US. Over 75% of them serve hospital and university campuses. Additionally, industrial zones can collaborate to create power parks to share waste heat streams. Many downtown cores of cities have steam utilities that complement the electric grid and natural gas systems. The remainder serve military and government functions (International District Energy Association, 2009).

The advantage of steam is that it can transmit a large quantity of high-quality heat. Historically, pressurized steam from a utility directly replaced central boilers or waterwheels that ran factories. Today, customers use it to heat buildings, run water systems, or generate electricity. A heat exchanger transfers the heat from the utility to a building's internal system. In this case, the heat exchanger is analogous to the electric meter that serves a building. It is the interface between the utility's network and the customer's systems. Once the steam condenses back to water and the water is cool enough, it is disposed down the drain. The central plant needs a constant supply of fresh water to deliver to its clients. In essence, most steam systems are open-loop systems, being unidirectional and radial.

Steam pipes need to be carefully managed for contaminants and condensates. Contaminants in the water can reduce the thermal load and efficiency of heat transfer. To minimize this, the water is usually filtered prior to converting to steam. Condensates are water droplets that form if the steam comes in contact with cold pipes. If condensates are not properly removed, they can damage the system. Hot water droplets traveling at the speed of gaseous steam can strike valves and curves in the pipes, causing catastrophic failures. Condensates within a steam system are also responsible for the knocking sound one sometimes hears when a steam system is ramping up to heat. Fast-moving water droplets are hitting the inside of the radiators and steam system. Removing these condensates results in approximately a 15–25% transmission loss. These factors constrain even modern district steam systems to within a 1- to 3-mile radius from the central plant.

District heating with water

District heating via hot water, or hydronics, has become a popular alternative. Hydronic systems are typically closed-loop systems, meaning that water is conserved. A customer receives hot water from the central plant. After a customer consumes heat from the water, the colder water is returned to the central plant. It is then reheated and sent out again. This requires a two-pipe system for water out and water return.

Hydronic systems are limited to delivering low-quality heat at 350 °F (175 °C), and it is more frequently transmitted at 250 °F (120 °C) (Phetteplace, 1995). Temperatures of 210 °F (99 °C) and lower are becoming more common, enabling innovative energy storage technologies. Many disadvantages of low quality heat are outweighed by the ease of management. If the temperatures are below 180 °F (82 °C) and at low pressure, the pipes can be made out of PVC or plastics. PVC pipes do not corrode easily, meaning the water needs less filtering for contaminants as compared to steam. This lowers maintenance costs. Transmission losses are lower than those for steam, around 3–5% (Gent, 2014). As a result, hydronic systems, such as the system that serves the city of Copenhagen, can be anywhere from 15–70 miles long and serve large swaths of regions and neighborhoods.

The amount of energy extracted by a customer is also easier to measure. One merely has to measure the temperature of the water entering a facility, the temperature of the water leaving the facility, and the flow rate of the water. The integral difference is the amount of energy the customer consumed. If a customer does not consume any energy, there is no flow. Water is not accepted into the return system until it meets its designed temperature difference in the heat exchanger.

One advantage of a hydronic system is that customers can both extract heat and transfer excess heat back into the system. This property makes water ideal for collecting and transporting waste heat. Factories can buy heat energy when needed and sell their waste heat to the district energy company. Commercial buildings can buy heat energy in the winter when there is a need. Solar hot water heaters can feed excess heat into the system. Even heat generated from pumping water around the system can be captured and fed into the transmission system, making the pumping costs less impactful. This means that the heat utility does not need as much generation capacity to meet client need. The utility could offset fuel needs by gathering low-quality waste heat from the customers it serves.

Finally, the heat energy in hot water can be stored. Since energy is measured based on the difference of temperatures, one can continually put energy in by raising the water temperature up to the safety limit of the system. Additional storage capacity can be gained by constructing hot water tanks. When there is an excess of heat generation, the water is heated. When there is a demand for heat, the hot water can be circulated. This allows for atmospheric water (energy) storage to become part of the district heating design.

8.3 The value proposition of managing risks

The primary benefit of managing heat is reducing resource consumption. Managing heat energy has also proved to increase resiliency against disasters. Some use heat management as a way to increase flexibility of fuel choice against increasing costs. In essence, heat management has become part of the larger energy risk management.

Resiliency against disasters

District energy systems improve resilience to natural disasters. Resilience is the ability to prepare for, withstand, and recover rapidly from major disruptions. Prior to September 11, 2001, disaster preparation for many of the organizations in Manhattan revolved around the response in the minutes to days after an event. After the destruction of the World Trade Center towers, disaster preparation also included resilience, or the ability to stay operational despite interruptions.

The importance of resiliency became apparent during Superstorm Sandy in 2012. After the storm, about 8 million customers were without electricity—some for several weeks. In spite of the damage, every district heating and CHP system that was designed to operate despite these emergencies functioned as expected, providing electricity and heat to customers. These systems ranged from a small 65-kW system designed to provide uninterrupted service to a small data center to a large 40-MW system serving Co-op City, a neighborhood in the Bronx with 60,000 residents. Hospitals and care facilities with CHP capabilities stayed open, admitted additional patients, and allowed staff to stay overnight (Hurricane Sandy Rebuilding Task Force, 2013). CHP demonstrated the value of a robust distributed generation network to mitigate against risks to larger-scale network disruptions within the local region, providing a bright spot in an otherwise bleak demonstration of regional disaster preparedness.

Earthquakes are a concern to underground infrastructures like steam systems. Underground pipes take longer to repair after a catastrophic event. The Great East Japan Earthquake and Tsunami of 2011 is an extreme example of potential damage. The 9.0 earthquake destroyed much of the surface infrastructure, including large numbers of electric power poles and roads, as well as underground infrastructure, including water and sewage pipes. The quake even destroyed power plants and precipitated the Fukushima nuclear disaster. While some areas of the country were able to restore power quickly, others suffered prolonged service disruptions. Water supply recovered slowly following the event, requiring nearly 1 month to reach 90% recovery (Sagara and Ishiwatari, 2012).

District energy has also demonstrated its resilience to earthquakes as well. In Seattle, the magnitude 6.8 Nisqually Earthquake in 2001 caused substantial surface damage. Approximately 217,000 customers were without electric power for several hours. The underground infrastructure, such as water, sewage, and steam pipes, experienced no damage or outage (ABS Group, 2001), allowing Seattle Steam to function as designed. San Francisco had the same experience in 1989 after the magnitude 6.9 Loma Prieta Earthquake. Although 1.4 million people lost electric power, the NRG Energy district heating system continued to serve the downtown core without any disruption to service (NRG Thermal, 2014).

There are several factors contributing to the success of CHP systems during disasters. First, a CHP system is more robust than a backup generator. CHPs are designed to be run and maintained regularly. Backup energy generators are commonly used infrequently and are prone to a number of human failures due to unfamiliarity by the staff or a lack of on-site fuel. The equipment may not function as expected or the fuel supply may not be planned properly. For example, over 50% of hospital backup generators failed during the 2003 New York City blackout—a marked contrast to the Superstorm Sandy experience where CHP plants were utilized (Hampson et al., 2013). Second, the transmission system for CHP tends to be underground. Since most disasters affect the surface, underground infrastructure often is better protected. Surface infrastructure, such as electric lines, bear the brunt of storm damage from fallen trees, flooded substations, and downed utility poles. Third, a CHP system can be islanded from the rest of the energy system. This allows it to function as a microgrid that is separate from the main electrical system. This is a key component of its resiliency, allowing properly designed systems to maintain functionality while the rest of the grid is down. This also reduces the load that local utilities have to fix, freeing up valuable skilled labor and resources to serve other sites during an emergency.

States in hurricane-prone regions have long acknowledged the advantages of CHP as part of resilience. In 2009, Texas enacted a law that required a CHP feasibility study as part of any construction or renovation of a government facility deemed to be part of the critical infrastructure (Texas House of Representatives, 2009–2010). In 2012, the Louisiana House of Representatives passed Resolution 167 modeled after the Texas law.

Reliability

Electric reliability is the ability to deliver electricity without failures. An average customer in the US will have access to electricity 99.98% of the year, experiencing only 1.7 hours of power outages (EPRI, 2005). Although this may be short, one study by the Electric Power Research Institute (EPRI) estimates that the US loses $120–$190 billion per year due to power outages and power quality issues (EPRI,

2001). The digital economy is hardest hit, with some companies losing anywhere from $50,000 to $30 million per minute during a power outage (EPA, 2007).

CHP has been identified as a way to improve electric reliability. Hospitals and health care facilities that utilize CHP have seen a decrease in power interruptions by up to 95% (McGee and Nussbaum, 2006). A well-designed data center with a CHP system and multiple electric grid connection points could theoretically reduce outage time to approximately 7 seconds per year (Preservation Green Lab, 2011). Industrial manufacturing uses CHP to help provide high-quality electricity and avoid factory downtime during outages.

Flexibility in integrating renewables

CHP has also been investigated as a solution to help integrate renewable electricity into the electric grid. Some renewable resources, such as geothermal energy, extract heat energy from the ground. The heat can be directly fed into a hydronic district heating system, eliminating conversion losses to electricity. CHP facilities can also improve the integration of variable renewable resources. Boilers can be configured to both produce and consume electricity in order to produce heat. If there is not enough electricity on the grid, such as when clouds are covering solar panels, the facility can produce electricity. When there is too much electricity, such as when there is a gust of wind, the facility can consume electricity to produce heat. This ability to produce or absorb electricity creates a virtual battery (IEA, 2011a).

Hedge against future fuel prices

District heating can also act as a hedge against future energy prices. Dual-fuel systems can be built to switch between the burning of wood and other biomass to natural gas, depending on the price and availability of the resource. This allows the utility to choose the most cost-effective fuel to deliver services to customers. Frequently, these alternative fuels are only cost effective when purchased in bulk, meaning that individual customers are unlikely to pursue such alternatives independently. Instead, a district heating system can make the switch to primary fuel on behalf of all customers. As an example, when District Energy St. Paul in Minnesota switched from coal to natural gas and waste wood, they dramatically lowered the carbon footprint of all buildings connected to the system (Preservation Green Lab, 2011).

Additional customer benefits

Customers also enjoy secondary and tertiary benefits from buying heat energy as a resource. Without a district heating system, customers would have to supply their own boilers to heat the buildings. Each owner would have to purchase capital equipment, manage additional facilities staff, operate fairly complex equipment, and purchase additional fuel. With a district heating system, the customer only needs a heat exchanger. The heat exchanger has a much smaller physical footprint than a central boiler. In a downtown core, where space is a premium, the value of the additional rental income can be significant. For Seattle Steam, which has nearly 200 customers, this space savings is worth approximately $2 million in annual rent in the downtown core, improving asset values by over $30 million.

8.4 Upfront costs to provide energy

The cost of building a district energy system is not insignificant. As an example, the city of Seattle has identified four neighborhoods that could benefit the most from district energy. They span a variety of sources, from Seattle Steam's facilities to factories in the industrial district. The system could serve customers such as the medical district, several university campuses, the downtown core, and a fast-growing residential neighborhood.

One option is to convert Seattle Steam into a co-generation facility. It currently operates a radial steam system and does not generate electricity. It could be upgraded to generate electricity and operate as an island in downtown Seattle. An 18-MW CHP facility (sized to meet the thermal demand and include the virtual battery feature) is estimated to cost $42 million. A CHP facility would be able to generate more revenue from both the sale of steam and electricity.

Another option would be to build a hot water system that connects the districts together. This would connect lower-cost waste heat from the industrial district, pool together resources from various power plants, and connect to neighborhoods that need heat. However, the amount of new pipes which would need to be laid to support this course of action is immense. The total length of the system would have to stretch for over 8 miles. Connecting to each individual building could require upwards of 30–50 miles of hydronic pipes. At approximately $1,000 per foot ($5.2 million per mile), this would cost up to $250 million.

Long-term benefits would be realized from these projects via lower operational costs. Fuel source selection is an important part of the decision. Natural gas prices are projected to be lower than coal over the long term. Agricultural biomass from chipped wood is cost competitive to natural gas and is a renewable resource. A hot water system could also utilize the low-to-no-cost waste heat from the industrial district. However, development of district energy is not without risk. Industrial waste heat is only available if the plant stays operational. As companies open and close factories, the source of heat may not always be reliable.

A third option would be for the city to forgo investments in heat energy altogether. It could expand the electrical infrastructure to meet the needs of the growing population. Additional substations cost anywhere from $110 to $240 million apiece, and distribution lines cost $1 million per mile. The benefit is that Seattle City Lights is the first carbon-neutral utility in the country, with 80% of its electricity generated by hydroelectric dams (City of Seattle, 2013). This would be an ideal system to expand. However, the hydro capacity is at its limits and may not be able to grow to meet the forecasted demand.

8.5 What is the value of well-managed heat?

Heat energy and electrical energy are frequently used interchangeably, as they can both provide the same benefit of thermal comfort. However similar some of the produced benefits might be, appropriate management techniques for these two forms of energy are quite different. In the US, utility commissions are mandated to regulate and provide the lowest-cost electricity. Such policies often neglect the additional value of heat energy in providing comfort and other energy outcomes.

The city of Copenhagen has pursued a policy that optimizes total energy consumed, not just electricity. Their power generators are

CHP facilities that approach 80% heat recovery. A second significant source of heat is from waste-to-energy plants, integrating their trash needs into their energy infrastructure. They distribute heat via a large system of steam and hot water pipes that serves 97% of the city's buildings (Thronton, 2009). Waste heat is collected from every possible source, including the industrial district, transportation trolleys, and crematorium.

The cost to implement the program was not insignificant—over $3 billion borne through tax incentives and loans. The benefits have been quite dramatic. By recycling waste heat, the city has reduced emissions by 655,000 tons of CO_2, the equivalent of diverting 1.4 million barrels of oil annually away from power plants. The Danish fuel mix for CHP has dramatically shifted towards renewable sources. By 2009, the total share of fossil fuels dropped to below 60% from a high of 85% in 1980. Both coal and oil consumption decreased (Danish Energy Agency, 2010). As a result of its efforts, the per capita carbon footprint in Denmark is 8.4 tons of CO_2. In comparison, the per capita carbon footprint in the US is 18.5 CO_2 per capita, over twice that of Denmark.

Critics may argue that Denmark also has some of the highest electricity rates in the world. The cost of electricity in Denmark is approximately $0.38 per kWh, over 50% of which is taxes (Eurostat, 2014). Yet high rates do not always translate to high bills. Remember that electricity is only one form of energy. Residents of Copenhagen purchase heat energy separately at approximately $0.12/kWh. Furthermore, Danish buildings are more energy efficient than their US counterparts, resulting in needing less energy to run them. In the end, the cost of energy for both countries constitutes approximately 3–5% of a typical household's monthly income.

What cannot be disputed is the successful decoupling of energy consumption from GDP. Since the 1970s, Denmark's GDP has grown by 80%, while energy consumption increased by a mere 5%. Its energy policies have also reduced CO_2 emissions by 18%. This shows that growth in the economy is not dependent on consuming more energy. Furthermore, its energy policies have transformed the country from a net importer to a net exporter of energy. In the 1970s and 1980s, Denmark imported nearly 80% of foreign oil for its energy resources. This cost the country nearly $5.2 billion to purchase, or about 7% of its GDP. By 2006, it exported $6.1 billion worth of energy to the rest of Europe (Thronton, 2009).

Heat management is slowly gaining in popularity. In 2012, President Barack Obama signed Executive Order 13624. In it, his administration recognized the importance of managing heat to reach climate goals. It directed accelerating investments in CHP technologies. Copenhagen has set itself an ambitious goal of being carbon neutral by 2025. The city of Seattle has also set its own goal of reaching zero net greenhouse gas emissions by 2050. For the two cities to reach these goals, district energy must be a carefully and comprehensively integrated component.

8.6 Chapter 8 Summary

- Heat and electricity are both forms of energy that can be used to satisfy common end-use needs. Their properties and management techniques are very different.
- Up to 60% of wasted energy is lost as heat. This indicates that there are large opportunities to recapture useful energy.
- Heat management can result in increased financial flexibility for many facility owner/operators and improve physical resilience.
- Denmark has transformed itself from a net importer of energy to a net exporter of energy by incorporating heat energy management into planning and operational strategies.
- CHP and district energy are important components of robust and resilient energy delivery systems in the built environment.

9 | *The collision of two critical infrastructures*

Resilience of the electric sector to cyber attacks should be no less resilient than to a tornado.

Cybersecurity for State Regulators (Keogh and Cody, 2012)

In November 2013, a cybersecurity breach at Target Corporation resulted in the theft of as many as 110 million customers' data, cost the company hundreds of millions of dollars, and triggered the first termination of a publicly traded company's CEO in response to a cyber event (Harris et al., 2014). The attack was one more indicator of a growing trend of cyber*in*security now challenging every public and private organization. The problem has been noticeably accelerating since the Estonian government and economy were crippled by a large and coordinated cyber attack (primarily from distributed denials of service) on government and private organizations in 2007. The Estonian attacks were a watershed event triggered by the government-led movement of a memorial to the Soviet-led liberation of the country from Nazi Germany to a less prominent location in the city of Tallinn. The ensuing riots and cyber-violence catalyzed a broader international discussion about cyberwarfare (Herzog, 2011). While the cyber attacks in Estonia and the breach at Target had vastly different objectives and used totally different approaches, they illustrate the unanticipated security consequences of embracing increasing connectivity.

In response to these new threats, US public companies are beginning to receive guidance from the Securities and Exchange Commission (SEC) on requirements for investor disclosures related to attacks and compromised networks (SEC, 2011). With a number of high-profile attacks in 2013 and 2014, more regulation is likely on the way (Cleary and Eades, 2014). Organizations such as the North American Electric Reliability Corporation (NERC)—which sets the standards for utility best practices in North America pending regulatory approval—are also conducting scenario-based assessments and issuing guidance about how to protect critical infrastructure from these new threats (FERC, 2013b; NERC, 2012a).[1] Society is now faced with how to balance the desire to benefit from the potential of a more connected world with the practical security risks that come from connecting our physical infrastructure with cyberspace.

While enhanced-connectivity between various organizations, people, and systems in the electric power system has made operations more efficient, it has also introduced new kinds of vulnerabilities. In the envisioned future grid, every car, home, and business will have some form of control system enabling its participation in the emerging "internet of things." Today's refrigerators, entertainment systems, security systems, and other devices can already be networked and remotely controlled by the homeowner. This continued increase in connectivity is fraught with unanticipated consequences of unknown, and potentially unknowable, magnitude due to the immense complexity of the systems when viewed as a whole. Experts like Dan Geer are rightfully concerned "because we are finding that there is no separate 'digital' economy but only a single economy where even the mundane comes to depend on distant digital perfection" (Zurich, 2014). For the historic centrally controlled electric system, the cyber-physical intersection offers unique new benefits at the cost of having created new, and sometimes critical, failure modes.

9.1 A centralized grid emerges

Thomas Edison's grid emerged from a decentralized group of small power producers competing to serve new regions, but today's grid ultimately became a highly centralized system with relatively few participants. At the "bulk" level of utility-scale generation and transmission, electricity primarily flows in one direction—away from the power plant and towards the consumer—largely based on the business model developed by Samuel Insull and the Chicago

---

[1] On November 22, 2013, the FERC issued Order No. 791, Version 5 Critical Infrastructure Protection Reliability Standards that adopted the NERC's proposed standards with some limited modifications. The Order also directed the NERC to propose an updated version by February 3, 2015. See FERC (2013b).

Edison Company from 1892 to 1914 (Hughes, 1988). Understanding
the historical evolution of the electric power system is crucial to
comprehending the current physical system, behavior of participants,
existing legal structures, and limits on the nature and rate of future
changes to the grid.

Samuel Insull originally served as Edison's personal secretary for
12 years but eventually assumed a role as a vice president in the
General Electric Company, which had been formed from the merger
of Edison General Electric (Hughes, 1988). In 1892, Insull departed
General Electric for the Chicago Edison Company which, over a
period of 20 years, became a dominant Chicago-area monopoly and a
technological leader. Insull's focus on superior margins and reliability
through large applications of the new steam turbine technology and
effective use of existing infrastructure was the foundation of today's
high-capital-cost, low-marginal-cost, exogenously insensitive utility
business model.[2]

As multiple load-serving entities (i.e. utilities which sold electricity to end users) emerged, connections between once-independent
networks grew to more cost-effectively serve the combined needs
of customers. In Edison's originally proposed distributed generation
model, a factory or community required a power plant capable of
satisfying the maximum possible load. Insull took advantage of the
fact that customers were not likely to simultaneously consume their
maximum capacity requirements to deliver cost savings. Over a period
of 22 years, the Chicago Edison Company worked to find exactly how
best to value this load diversification phenomenon in order to offer
better service and lower prices and generate increased profits.

In 1914, Insull graphically described his managerial approach
to load management, which became the basis for much of the
twentieth-century utility business model. He explained the virtues of

[2] Insull's growth during this period of consolidation was extremely rapid. He
deployed the world's largest steam turbines, rated at 5,000 kW each, which
began entering service on October 2, 1903, and by 1911 he had upgraded
to ten 12,000-kW units at Fisk Street Station in Chicago (CEC, 1911). Insull
admitted that early turbines were likely not as economical as alternative
technologies initially, but the decision ultimately proved to be valuable and
gave Insull the confidence to pursue other ambitious projects like the world's
largest generating unit at the "State Line" generating unit, which included a
208,000-kW (278,820-hp) turbine when it opened in 1929 (Hughes, 1988;
ASME, 1977).

diversified load, referencing 11 separate classes of consumers, which helped create a diversity factor for each region, allowing for a more scientific estimation of maximum simultaneous demand: also known as the peak load required by the system (Hughes, 1988). He used a case study from a Chicago apartment block with 193 customers possessing a separate maxima of 92 kW but an observed transformer maximum of only 29 kW (Insull, 1914). The analysis proved that Insull could serve his entire customer base without building as much generation, transmission, or distribution capacity as they would require separately and meet reliability metrics.

Insull's focus on using empirical data and statistical analysis also led to the identification of seasonal loading phenomena and the formalization of capacity factor as a central metric for plant operation economics in business processes (Insull, 1914; Hughes, 1988). The net capacity factor of a power plant is the ratio of its actual output to its theoretical, or nameplate, capacity (Tidball et al., 2010). The same concept applies to power lines, transformers, or other capital-intensive assets. This method allowed utilities to optimize the utilization of invested capital and maximize return on investment. With these tools at his disposal, Insull actively sought out both generation and customers to raise capacity factors and improve load diversity. His approach to customer acquisition, which used targeted business development efforts to simultaneously increase the profitability and reliability of the electric service received by his customers, is still relevant today—even within modern market structures.

The emergence of a strict hierarchy to govern relationships between multiple entities seeking to take advantage of Insull's demonstrated economies of scale is unsurprising. From 1907 to 1920, most states agreed that electric utilities should be legally protected monopolies with geographically defined service territories (Kwoka, 1996). At the time, coordinating multiple generators on the early grid was a tedious and deliberate process that was built around these early service territories, or balancing areas. Utilities were equipped with telegraphs—and, later, telephones—but the amount of data that could be practically managed over such devices was limited. Detailed schedules were produced, and utilities scheduled *interchange* across the geographic boundaries at the connection points between their networks, when they existed. With a small number of participants and advanced scheduling with a relatively predictable demand, this proved to be sufficiently reliable and controllable for both utilities and regulators—cost and reliability were significantly better than what privately owned generation and distribution could provide for most factories or communities.

However adequate these processes were for much of the twentieth century, they remained inefficient. Overall system stability was maintained via large amounts of (primarily) excess electric generation capacity—only brought to bear if unexpected demands occurred from within, or across, a utility's territory. These *ancillary services* were largely required to account for information disparities between the individual generation and transmission facilities and the rapidly growing number of customers.[3] Today, the falling cost of data processing, storage, and transport technologies can result in more efficiencies than Insull could have imagined by maintaining system stability with brute force. This is only possible if utilities, consumers, and generators embrace modern information technology and communications infrastructure, making the "smart grid" capable of enabling bidirectional flows of information and electricity.[4]

9.2 A "smarter" grid

The term "smart grid" gained popularity as a way to describe the merger of the physical grid with cyberspace—as power networks and telecommunications became increasingly intertwined. Proponents such as Prof. Massoud Amin argued that embracing modern communications and data processing technology was required to provide a grid which was secure, agile, and robust enough to be able to survive threats and unanticipated conditions in the future (Amin and Wollenberg, 2005). Others, including leading security expert Dan Geer, have argued that we should be more circumspect in chasing these potential benefits until we better understand how to deal with the new risks associated with the merger of these critical infrastructures. Geer asserts that "Electronic Health Records and the Smart Grid [...] are at the stage of 'last chance to get it right without breakage'" (Geer, 2014).

---

[3] Common ancillary services include load-following, spinning reserves, power for voltage control, and devices that provide no additional power but help with the maintenance of power quality (e.g. flexible AC transmission devices or shunt capacitors).

[4] See e.g. Yan et al. (2012).

For large producers or consumers of energy, the decision to generate or consume power from the grid has always been one of economics—while regulators are charged with ensuring that it remains in the economic interest of utilities to provide reliable electric service. Insull's methodology still applies. Balancing supply and demand more efficiently is the *physical* problem addressed by the smart grid; however, facilitating information exchange to enable a larger and more diverse set of participants to engage in *economic* transactions without undermining the reliability of the system is the actual implementation challenge. Without this, no economic benefit is conferred to utilities or consumers sufficient to justify the liberation of capital required for modernization. Any efforts to modernize (e.g. enabling active supply- and demand-side participation, scaling the use of less controllable utility-scale generation such as wind and solar, or increasing the use of flexible distributed energy resources) must simultaneously address economic and "control system implications to ensure grid reliability" (GWAC, 2013).

Most of the existing efforts from utilities and smart grid companies focus on evolutionary improvements to the centralized control approach to modulate supply in response to actual system load, effectively promulgated by Insull over 100 years ago. The Pacific Northwest National Laboratory's (PNNL) GridWise Testbed Demonstration Project is a notable exception which focused on enabling *transactive energy* interactions (Hammerstrom et al., 2007).[5] During the pilot program, electrical devices, including distributed generators, water pumps, and 112 homes, "bid" into a micromarket for energy. Prices fluctuated every 5 minutes based on grid conditions, and trading software automatically controlled energy devices (e.g. thermostats and hot water heaters) on behalf of participants (Hammerstrom et al., 2007).

Many studies had demonstrated that consumers would respond to price signals,[6] even in the residential market, but the PNNL event proved that near-real-time bidding could successfully balance a local distribution grid in a more sensitive implementation of the same

---

[5] See the Gridwise Architecture Council for a more thorough overview of transactive energy concepts from a multistakeholder public–private consortium—originally sponsored by the U.S. Department of Energy—dedicated exclusively to this challenge.

[6] See e.g. an Oxford University report on the Effectiveness of Feedback on Energy Consumption by Sara Darby (2006).

way bulk power was managed between large utilities by RTO/ISOs. Embracing this approach would mean growing the total number of market participants from about 7,000 total electricity market participants (including generators and transmission entities) to well over 150 million in the US alone.[7]

9.3 An introduction to cyber-physical systems

Today's grid increasingly relies on information and communication technologies to meet pressing operational requirements. Because of these capabilities, which have become increasingly essential over time, the grid has truly become a cyber-physical system. Cyber-physical systems generally refer to machines which can interact with the physical world (i.e. control or sense), make decisions based on exogenous and internal data, and then take actions in the physical world. Actions could occur directly, such as sending a control signal triggering an action or alert in an industrial control system (ICS) or supervisory control and data acquisition (SCADA) system, or indirectly—i.e. alerting a human actor.

Although terms like "smart grid" and "cyber-physical system" are relatively new, utilities have been utilizing data and improved computational power for over half a century (NCS, 2004). It is important to understand the creation of control systems to help manage complex engineering and economic problems which are inherently part of reliably and effectively operating energy delivery networks. Cyber-physical systems are instrumental to the day-to-day operation of the infrastructure that manages the energy, water, sewerage, and transportation infrastructure silently supporting modern society in industrialized nations. Since it is unreasonable to protect all information, or information systems, against all attacks, information security is generally considered to involve tradeoffs between confidentiality, integrity, availability, and accountability (Blakley et al., 2001). For ICS/SCADA systems, availability is generally prioritized above all other considerations. As the "internet of things" further develops, the same systems that once applied only to large utilities and infrastructure operators are increasingly part of the common

---

[7] According to the U.S. Census Bureau (2013), there are 132,802,859 identified housing units and 27,092,908 total firms (2007).

building or home—e.g. both Apple and Microsoft unveiled software frameworks for controlling devices in the connected home in 2014, called HomeKit and HomeOS, respectively.

In the early 1960s, ICS/SCADA systems helped factories and large industrial facilities control and optimize production within the confines of factories and power plants with mainframe-based systems (NCS, 2004). Later, these same facilities would use wide-area networks (WANs) that leveraged communication lines spanning cities, states, or countries. This further enhanced the connections common to SCADA systems and related business systems, both of which had evolved to use more modern client-server architectures and common communications protocols, like Ethernet and TCP/IP (Anderson, 2002).

Prior to the widespread deployment of information and communications technologies, power plant operations remained virtually indistinguishable from Insull's early centralized plants. Expert operators monitored systems and made decisions seeking to balance the grid, accounting for the limited availability of information with excess capacity under their control. With SCADA controls and WANs, operators could tightly couple operations to reduce redundancies without compromising the system and even control power plants from remote locations if needed.

Understanding automatic generation control

The deployment of SCADA systems enabled numerous operational advancements within industrial systems and across the energy sector, including automatic generation control (AGC). AGC allowed system operators to more effectively adjust the power output of generators in response to system conditions within isolated, and eventually in multiple, control areas.[8] Initially, controllable generation units used physical systems, called *governors*, capable of sensing changes in

---

[8] AGC implementation varies widely depending on the kind of generating unit (e.g. hydro versus coal versus natural gas), power plant type (e.g. drum-type or once-through boiler), plant control system model (i.e. boiler-follow, turbine-follow, or coordinated), operating point (the ability of a plant to respond to control signals can vary widely across its viable operational range), and operator actions (including the process and conditions under which AGC will be disabled) (VanSlyck et al., 1992).

frequency caused by the reduction of output from other generators, or from the addition of load, to implement follow-up control units. As a generator is placed under increased load, the speed of the generator decreases until the frequency stabilizes.[9] If the disparity is large enough, frequency will remain outside of a tolerable set of bounds, known as the *governor deadband*, which stabilize generating unit speed to achieve equilibrium within 12 seconds (Ewart, 1975). In small isolated systems, speed stabilization is more commonplace. Large interconnects may require thousands of megawatts (MWs) of instantaneous capacity (supply or demand) changes to trigger frequency anomalies large enough to trigger unit speed stabilizers (Concordia, 1990).

AGC was developed to reduce portions of this manual control to reduce fuel and maintenance costs, improve power quality (i.e. voltage and frequency stability), and to reduce the amount of energy being unintentionally transferred to neighboring utilities as a result of over- or underproduction within each control area—known as *area control error* (ACE). The adoption of AGC was necessitated as the integration of diverse power systems across geographically defined balancing areas became important. Larger systems required higher degrees of coordination to maintain stability and increase derived benefits.

Utilities begin network expansion

Utilities leverage data sampled from sensors on *tie lines*, which connect different balancing areas within each interconnect. This data must be captured, sampled, filtered, and transported to a centralized location before it can be used to compute the ACE from which control instructions are drawn and, ultimately, sent back to generators. To accomplish these types of time-sensitive tasks without breaking the bank, utilities adopted a wide range of hybrid communications networks leveraging existing available infrastructure when possible (e.g.

---

[9] When new loads are added, a power mismatch extracts kinetic energy from the inertial storage of the generator (i.e. a large flywheel on the spinning generator) until frequency has declined to the point where kinetic energy extraction has ceased. If the added load is small enough to be supported by the system, the reduced power requirement caused by declining system frequency will match the output power of the supporting generator within less than two seconds (VanSlyck et al., 1992).

fiber optic packet-switched networks, copper wire lines); power line communications if appropriate; and a wide range of wireless technologies such as GSM, GPRS, WiMax, and WLAN (Yan et al., 2012).[10]

As these networks were deployed, SCADA system-collected data from single sites was used to improve state estimation of the whole system. State estimation (SE) and energy management systems (EMSs) rapidly developed to provide control-area and regional-level control signals across multiple generating plants, tie lines, FACTS devices, etc.[11] Distribution management systems also emerged to help improve the stability of the radial lower-voltage networks that help provide the last-mile connections from the interconnected grid to typical individual customers.[12]

By and large, second-generation SCADA systems were more distributed and could integrate many more sensors and devices. The proliferation of personal workstations drove the creation of local area networks (LANs) within small geographic areas or buildings. These LANs were increasingly stitched together across wide-area networks, later via virtual private network (VPN) services across an internet, to enable even more individuals and teams to expeditiously access and analyze data.[13]

As computers and networks became more common, other network-enabled technologies began to complement SCADA systems across other parts of the utility business. For example, automatic meter reading (AMR) and time of use (TOU) meters that leveraged solid-state sensors were introduced to compete with traditional electromechanical meters. By the end of the 1980s, AMR and TOU meters (significantly cheaper, largely thanks to general cost reductions in electronics and supporting technologies)

---

[10] Power line-based communications on AC networks is an old concept, with at least two patents issued in the 1920s (e.g. 1,607,668 and 1,672,940)—for a more thorough treatment of challenges and technologies, see e.g. Sutterlin and Downey (1999).

[11] FACTS stands for flexible AC transmission system. For more details on SE and EMSs, see e.g. Schweppe and Wildes (1970), Huang et al. (2012), Abur and Exposito (2004).

[12] See e.g. Kersting (2007) for a more detailed look at distribution system modeling and analysis.

[13] A virtual private network, or VPN, creates a private, shared logical network between multiple LANs connected over public networks like the internet via a series of dedicated connections, tunneling protocols, and encryption.

had been validated operationally—in part due to Federal Energy Administration—sponsored research into the system-wide benefits of TOU rates for consumers (EPRI, 2010).

Utilities also benefited from the cost savings from reducing manual labor used to physically inspect meters and record how much electricity a household or business had consumed in the last billing period. Many early AMR systems had a small radio that broadcasted this information in its immediate vicinity. A van could drive through neighborhoods with a receiver and rapidly gather data, resulting in more accurate billing information and lower transaction costs per customer. With new technologies, this experimentation delivered results—Pacific Gas & Electric implemented their first voluntary TOU rate program in 1982, and 80% of their customers saved $240 per year by the early 1990s (King, 2001).

The current generation of common SCADA systems is fully networked, bidirectional, and utilizes both wired and wireless wide-area networks via secure VPN connections. Figure 9.1 shows an exemplary modern enterprise network with remote access capability, firewalls, intrusion detection systems, and industrial control hardening devices. It allows for even more devices to be integrated and enables even more people to gather, view, and analyze the data. One-way meter readers have evolved into two-way systems called automated metering infrastructure (AMI). These not only transmit information back to the utility but can also receive information, such as real-time costs for demand-side management solutions.

Cursory introduction to industrial control systems

Today's ICSs are multi-tiered systems that have numerous components which could be utilized to impact their operation. Modern SCADA systems typically refer to centralized control systems, while the more general term of ICS may be used to refer to a SCADA system or a distributed control system (DCS). DCSs typically have a controller element located with each component where each discrete subsystem has a separate controller and sequential processes are typically the focus—generally processes with less sensitivity or coordination required.[14]

---

[14] For a more thorough treatment of ICS fundamentals from a cybersecurity perspective, see e.g. Knapp (2011) and Macaulay and Singer (2011).

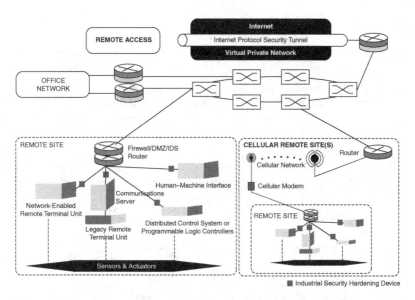

Figure 9.1 Exemplary modern SCADA system with legacy, network-enabled, and cellular system components

Remote terminal units (RTUs) and programmable logic controllers (PLCs) are generally the most numerous single components within most control system networks and are located in close proximity to sensors and actuators within the physical system being controlled. RTUs convert and relay telemetry information and other data. In other words, they collect data from analogue sensing devices and convert captured signals into a digital format or take digital signals and deliver analogue stimuli to actuators. Depending on the system, PLCs may be used with RTUs or in lieu of them, interacting directly with a controller. PLCs are designed to position logic directly adjacent to controlled systems to minimize the chance of disruption, degradation, or destruction—even in the event of temporary or permanent loss of connectivity to higher-level controllers. Like most engineering solutions, most PLCs and control systems were designed to be resistant to random failures from normal operations, not targeted failures from malicious actors. As a result, most have minimal, if any, embedded security protections or hardening.

RTUs and PLCs connect with a SCADA master or a higher-level controller via some form of wired or wireless network—increasingly

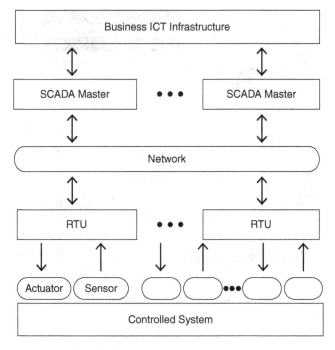

Figure 9.2 A simplified SCADA testbed

converging onto standard internet protocol-based (IP) infrastructure.[15] These higher-level controllers are typically additional kinds
of custom software which are installed on top of common Windows
or Linux variant operating systems that facilitate human–computer
interaction to set system objectives, visualize performance, etc. Thus,
many top-level controllers can be manipulated via the compromise of
standard operating systems.[16] Figure 9.2 shows a simple lower-level
representation of a typical SCADA system testbed—based on the
model used in the TRUST SCADA Testbed developed by UC Berkeley

[15] TCP/IP is the IP routing protocol, defined by IETF RFC791, that uses
Transport Control Protocol as the primary error-correcting and delivery
assurance mechanism to deliver information across packet-switched networks
common to all modern business networks and the internet. Broad convergence
of ICS, physical security, banking, facilities management, and metering onto
TCP/IP is ongoing and accelerating (Macaulay and Singer, 2011).
[16] More discussion of monocultures as a security risk within key systems is
provided in Section 12.2 and Section 11.3.

and Vanderbilt University (Cardenas et al., 2009). Simulators and emulators are a crucial component of understanding potential security concerns and controls for decision-makers in organizations that own or operate cyber-physical systems.

9.4 Conflicting values and design criteria

Cyber-physical security research has only recently come into vogue as industry, government, and academia attempt to determine which long-standing information and network security principles apply to the converging cyber-physical world (Mo et al., 2012). The effort is made particularly challenging as new innovations, techniques, and technologies are being rapidly developed to meet ever more stringent operational requirements that generally emphasize efficiency gains to achieve cost savings. Most engineered systems, including common industrial control systems and the internet, were designed with the intention of being resistent to random faults and common-mode failure (Branigan and Cheswick, 1999). Since these systems were designed by engineers for the purpose of accomplishing something positive, few were designed with security in mind from inception. Security to protect against adversaries who might attempt to create targeted failures is normally bolted on later. Unfortunately, a growing body of work demonstrates that resistance to targeted and random failures cannot be accomplished simultaneously (Barabasi and Albert, 1999). This is of immense concern for both control systems themselves and the underlying systems that form the current internet.

For electric power systems, almost every aspect of state estimation and control must be re-evaluated under the assumption that both random faults and nefarious actions by adversaries are indeed possible, or even likely. This is a major shift from the status quo. At a minimum, the increasingly interconnected and interoperable ICS and SCADA systems must source, transport, analyze, present data to system operators, and disseminate control instructions. System operators must consider the potential consequences of compromised systems or data at each of these levels.[17] Answers are far from clear, and sound policy choices must be rooted in assessments of

---

[17] See e.g. Fawzi et al. (2011) for a discussion of power system state estimation when sensor measurements are compromised by an attacker.

risk—which by definition requires an estimation of consequences stemming from the non-performance of particular assets such as specific control systems (e.g. at generators) or systems which support state estimation within particular balancing areas of the grid. A more thorough discussion of this challenge is provided in Chapter 12 where we evaluate the question posed by noted internet security expert Dan Geer in his thoughtful paper, "Resolved: The Internet Is No Place for Critical Infrastructure": "Is the creative destruction that is free enterprise now to be focused on remaking what are normally the steadying flywheels of American society, by which I mean government and other capital-intensive industries?"

Much like the internet, the currently envisaged grid will allow an arbitrary number of devices, sensors, and actors that gather, analyze, and interact with the information to make better decisions about their consumption (NCS, 2004). Manufacturers have begun to sell "smart grid-ready" appliances with some built-in energy management solutions. Electric vehicles and solar panels have built-in electronics and energy software—often integrated with more sophisticated telematics systems. Many standard building energy management systems can interface with and control common devices within homes and businesses to help limit energy expenditures or meet other goals such as environmental impact. This kind of connectivity can offer significant advantages.

The same tools required for responsible testing or design can also be used maliciously. All security technology is dual use. The same connectivity which enables these networked appliances to contribute to grid balancing via intelligent scheduling could be easily used to cause large and sudden shifts in demand to destabilize the system. Individually, these small spikes are manageable. At scale, they are much more threatening. For example, some household appliances, such as refrigerators, use inductive motors. Inductive motors consume different amounts of power during normal operation than during start-up. For example, if a common refrigerator requires on the order of 0.6 kW to maintain steady-state operations, it may require up to three times more power during start-up. With the new kind of connectivity enabled via the internet, the start-up of refrigerators could be coordinated to happen simultaneously. Considering the numerous disclosures of major vulnerabilities in typical household internet-facing home routers, this is a plausible scenario.

If a third of the 120 million US households all turned their refrigerators on at the same time, the surge in demand would be equal to roughly 216 GW before stabilizing at approximately 72 GW. The instantaneous surge is equivalent to the output of more than all the California-based natural gas–fired power plants operating in 2011, which had a combined capacity of 45.85 GW (Nyberg, 2012). At scale, this connectivity can expose the grid—and, by extension, society—to new kinds of threats regardless of whether or not existing SCADA vulnerabilities are addressed or how well utility control networks are secured and managed.[18]

9.5 Evolving and disruptive security requirements

The cybersecurity needs of the electric power system will continue to evolve more rapidly as commonly used utility networks continue to converge on standard internet communications technologies and domain-specific protocols (e.g. TCP/IP-based communications and standards such as IEEE 1547 for distributed energy resource interconnections and DNP3 for SCADA protocols) (IEEE, 2003; Clarke et al., 2004). The historic, haphazard proliferation of devices from many different vendors has created an urgent business need to standardize. As of 2010, the National Institute of Standards and Technology (NIST) has compiled a list of 75 different standards that affect the smart grid in some manner (NIST, 2010b).

The once-diverse set of protocols and devices provided some degree of security simply because developing exploits that would affect a large number of devices or systems was difficult, costly and time consuming. This movement towards standardization and interoperability has reduced this natural impediment to attack while also subjecting utilities to new vulnerabilities common to off-the-shelf technologies used more broadly in modern communications networks. Improved interoperability and standardization also increases the likelihood that previously isolated errors may propagate across more parts of the system and cause broad disruptions as opposed to isolated faults.

This tension among cost-effectiveness, ease of use, and heterogeneity is a good example of the need to develop priorities and acknowledge tradeoffs in cybersecurity. Providing lasting security in such an environment is particularly challenging because of the

---

[18] See e.g. Igure et al. (2006) for a discussion of common SCADA vulnerabilities.

requirement to remain responsive to both evolutionary and disruptive security needs simultaneously. Many, if not all, strategies or technologies for cybersecurity only remain useful until a particular threat changes their approach. Retooling for a new offensive action remains much cheaper and easier than defending against a broad set of potential threats and vulnerabilities.

Following the development of formative technologies, such as the solid state transistor, and concepts (e.g. Claude Shannon's Information Theory in 1948), a major explosion in electronic data processing and communications technology quickly occurred (Shannon, 1948). The growth of information assurance and security followed as a natural consequence of this technology's increasingly central role in modern systems and organizations. Initially, vulnerability identification and exploitation started with technology experts who were mostly troubleshooting and experimenting with their own creations. As devices grew more numerous, complex, and diverse, the system vulnerabilities increased. Experts soon formed collaborative groups to take advantage of increasingly specialized skill sets. The "development of additional testing tools correlated with the expansion of human teams [...] increasingly enabled less sophisticated technologists to take advanced actions," contributing to the democratization of hacking tools and services.

One recent example of increasingly sophisticated dual-use technology becoming freely available on Github is masscan (Graham, 2014). Created by Robert Graham, who also created one of the first popular desktop firewalls (BlackICE Defender) in 1998 and the first intrusion protection system (IPS) known as BlackICE Guard, masscan can be used to scan the entire IPv4 address space, which contains 4,294,967,296 IPv4 addresses, in under 6 minutes, at 10 million packets per second on a single machine (Graham and Johnson, 2014). To demonstrate this, Paul McMillan scanned all accessible IPv4 addresses on port 5900, commonly used for virtual network computing systems connections—returning screen captures from insecure machines that did not use any authentication, including industrial control systems, security systems, food production systems, and many others (Zetter, 2013).[19] Since scanning with masscan is an order of magnitude faster

---

[19]   There are actually 65,535 TCP and UDP ports which are accessible for each IP address, so selecting one of the useful and typically vulnerable ports is important for security researchers or hackers.

than other popular "consumer grade" scanners such as Nmap (Nmap, 2014) and Zmap (Durumeric et al., 2013), user capabilities have been amplified. This is a particularly powerful illustration of the larger trend of search tools such as Shodan—a search engine designed by John Matherly focused on enabling searches for control systems and the industrial internet by common services, ports, geographic regions, etc. (Shearer, 2010)—and Google, making vulnerabilities much more visible and accessible to would-be mal-actors or curious individuals.

While the increased visibility of these systems cannot always be mitigated, infrastructure operators can ensure that basic security and authentication are utilized to prevent unintended users like McMillan from viewing control interfaces or traffic. Beyond basic configuration, ICS-specific firewalls (e.g. Tofino), intrusion detection systems (e.g. Snort), and network traffic analysis tools (e.g. ntop) can harden networks against intruders and reduce the likelihood of unknown compromises. While the most common issues can be quickly and easily addressed with a combination of data traffic segmentation, data logging, pattern analysis, policy definition, and enforcement against the baseline network and application activity, new approaches are required to manage and secure the networks and devices in the grid and other critical energy infrastructures.

Many of today's primary approaches to cyberspace defense involve signature-based techniques—including patching, anti-virus scanning, firewalls, and intrusion detection systems—which focus on already-known vulnerabilities and impose a cost on the functional operation of the systems by introducing storage, computation requirements, and additional latency. There are practical limits to the tolerable number and costs of these systems, and many of their operational processes are not well suited to use within real-world ICS/SCADA systems.

Cyber-physical systems can also benefit from actual reductions in the number of potential vulnerabilities in common network components and even control systems themselves. Security expert Felix Lindner argued in *Software Security Is Software Reliability* that "if software always worked as specified or intended by its makers, only a small subset would be vulnerable to attack, and defenses would be much easier to implement" (Lindner, 2006). There are a number of research efforts seeking to address this challenge via investments—to include funds from the Defense Advanced Research

Projects Agency (DARPA)—in interactive software synthesis systems; verification tools such as theorem provers and model checkers; and specification languages that can result in software which is formally proven to meet functional, security, and safety requirements when run on satisfactory hardware in satisfactory operational conditions (DARPA, 2012). Since legacy systems, remote access points, and high-availability requirements of systems all mean that "patching and frequent updates are not well suited for control systems," this work may eventually prove extremely important (Cardenas et al., 2009).

Increasing costs and complexities of information technology and large-scale computing used in the everyday operation of critical infrastructure systems will require increased technical and managerial expertise within utility and energy companies. Due to the high capital costs and low marginal costs for computing resources, the need for highly resilient data centers with extensive physical backup and failover protection services, increased federal and international cybersecurity standards, and cost advantages possible for most organizations leveraging cloud computing, "utility scale" Infrastructure-as-a-Service (IaaS) and Platform-as-a-Service (PaaS) offerings will play an increasingly central role for most utilities.[20]

While limited investor-owned utilities (IOUs) with sufficient scale may embrace private data centers, most of the 3,000 US electric public utility districts (PUDs), rural electric cooperatives, and numerous other entities have much to gain from application deployments on virtual machines from IaaS/PaaS offerings such as Microsoft Azure or Amazon Web Services. Executives must keep in mind that these platforms still bring additional dependencies that come with their incorporation; the "cloud" still relies on physical systems which require power, cooling, and connectivity to perform their intended business functions. More comprehensive discussion of threats and vulnerabilities for cyber-physical systems is provided in Chapter 11.

9.6 Stuxnet and the debunking of ICS/SCADA security myths

From a security perspective, many early control system network designs assumed (usually incorrectly) that many ICS/SCADA

---

[20] See e.g. the National Institute of Standards and Technology release of Special Publication 800-53, Security and Privacy Controls for Federal Information Systems and Organizations.

systems—like those common to major power plants—were unreachable as a result of an "air gap" between business systems that would later connect to the internet (Leverett, 2011). While ICS/SCADA networks generally try to minimize connections with business systems, many within the industry actually assumed that isolation had been achieved: that a 'gap' protected them from would-be malactors. This kind of assumed lack of vulnerability from particular vectors would later prove dangerous as increased interaction between SCADA systems and other equipment (e.g. thumb drives, laptops that were used to access the internet and provide updates, etc.) became more common (Byres, 2013; Lemay, 2013). Stefan Woronska, Siemens' director of Industrial Security Services, advises today's electric power systems professionals to "forget the myth of the air gap—the control system that is completely isolated is history" (Byres, 2013). US government agencies, including the Department of Homeland Security's (DHS) ICS-CERT (Computer Emergency Response Team), also now acknowledge that true air gaps are effectively impossible in multiple published advisories. Seán McGurk, the former director of the National Cybersecurity and Communications Integration Center at the DHS, testified to Congress that "in our experience in conducting hundreds of vulnerability assessments in the private sector, in no case have we ever found the operations network, the SCADA system, or energy management system separated from the enterprise network" (McGurk, 2011).

In 2010, the discovery of the Stuxnet malware definitively proved what Woronska later helped evangelize to the industrial controls community—it was time to start thinking differently about security (Karnouskos, 2011). Widely considered to be the first fielded cyber-physical weapons technology known to have been utilized (Langer, 2013; Matrosov et al., 2010; Falliere et al., 2011; McDonald et al., 2013), the 500-kilobyte computer worm is known to have infected the software of at least 14 industrial sites in Iran (Kushner, 2013) but would later be found spreading across the globe in other systems which shared components with the Iranian systems. Sergey Ulassen discovered Stuxnet while working for VirusBlokAda, a Belarusian cybersecurity firm, whose Iranian clients reported a number of Blue Screens of Death (BSoDs) on their Windows machines (Kaspersky, 2011). Nearly 60% of the infected Windows hosts were in Iran, but infected hosts were in 155 countries from over 40,000 unique external IP addresses (McDonald et al., 2013). Soon after Stuxnet, Flame and Duqu—a piece of cyber-espionage malware that shares "striking similarities with Stuxnet" (Bencsath et al., 2012)—were also found throughout Middle Eastern countries (Kushner, 2013). Stuxnet targeted the Iranian uranium enrichment program at Natanz—causing premature failures of P-1 centrifuges and reducing the efficiency of enrichment efforts to delay the program by several years (Broad et al., 2011).

Stuxnet compromised systems in three primary phases. First, it targeted Microsoft Windows machines and networks utilizing multiple 0-day vulnerabilities to accomplish a series of tasks. One of these vulnerabilities, known as the win32k.sys 0-day, was used to escalate local privileges before completing payload malware installation and contacting command and control servers to report progress and seek guidance from the attacker (McDonald et al., 2013). Stuxnet spread using a self-replicating capability which allowed it to hide copies of itself on removable drives without user knowledge. Later versions allowed for network-based exploitation to infect additional machines. Once on a system, the malware sought out Siemens Step7 software, a common Windows-based control systems interface used at Natanz. While Stuxnet was highly automated, this general process is typical of many network intrusions.

Stuxnet had to accomplish two tasks to manipulate controls without detection—both the production system, which controlled the operation of centrifuges, and the protection systems, intended to detect abnormal behavior, needed to be compromised. The larger payload focused on compromising the top-quality Siemens S7-417 PLCs used to control the valves and pressure sensors of up to six cascades, or 984 centrifuges (Langer, 2013). The smaller and more obvious attack payload focused on the smaller S7-315 controllers believed to be connected to the Siemens WinCC SCADA system for monitoring centrifuges—likely one WinCC controller per six cascades with 164 drives each. Both attacks appear to have been designed to execute periodically and relatively infrequently—on the order of once per month—against very specific PLC logic using what is essentially a "man-in-the-middle" attack where the attacker makes independent connections to the victims and relays connections unbeknownst to either party. The *New York Times* remarked that "the attackers took great care to make sure that only their designated targets were hit. . . .

It was a marksman's job" (Broad et al., 2011). The malware ensured that the identified frequency converter drives were manufactured by the Iranian company *Fararo Paya* or Finnish company *Vacon* before altering frequency values expected to range from 807 to 1,210 Hz to new values between 2 and 1,410 Hz (Falliere et al., 2011). This wider frequency range was destructive to the equipment.

In addition to ending the myth of the isolated SCADA system which could be protected by tight control of personnel and limited external connectivity, Stuxnet also invalidated a trusted model of software legitimacy assurance (Shakarian, 2011)—the worm used certificates by Realtek Semiconductor systems and JMicron Technology Corp. to appear legitimate to Microsoft Windows (Matrosov et al., 2010). Since many industrial control system components such as the PLCs targeted by Stuxnet are used more broadly, there are also concerns about how the development of such capabilities by nation-states may lead to the proliferation of software which can be co-opted by non-state actors who would have otherwise been unlikely to develop such technologies—ultimately threatening governments' ability to safeguard national security interests (Kerr et al., 2010).

9.7 Chapter 9 Summary

- Samuel Insull's strategy of load diversification enabled utilities to optimize their infrastructure and resources. They could now serve more clients with a common infrastructure and lower levels of capital investment by managing disparate load profiles.
- Combinations of physical machines and computational control systems are known as cyber-physical systems. Numerous examples, such as automatic generation control, exist in infrastructure systems.
- Utilities and consumers should be aware that *any* security measure can be breached with sufficient effort. Concepts like "air gaps" are myths which have been thoroughly disproved by real-world events like Stuxnet.
- Standardization and interoperability increase the adoption of information technology into the grid. This also subjects utilities to new vulnerabilities that they have never faced before.
- Many "smart grid" concepts are computationally enhanced variants of proven methods for managing electric power systems. Increased

information sharing from a more instrumented grid is critical to increased integration of balancing areas and improving efficiency over the long term.

- Growing interdependence between electric power systems and telecommunications infrastructure poses new risks to both systems. Implications of these dependencies and associated risks are not suited to existing utility risk management or decision processes.

PART IV

Energy futurism

10 | *Towards better management of energy infrastructures*

The only thing we can be sure of about the future is that it will be absolutely fantastic. So, if what I say now seems to you to be very reasonable, then I will have failed completely. Only if what I tell you appears absolutely unbelievable have we any chance of visualizing the future as it really will happen.

Arthur C. Clarke (1964)

Modern societies currently face two distinct, but related, major challenges with core energy infrastructure(s):

- ensuring the reliable provision of energy; and
- managing rapid and continuing increases in infrastructure variability and uncertainty stemming from new technologies, services, global population growth, and urbanization.

Pragmatic approaches to these challenges require broad-based cooperation between numerous vested parties in addition to an understanding of the physical, informational, financial, and regulatory constraints and nuances.

Simulation and modeling will become an increasingly key component of grappling with this complex systems problem—it can aid us in gleaning knowledge about how, why, and how quickly we can, and should, attempt large-scale energy transitions. Shepherding society through a transition from today's fossil fuel–dominated energy industry poses numerous challenges, not the least of which is quantifying the cumulative effects of any shifts to the structure of our primary energy supply. The massive existing investments in infrastructure across the entire energy exploration, development, extraction, processing, and delivery pipeline preclude many of the breathless proclamations for near-term prime energy source reinvention. Said differently, with the "proliferation of inanimate prime movers"— i.e. robots, vehicles, and other machines that provide mechanical power from converted primary energy sources—transitioning to new mixes of primary energy sources continues to become increasingly complex and costly (Smil, 2010b). Thus, the confluence of the technical and infrastructure imperatives, economic and social changes, and behavioral adoption realities for large economies will result in a protracted period of transition where considerable economic opportunities and pitfalls await (Smil, 2010b). This human-imposed reality tempers Arthur C. Clarke's hopeful and unbounded view of transformation technological change used to open this chapter; however, human challenges frequently dwarf technological impediments to actual change.

Any potential efforts to accelerate this transition require realistic models at both the component and system level that can aid in discussion, adoption, and regulatory modernization using quantitative assessments, which can be periodically validated against actual real-world cases as the transition progresses. It is now an industry imperative to embrace multilevel, distributed decision-making capabilities that support wide-area energy system management for energy delivery, quality of service, and resilience.

This chapter provides some context for pragmatically assessing future transformations to energy systems, including a discussion of:

- complex versus complicated problems and systems;
- the current command and control paradigm for electric power systems operations; and
- an exploration of the transition towards an open-access electric power system.

10.1 Complex versus complicated

Our power system, largely ignored and forgotten by most today, is inextricably linked to the rise of industrial societies and the more recent development of the information economy. The inherent coupling of the physical and economic aspects of the system—electric power systems must balance supply and demand to maintain voltage and frequency stability—is a perfect example of a complex human–machine system. This balancing requirement has driven the power system towards the use of markets to enable multiple counterparties to participate in balancing the whole system. Today, the grid is on the verge of a major transformation which will touch every facet of modern life. Yet we have very little knowledge—and have made very little effort—to understand the effects of proposed changes when applied at scale, where emergent system-level behavior occurs.

If the true measure of a science is best obtained from the quality of its tools, the science (and art) of power systems operation leaves much to be desired. As a result, billions of dollars—forecasted by McKinsey and company as more than $130 billion of value annually (Booth et al., 2010) and at a total cost estimated between $338 billion and $476 billion within the US over 20 years (Gellings, 2011)—are being spent to upgrade components of a complex system with almost no input from the society the system is meant to serve. Huge amounts of money are being deployed around the globe, with the European Commission's Joint Research Centre reporting more than € 3.15 billion ($3.4 billion USD in 2015) spent or budgeted for more than 450 projects within the EU (Covrig et al., 2014). The lack of realistic data at the system level is astounding—especially because electric power systems have a number of peculiar behaviors. One example is known as Braess' paradox—a counterintuitive phenomenon best known for traffic networks. Adding new lines within transmission or distribution networks can actually increase congestion and destabilize synchronization of the grid in certain cases (Witthaut and Timme, 2012; Bjorndal and Jornsten, 2000, 2007). More investment can reduce power quality and efficiency in the absence of proper analysis.

It is not that any particular set of data or derivative models will be correct or provide perfect *a priori* knowledge of what the smart grid will do. But, as Bill Gates told journalist Ina Fried, simulation modeling is almost always a logical place to begin. He noted that "you are a hundred times smarter doing it that way than just going out there [without such models]" and personally funded disease-modeling software and other software model development to better understand human health concerns in the developing world as part of his work with the Gates Foundation (Fried, 2010). Reasonable models provide a point of departure from which we can iteratively make better decisions when solving complex problems. Simulation and modeling also provides an opportunity to allow non-experts to engage in "what-if" discussion and analysis leveraging science. This is aligned with Gates' related comment that "we do have a problem if we don't draw a large part of society into at least some understanding of science and the tools of science" (Fried, 2009).

Unfortunately, modeling remains drastically underfunded while investors pursue buzzwords like "geospatial analytics" which are often primarily focused on visualization instead of transformative technologies like real-time decision support systems fusing historical data with analytical models to move as far as possible towards establishing a causal relationship. As a society, we often remain fixed on adding widgets, feeling that the sum of the parts must make a whole or that simplicity is a universally appropriate goal. When this kind of reductionist thinking is applied to complex systems like electric power, it can be dangerous. Almost all of modern science (e.g. twentieth-century research) has been based on the concept of distilling information and interactions into much smaller—and ideally more understandable—models and equations. Throughout the industrial era, this approach of attempting to reduce complicated problems into simplistic models has been immensely successful. No longer. We have been spoiled by opportunities for large-scale economic growth which could be captured without looking at systems in a more holistic fashion.

There are dozens of apposite examples of industrial-era approaches delivering massive value to society and impacting human problems such as the spectacular advances in manufacturing silicon and mass productions of vaccines and antibiotics. For a less obvious example, consider Louise Fresco's work on the changes to global food production in the industrial era, which celebrates mass-produced agriculture and runs directly counter to well-known books that advocate for local food production, including Michael Pollan's *The Omnivore's Dilemma* (Fresco, 2011, 2014; Pollan, 2007). In each of these cases, the typical examples are based upon major improvements to the building blocks of industry and commerce. We have built a system which is designed for, and is successful at, cost-effectively producing widgets to increase marginal efficiency, not at designing new systems for large-scale gains and long-term sustainability. For example, the transition from agrarian society to industrial farming was easy when compared to the challenges we are soon to face: becoming sustainable in the long term with rising demand, increasing costs, and considerable uncertainty around issues such as weather and disease.

Although there will certainly be more advances at the "component level," we have exhausted many of the complicated problems. For better or worse, especially when evaluating approaches to large-scale energy problems, we have to embrace complexity. This is the difference:

- A *complicated system*—many of the problems we have solved in the twentieth century—can be simplified effectively and universally, often by viewing the constituent parts in isolation. The system's simplified form may not be as accurate, but the model created remains sufficient for its intended purpose.
- A *complex system*—many of the problems we are facing cannot be reduced by removing components. Removing a key element of the system causes it to collapse or to behave in such a substantively different manner that a simplified model has no utility. Components of the model can often be simplified, but the interactions between components, often involving interacting agents, must remain. The key characteristics or qualities of the system emerge from these interactions.

In many ways, complex-systems problems are the scientific equivalents of the "wicked" strategy problems introduced in Chapter 4. We are building new wind farms without sufficient shaping and storage capacity to integrate the highly variable and uncertain resource into regional power systems—such as excess energy events (NWC, 2011; BPA, 2012).[1] Thanks to major improvements in technology and attractiveness by companies like Tesla Motors, we are beginning to sell electric vehicles at unprecedented scale—the Electric Power Research Institute's US-focused smart grid report from 2011 uses 10 million electric cars in 2030 for a planning factor (Gellings, 2011)—but are doing so without a coherent regulatory framework or realistic estimations of the effect on the larger system. We are often overloading consumers with information about the number of kilowatt-hours they have consumed, a meaningless metric, while remaining blissfully ignorant of the fact that consumers are not, and never have been, interested in kilowatt-hours.[2]

---

[1] Usually resulting from a combination of high springtime runoff and low electrical loads when hydrostatic power generation (i.e. power from dams) is high and loads are low.

[2] For a detailed analysis of how information impacts energy consumption, see e.g. Darby (2006).

In short, we have applied lessons from previous technological revolutions without adequate consideration of their appropriateness for today. Smart meters, distribution automation, synchrophasors, etc. will all be part of an increasingly information-rich grid. However, as the internet has so clearly demonstrated on numerous occasions: information-rich does not mean "smart." Real intelligence will come from making sense of the data and leveraging it within actual operational processes to realize new overall efficiencies, increase power quality, or improve system resilience. Smart investments in infrastructure will similarly depend on the right mix of a wide variety of energy generation and storage technologies, however, the real performance gains and economic opportunities exist in the "glue" where coordination and transactions occur between vested parties.

Success, and perhaps even survival, of many modern-day systems and organizations increasingly depends on the ability to manage highly dimensional decision spaces where small changes may cause substantial shifts in outcome.[3] In many ways, complexity science typifies the convergence of scientific disciplines as we seek more holistic approaches to research and development of technologies. This is the direct result of our previously unimaginable computational power which "as a research instrument provides us with a new way of seeing reality, and the architectonic of the sciences must change accordingly" (Pagels, 1988).

China's attempt to rid itself of "the four great pests"—rats, flies, mosquitoes, and sparrows—during the Great Leap Forward is among the most tragic twentieth-century examples of myopic thinking applied to complex systems. After an immensely successful campaign against sparrows, crops were savaged by a plague of locusts (normally eaten by sparrows) with devastating consequences for rural farmers (Shapiro, 2001). A lack of understanding of relationships at the system level directly contributed to the millions that died of starvation between 1958 and 1961 (Dikötter, 2010). Total effects of the campaign are hard to quantify, but the lasting ecological imbalance caused by the campaign helped to underscore the importance of diversity within complex systems. As we have gained additional abilities to associate and correlate large data sets, we have been

[3] Generally due to non-linear characteristics of dynamical systems; see e.g. Lee (1992) for a technical description.

better able to validate that real-world ecosystems function "through the interaction of complex processes operating at different scales and time frames" (Swanstrom, 2008), that exogenous shocks to these systems coupled with diversity loss can trigger sudden system collapse (MacDougall et al., 2013; Ives and Carpenter, 2007; Hooper et al., 2012; Loreau, 2010; Barnosky et al., 2012; Rietkerk et al., 2004), and that species diversity can help provide buffering effects for environmental changes (Ives and Carpenter, 2007; Hooper et al., 2012; Loreau, 2010).

As our infrastructure, our services, and indeed the earth continue to progress towards some maximum capacity, overall resilience decreases—fewer buffers exist. Despite learning that socioecological systems in the natural world are undermined by a lack of diversity, we have continued to seek monocultures in our infrastructure systems, including electric power—this idea is explored more deeply in Chapter 12. We can choose to see complexity as an opportunity or as a precipice. Addressing challenges, particularly in electric power systems, demands simultaneous efforts at the nexus of energy, information, and finance. Stephen Hawking was right: the twenty-first century will be the "century of complexity" (Hawking, 2000)—the greatest opportunities lie precisely where systems, industries, and actors converge.

10.2 The future of the grid

Two dominant competing paradigms exist for the electric power system: command and control versus open networks. This is an oversimplification—a false choice—but the construct is useful enough for the purposes of highlighting major philosophical differences. Returning to the principles of today's bulk electric power system—the wholesale market between major generators, consumers, and resellers—at the RTO/ISO level helps to frame our current societal approach:

1. The electric grid requires supply and demand to be in balance, always.[4]

---

[4] This is a physical principle based on the First Law of Thermodynamics. Electricity travels at the speed of light and takes the path of least resistance, moving in all directions through conductors until balanced.

2. Grids maintain reliability by continuously varying generation to equal demand in real time.
3. Utilities fundamentally manage cost and risk to serve load.[5] Controllable, certain, and constant loads (over a given time period) are valued at a premium.
4. Wholesale markets are used to balance the system, with bilateral contracts relating a few hundred counterparties at a time. Pricing is determined using algorithms which determine the value of energy at each considered *location* and *time*.
5. User control at the retail level is equated with system instability, and utility control over users is seen as the solution. Utilities seek to utilize load as a resource, but generally one that they control.

Command and control networks at the utility level are perhaps best represented by the classic AI computer "HAL 9000" (Heuristically programmed ALgorithmic computer) made famous in *2001: A Space Odyssey* (Clarke, 1968). A supremely powerful version of HAL with access to numerous sensors and control over a wide range of devices ranging from electric cars to air conditioners would have the ability to significantly improve the efficiency of electric power system operations. Proponents of this general view point to concepts of "utility computing" as described in the *Big Switch: Rewiring the World from Edison to Google* (Carr, 2013) and the increasing level of interconnectedness between energy-hungry telecommunications networks and power systems. While the intertwining of energy and information is clear, the idea that centralizing intelligence within a command and control center to expand the reach of today's utilities is both short-sighted and dangerous.

Balancing the grid

Command and control approaches to electric power systems will ultimately fail because they are not long-term economically optimal at the system level. Simply put, this approach to balancing the grid does not scale to meet the future needs of the power system—especially if distributed energy resources (e.g. residential solar) or significant amounts of renewable energy are desired.

[5] Load is synonymous with end-user demand in this context.

To enable equitable participation in the bulk power markets, utilities gain access to common data sets via the Open Access Same-Time Information System (OASIS). This data enables them to make decisions about whether or not to submit an offer to produce electricity in a future time period based on a combination of factors which include start-up costs, running costs versus output, maintenance costs, and environmental costs associated with a particular plant.[6] Generally, power producers provide a *composite bid-curve* which is used by the economic dispatch and unit commitment processes for scheduling. When submitting offers, power plant cycling costs (i.e. ramping up or down output) must be considered. These costs are amplified with increasing system variability and uncertainty resulting from renewable energy portfolio increases such as wind (Jordan and Venkataraman, 2012; Kumar et al., 2012; Lew et al., 2012). Many current models use a concept of *net load* which treats wind and solar energy as negative load since these assets are not directly controllable. This approach contributes to increased system volatility and increases costs because thermal generators such as steam coal, nuclear, and combined-cycle plants are not well suited to short-term forecast errors requiring near-term corrections to plant output. At low penetration levels, effects are minimal, but at higher penetration levels, they become unsustainably expensive and contribute to system instability.

The National Renewable Energy Laboratory's (NREL) Western Wind and Solar Integration Study analyzed the worst week in a mid-April operational scenario without wind or solar and for a case with 35% of capacity from wind and solar (Milligan, 2011). At higher levels of penetration, short-term forecast errors are significantly disruptive to baseload generators using current approaches to centralized unit commitment where the constraints are explicitly managed at the ISO level. At this level, decision-making becomes computationally infeasible (Ilic, 2010b). If demand response, and distributed energy are considered, there are far too many decision variables to consider, whereas if demand is assumed to be inelastic and only centralized generation is considered, then the number of potential future system states to be considered is huge.

---

[6] See e.g. Lee (1989) and Mendes et al. (2004) for additional information on environmental constraints—emissions and cooling—on thermal units.

As an alternative, an approach with substantial amounts of distributed energy using a model predictive control-based distributed interactive unit commitment (DIUC) process can help supply peak loads with demand response, providing a buffer to resource output variability (Joo and Ilic, 2010; Ilic et al., 2011; Xie and Ilic, 2009). For instance, consumers can consume more or less energy based on local availability of power from distributed resources and renewables without requiring the centralized baseload generators to experience the major volatility demonstrated by the NREL Western Wind and Solar Integration Study (Milligan, 2011). In the centralized generation and centralized control model, the system "must have reserve power equal to the load in the area" where power outages from non-performing renewable or distributed resources occur (Ilic, 2012). Ultimately, whether or not distributed generation makes sense depends on the costs experienced by consumers if power disruptions occur, but distributed generation can be effective in increasing system reliability and freeing centralized capacity to serve priority customers during periods of stress (Nazari, 2012).

Practically merging these diverse and comparatively unpredictable resources into wholesale markets requires an intermediate structure capable of bearing responsibility for aggregate performance—i.e. an entity must schedule forecasted participation within the bulk power markets and bear responsibility if a large number of distributed resources such as solar, wind, and demand response do not participate as promised. The wholesale markets deal in firm contracts. Here, the intermediate market structure can allow groups, or aggregators, to represent these resource pools in wholesale markets.[7] In overly centralized approaches, the incentives do not support integrating diverse resources at value to the system as a whole due to a lack of alignment between physical, information, and economic incentives (Ilic, 2013).

Managing the data

Even if some new mathematical technique for centralized unit commitment and security-constrained economic dispatch were to be developed, the practical challenges of capturing, moving, associating,

---

[7] See e.g. Joo and Ilic (2010, 2011, 2012).

and correlating data to support fully centralized control is impractical. Managing large-scale distributed computing platforms across broad geographic regions is complex, difficult to do well in practice, and subject to real-world limitations as to the degree of coordination of activities in time and space. The eight fallacies of distributed computing were developed by Peter Deutsch and James Gosling to aid software engineers in building complex data-processing systems when evaluating system architecture (Rotem-Gal-Oz, 2006):

1. The network is reliable.
2. Latency is zero.
3. Bandwidth is infinite.
4. The network is secure.
5. Topology does not change.
6. There is one administrator.
7. Transport cost is zero.
8. The network is homogeneous.

These fallacies underscore why command and control approaches—while attractive to utilities and regulators in the near term—are doomed to fail. Since all sensor data is collected and stored, at least initially, at the location of the energy generation, storage, or transmission assets all of which are distributed across the world, information must be moved to support analysis. Not only does this mean that the communications infrastructure connecting each asset to the central control node must function in an uninterrupted fashion, but the total distance traveled must also be within an acceptable limit to avoid unacceptable levels of latency.

When thinking about moving information over optical transport networks (i.e. fiber optic cables) to help facilitate maintaining the balance of the grid at large distances, continental-scale systems like electric power must account for the time it takes to traverse distances, even at the speed of light. It takes a photon more than 67 milliseconds to get halfway around the world (i.e. an antipodal point on the Earth's surface). This means that, similar to high-frequency trading which is extremely time sensitive, there are optimal intermediate locations to site information-processing infrastructure seeking to coordinate control signals for electric power systems (Bhupathi, 2010; Gomber and Haferkorn, 2013; Wissner-Gross and Freer, 2010). There are calculable limits to the degree of spatial and temporal coordination that can occur within our current understanding of physics. For power systems operating at continental scales, these considerations become immensely important.

Privacy concerns

Even without the physical limits to coordinated action, there are numerous privacy concerns in a centralized decision-making model where utilities, the government, or any third party stores large amounts of high-fidelity consumption data on behalf, or in addition to, the consumer (Department of Energy, 2012). In 2010, the U.S. Department of Energy acknowledged that the "long-term success of Smart Grid technologies depends upon understanding and respecting consumers' reasonable expectations of privacy, security, and control over who has access to potentially revealing energy-usage data" (DOE, 2010). To better frame the issues, Caroline Winn of San Diego Gas & Electric and Dr. Ann Cavoukian, Information and Privacy Commissioner of Ontario, published a paper describing how privacy-by-design could address consumer concerns about data—it was unveiled at the Grid ComForum West Conference on March 8, 2012, and set forth the following foundational principles (Winn and Cavoukian, 2012):

- embedding privacy requirements into smart grid designs and overall project framework;
- ensuring that privacy is the default stance of all programs;
- making privacy an essential design feature in smart grid systems and practices;
- solidifying privacy as a core objective of all smart grid projects;
- incorporating privacy end to end throughout the entire lifecycle of any personal information;
- bolstering visibility and transparency for smart grid efforts with consumers; and
- focusing on consumer privacy as a core foundational requirement.

Any centralized model where data about energy utilization is stored *en masse* will require careful protection—especially given the well-documented use of metadata by governments around the world. This issue affects all consumers, to include those in the US, because metadata is generally not subject to the same protections provided to other types of sensitive personal records such as financial or health information, or the content of personal communications. Now that it is possible to cheaply, quickly and easily collect detailed information about the utilization of energy, and other utility services or commodities, consumers are understandably nervous. The "distinction between content and metadata might have made sense decades ago when technology to collect and analyze data was virtually nonexistent" (ACLU, 2014), primarily because "metadata can actually be more revealing than content" (Cavoukian, 2013).[8]

10.3 Towards open-access transactive energy networks

Open networks have a better chance of improving the overall performance and profitability of electric power systems in an era of increasing connectivity and interoperability.[9] Open-access transactive energy networks focus on reducing transaction costs between producers and consumers of electricity who rely on the physical power system as a platform to facilitate and clear their commercial interactions. This requires embedding "intelligence and active control, management and interaction capabilities into the existing grid using advanced technologies sensors and controls" (Knight et al., 2013).

A well-designed system will be fundamentally about transparency, replacing the opaque structures and lack of access to network information that enabled Enron's infamous acts of corporate fraud and illicit business practices within opaque portions of the energy sector (Munson, 2005). This means operating at the nexus of energy, information, and finance to save money for all system participants while maintaining a continuous instantaneous balance between supply and the total system demand and losses as the result "of interacting, thoughtful (but perhaps not brilliant) agents" (Miller and Page, 2007). Richard Tabors of Charles River Associates wrote that "it is the Smart Market that will define and provide the market information—specifically the price, quantity, and control signals—by which both the Smart Customer and the Smart Utility will find

---

[8] For a detailed deconstruction of the often falsely asserted tradeoffs between privacy and security, see e.g. Solove (2011).

[9] Three general categories of interoperability—technical, informational, and organizational (Tolk and Muguira, 2003; Tolk et al., 2013; Knight et al., 2013)—have been developed for the smart grid, each of which carry their own specific benefits and risks.

mutually beneficial and economically efficient solutions" (Tabors, 2009).

It is through the "Smart Market"—which we refer to as intermediate market structures—that the meaningful opportunities to reduce costs, streamline operations, increase efficiencies, hedge risks, and expand into new markets and services arise. This parallels the Internet's evolution from an initially passive tool focused on sharing data into an active network of collaborators, eventually becoming an interactive content delivery platform that further matured into a major platform for business and consumer transactions.

Open networks are more analogous to *cybernetic* systems where intelligence emerges from the relationships, not from a single, omniscient entity. Thus, understanding—and sometimes selecting—the right rules and methods for obtaining and exchanging information facilitating economic transactions is paramount because of the cumulative effects of individual agents' behavior which produce a cybernetic control system. This is perhaps most eloquently described by Adam Smith in his treatise when he stated that emergent control without central planning was the result of individuals "led by an invisible hand to promote an end which was no part of his intention" (Smith, 1970). In this case, system participants create value for the system through well-defined mechanisms to ensure reliable system operation while remaining more conducive to privacy-by-design concepts.

The emergent principles of the electric power system are as follows:

1. The electric grid requires supply and demand to be in balance, always.
2. Grids maintain reliability using economic incentives to control energy transformations and interactions with the system.[10]
3. Intermediate markets are used to create liquidity within the constraints of the highly regulated wholesale market systems. Consumers interact with intermediate markets which facilitates energy transformation aggregation of diverse microevents into products which are similar to current wholesale market agreements.
4. Microtransactions can occur between participants via intermediate markets across an open-access grid.

---

[10] Energy transformations refer to any energy consumed or injected into the electric power system.

5. Users gain increased control without sacrificing system reliability.
6. Multiple types of energy-related services are available to consumers. Few will sell a bulk commodity. Consumers will choose between offers of long-term fixed-price service contracts, purchasing "cold air," or other energy outcomes.
7. Energy efficiency improvements and increased control technologies are financed by a rich ecosystem of investors and service companies.

Notice that we still emphasize meeting the fundamental physical constraints but embrace a new paradigm for the electric power system: equal and open access for all consumers, achieving the value of least-cost service via security-constrained economic dispatch (SCED), optimized investment, and maximized operations, which will result in increased energy independence, power sector reliability, and long-term sustainability.

A sense of where we are

Simulation and modeling is only now becoming capable enough to aid use in designing and operating better complex systems like energy delivery networks where very large numbers of market participants can exercise independent action and make their own choices. The new tools which are able to evaluate these large and complex systems are now emerging, enabled by the same distributed computing concepts that have helped give rise to modern-day internet giants like Google and Amazon. In the UK, a major smart meter simulation project was used to help the Office of Gas and Electricity Markets (OFGEM) make regulatory decisions about national smart meter implementations. Just as we are increasingly using these tools for understanding the future, the deployment of increasingly ubiquitous sensors has led Oak Ridge National Laboratory to create the Visualizing Energy Resources Dynamically on Earth (VERDE) tool for large-scale situational awareness of energy delivery networks (King, 2014) and analysis of potential policy options—such as state policy incentives for renewable energy generation (Sukumar et al., 2010).

At a high level, power grid models can be classified along three dimensions—operations, planning and evolution, and disruptions—as illustrated in Figure 10.1.[11]

---

[11] This figure is an extension of a depiction from DHS (2008).

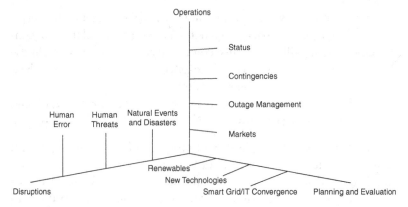

Figure 10.1 Three axes of power grid models

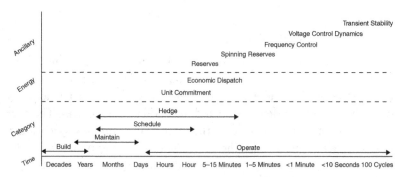

Figure 10.2 Temporal complexity of power system operations and modeling

Current state of the art in electric power system modeling is limited in its ability to solve realistic instances of unit commitment and economic dispatch problems in near real time, doing so with non-linear transmission constraints, incorporating stochastic analysis, and convergence of solutions for large-scale system optimization (Acar et al., 2011). The temporal complexity of modeling must be highly variable to enable the core modeling engines to aid in decision-making and analysis across the multiples time scales required to build, operate, and upgrade electric power systems. Different types of operations with different temporal characteristics are shown in Figure 10.2.

The representative types of electric power systems are also helpful to analyze. Five representative types which are explored throughout the book are:

- Regulated bulk systems
- Restructured bulk systems
- Hybrid systems
- Fully distributed systems (developed countries)
- Fully distributed systems (developing countries)

(Ilic et al., 2011; Ilic, 2007)

To get an idea of just how inaccurate some of the current control paradigms are, consider some of these examples. The following examples use root-mean-square (RMS) error. This is a frequently used metric of the differences between values predicted by a model or an estimator and the values actually observed.

- **Load forecasting errors:** The RMS errors for day-ahead load forecasts are usually 3–5% and reach as high as 10%. The errors will increase as more electric vehicles, distributed generation, and larger amounts of demand response are present. Similarly, time-of-use and dynamic pricing will clearly stress utility control centers, which have plenty of forecasting issues already with much more stable loads. Errors in forecasting make unit commitment and economic dispatch scheduling processes less efficient, increasing costs to the consumer. Even the time resolution of the load data collected from consumers and used for estimation can affect normal distribution loss estimates required for accurate scheduling and purchasing in the bulk power system (Urquhart and Thompson, 2014).
- **Wind forecasting errors:** The RMS errors for day-ahead wind energy forecasts are commonly as high as 30% of total production. The uncertainty (whether or not the wind is blowing) and the variability (is it blowing constantly) limit the practical amount of wind which can be integrated into today's grid. Wind forecasting relies on numerical weather prediction models to forecast wind speed and direction, which is then fed into analytics to estimate turbine output for day-ahead unit commitment and (usually) hour-ahead economic dispatch scheduling processes (Hodge et al., 2012). Developing confidence intervals (e.g. 95% confidence that a particular wind

turbine will generate 10 MW and 90% confidence that output will be more than 12 MW) is critical, as any shortfalls in scheduled production must be accounted for via marginal producers like natural gas (Hodge and Milligan, 2011). These errors are costly; one study of the Spanish electricity market found that forecast errors for each MW of installed capacity averaged between €15,000 and €18,000 per year (Fabbri et al., 2005).

- **Loop flow:** The RMS errors of unscheduled electricity flows between control areas—each utility or balancing authority's area of responsibility—commonly exceed 100%. Little progress has been made to properly model and eventually account for these flows, which amount to unintentional commercial exchanges. Electricity, in accordance with Ohm's law, takes the path of least resistance. It cannot be "routed" in the same manner as data transiting modern packet-switched telecommunications networks. A combination of "network effects," non-linear effects of ambient temperature, and non-thermal transmission constraints are responsible for the challenges. For example, from June 2007 until the end of December 2008, the Michigan Electric Coordinated and PJM Interconnection boasted a 482% RMS error of inadvertent power flow relative to the mean of absolute scheduled flow (Forbes and Cyr, 2011). Other problems include data availability and the size of the mathematical problems to be solved. Loop flow can only be partially addressed by technologies such as phase angle regulators which can help control power flows—but in an inflexible manner that is strongly linked to their designed purpose (e.g. facilitating energy flow into New York City via the New Jersey transmission grid) (Champ, 2010; Midwest ISO and PJM, 2007).

Clearly, substantial opportunities for improving our models and control systems remain.

Towards an open-access power system

The DYMONDS-enabled physical grid proposed by researchers at Carnegie Mellon is shown in Figure 10.3. It is based on just-in-time (JIT), just-in-place (JIP), and just-in-context (JIC) delivery models with extensive use of information technology for decision-making (Ilic et al., 2011). The proponents of DYMONDS go as far to

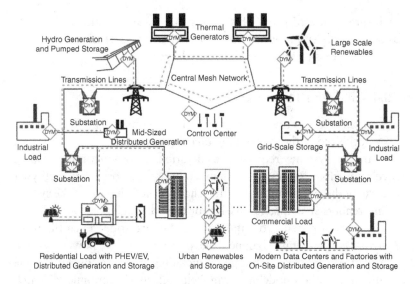

Figure 10.3 DYMONDS-enabled physical grid

state that "it is indeed not possible to reach energy and environmental objectives expected by society without beginning to rely on [Information Communications Technology] automation" (Ilic et al., 2011). DYMONDS, as proposed, is a logical evolution of existing energy management systems (EMSs)—or utility control systems—and the supporting supervisory control and data acquisition (SCADA) systems. These systems work in concert to obtain data required for decision-making and coordinating control signals at the bulk electric power system level within single utility power systems (Ilic, 2010a).

Notice that the diagram shows multiple DYMONDS data flows complementing the limited number of existing SCADA data flows within the bulk transmission system today—each DYMONDS emblem, shown as "DYM" in the figure, represents additional intermediate markets, as previously described. Microtransactions can occur within each of these hierarchical systems, getting larger towards the bulk power system and getting smaller towards the individual homes or cars at the bottom of the figure. Consistent with privacy-by-design concepts, these hierarchical markets can, in aggregate or via participants who trade in higher-level markets, represent the aggregate position of the energy resources being consumed and generated at that location on the grid for each current or future period of time. This means that sensitive data can be kept as close to the end user as possible and that higher-level systems need not have access to lower-level data with high fidelity.

10.4 Chapter 10 Summary

- Complicated systems can be reduced to components and evaluated or optimized without changing their behavior.
- Complex systems require consideration of interactions between components. Removing system components, or oversimplifying parts of the system meaningfully, changes the behavior of the system and renders such analysis useless or even damaging.
- The laws of physics still apply to the "smart grid." Even the most advanced grid must balance the physical electric power system. Advanced control techniques, including the use of large-scale simulation and modeling, can help improve economic dispatch and unit commitment processes used to cost-effectively balance the system.
- With advanced information and communications technologies, open-access electric power systems that provide just-in-time, just-in-place, and just-in-context electricity services are possible.
- Substantial work remains to be done to manage end-user privacy concerns while still ensuring that the data required to operate maximally efficient electricity networks is available to market participants.

11 | *Risk management in energy*

In the business world, the rear view mirror is always clearer than the windshield.

Warren Buffet

Modern societies' ability—or, more realistically, their perceived ability—to model and quantify risk remains the central and revolutionary idea that defines the boundary between modern human societies and our precursors (Bernstein, 2007). In the business world, most discussions of risk begin with establishing the likelihood of something occurring in the future, because risk and uncertainty are separated by the notion of an undesirable outcome. The risk stemming from an undesirable or catastrophic event implies an understanding of both the certainty of the event and the cost or impact of the occurrence itself. This is more easily asserted than accomplished in practice. The widely used paraphrase of famous baseball manager Yogi Berra or physicist Neils Bohr is applicable here: "It's tough to make predictions, especially about the future."[1]

Humans tend to struggle with reasoning about randomness or uncertainty. Many mistakenly overly attribute success to personal accomplishments or skills while assuming that failures were the products of chance, as described at length in Nassim Nicholas Taleb's book *Fooled by Randomness* (Taleb, 2001). Misuse of heuristics and projection of pre-existing human biases via data is also well documented (Tversky and Kahneman, 1974). Forecasters also commonly underestimate the likelihood of impactful but rare events which Taleb refers to as "black swans" (Taleb, 2007) and

---

[1] Versions of these quotes are widely attributed to both Neils Bohr and Yogi Berra. A fun and thorough frequency analysis of attribution was published by blogger Peter Patau in December 2006 after reading a piece by Paul Krugman entitled "Economic Storm Signals" in *The New York Times*.

others have called "dragon king" events (Gauthier et al., 2013). These low-frequency, high-impact events are hard to predict, although some work has been done with the intent of elucidating hidden organizing principles (e.g. Barnosky et al., 2012; Comfort et al., 2010; Field et al., 2012; Nott, 2006; Gauthier et al., 2013). The dragon king–type statistical distribution closely matches a power law distribution until the "bubbling events" within the complex system cause a sudden increase in high-consequence events.

In general, estimating the impact of rare events is even more difficult than assessing likelihood. Often, little to no actuarial data or experience exists to inform estimates or response planning. Miscalculations about risk or likelihood frequently require managers to make numerous simplifying assumptions. The reasonableness of these assumptions, commonly made more optimistic by management and regulators, can lead to significant failures in investment or policy. Fukushima, described in detail in Section 13.3, is a major example affecting energy and security. Earthquakes, volcanic eruptions, tsunamis, and solar storms are some of the rare but extreme geophysical events with sufficient magnitude but insufficient frequency to estimate accurate probabilities of occurrence (Love, 2012). A less dramatic example of this kind of failure in strategic judgment is the emergence of the personal computer. In 1977, Kenneth Olsen, then-president of Digital Equipment Corporation (DEC), announced that "there is no reason for any individual to have a computer in their home" (Courtney et al., 1997). At the time, performance improvements and cost reductions in mainframe computers indicated that gradual evolution was most likely, but the rapid emergence of the PC was openly discussed. If impact to DEC and likelihood had been plotted at the time, it would likely be reminiscent of the same chart used to depict the impact of earthquake events and tsunamis at Fukushima.

Assuming certainty is tempting, especially for planners and executives seeking to make precise predictions about future earnings, profitability, or returns on investment. Business processes that require overly specific forecasts about the future are dangerous because they "force managers to bury underlying uncertainties in their cash flows [...] in order to make a compelling case for their strategy" (Courtney et al., 1997). However, the assumption that assuming that everything about the future is uncertain creates the conditions for inaction; if nothing is assured, then modern concepts of investment to attain greater long-term value are invalid. Real risk management is about making better decisions to manage this inherent tension. The science and art of doing so are central to making decisions about energy—for individual consumers, energy companies of all types, and governments alike.

11.1 Security and risk

The ability to estimate risk over a variety of timelines and to make associated investment decisions as part of an intentional strategy is core to success in the modern enterprise, especially as our world has been made increasingly volatile by numerous assumptions to seek efficiencies by optimizing business processes and technologies for a narrow range of expected system states. Deviations from future expectations undermine the suitability of previous resource investments, which are the most tangible and direct manifestation of business strategy. This *surprise* is the source of insecurity. Some surprise—whether driven by technological change or by innovations that create new business or organizational processes—is inevitable, but *mitigable* surprise is avoidable. Therefore, costs which result from mitigable forms of surprise are the focus of our efforts as risk managers.

Mitigating surprise means that a range of likely futures must be considered. This is generally a balance between some level of a "good enough" estimation of a likely vector, or set of trends, and specific scenarios. McKinsey and Co. partner Hugh Courtney suggests a model with four levels of uncertainty to aid decision-makers in grappling with the challenge of deciding which approach is most appropriate. The lowest level results in describing "a clear and single view of the future," the second level is a finite "set of outcomes, one of which will occur," the third is a "range of possible future outcomes," and the fourth is a "limitless range of possible future outcomes" (Courtney, 2001). The appropriateness of the approach is closely linked to the time scales being considered—prediction in the short term is more likely to be accurate than that in the long term. The more time expected to transpire, the more likely it is that technological changes, environmental changes, or social changes with significant impact on both the consequences and probability of small events will occur.

Finally, decisions about the future are commonly viewed from the lens of how changes might pose new risks for the system, a business, or society. Risk analysts who treat the world as if it is in stasis at their own peril. Inaction is not risk-less or effective at delaying real change, however comforting. Changes to the relationship between systems and their environment progress continuously, whether they are intended or not.

Understanding operational risk for electric power systems

Before broadening our discussion to include risk management for long-term investment, we explore near-term operational risk management for electric power systems from the perspective of the smart grid. Professionals seeking to manage risk within the integrated energy supply chains or delivery systems, especially the sensitive electric power system, can employ a simple framework for initial estimation of operational risk in the context of their core organizational functions. The process begins with the identification of risks and potential material or organizational process solutions which provide some degree of mitigation. Fundamentally, "risk is a consequence of dependence," and security is a subset of reliability. A practical security model must be focused on the *raison d'être* for the system being managed, what Jim Collins calls the "economic engine" of the business in *Good to Great* (Collins, 2001). For today's electric power systems, this means maintaining system balance while minimizing system cost to consumers.

Risk management is entirely about decision-making—especially impacting the future by focusing limited time and resources on actionable factors which have a causal relationship with future system states. Reliably enabling rational decision-making requires ensuring both the availability and the quality of information required. Thus, information flows relevant to actual system state estimation are required at both the component level and for situational awareness to maintain system stability. Control signals used to coordinate future action and coordination signals that facilitate economic transactions between system participants must also be managed.

The following security model is consistent with the major principles of the Transactive Energy framework published by the Gridwise Architecture Council (GWAC) and is provided to better elucidate the full range of information which must be secured (GWAC, 2014):

- Transactive energy systems should provide for non-discriminatory participation by qualified participants.

- Transactive energy systems should be observable and auditable at interfaces.
- Transactive energy systems should be scalable, adaptable, and extensible across a number of devices, participants, and geographic extent.
- Transactive energy systems should use highly automated coordinated self-optimization.
- In transactive energy systems, transacting parties must be accountable for standards of performance.
- Transactive energy systems must maintain system reliability and control while enabling optimal integration of renewable and distributed energy resources.

Disruptions to the broad transactive electric power systems can stem from a wide range of potential events. In addition to the cyber-specific risks, which are subsequently discussed in more detail, major physical risks to the electric power system include: space weather; earthquakes; volcanic eruptions; tsunamis; tornadoes; hurricanes, monsoons, typhoons; and the targeted destruction of control infrastructure, transmission, distribution, storage or generation assets. These principles can be extended into a more descriptive, functional security model which is appropriate for any of these cases as follows:

- **Information exchange:** Transactive energy is about enabling the development of a robust energy economy. Like all other marketplaces, electronically enabled platforms become more efficient when interactions can occur between all qualified participants via a robust and standardized information exchange. The Transactive Energy Market Information Exchange (TeMIX) model has been proposed as a foundational protocol that is analogous to the Financial Industry Exchange (FIX) Protocol that enables global, high-speed, high-volume messaging for financial transactions (Cazalet, 2010).
- **Transaction authorization:** Transaction authorization mechanisms must reflect current best practices for unit commitment. This means that consumers' decisions to consume or produce energy must be in response to a transparently communicated price for a specific time at their particular location on the system. End users or their agents, must be able to easily interact with utilities in the same manner that utilities interact with one another—potentially as a group or via a third party.

- **Transaction management:** Real-time *verification* of transactions and performance records is necessary. All transactive platforms must allow counterparties to manage the risk of non-performance—for direct and indirect effects (e.g. non-availability of energy they've procured and needed for a business process). Since the practical availability or desirability of all potential energy sources or consumption is dependent on the economics of participation, transparent energy prices enable customers of all sizes to participate in the electric power system.
- **System performance and reliability:** Management of emergent phenomena (i.e. overall system state and performance) and broader equities of all market participants often includes aspects of all three of the previously introduced areas. However, here we focus on those specific attributes which transcend typical individual or organizational bounds.

The enabling information flows move along logical and physical information and communications technology infrastructure. Mapping the information flows is foundational to bound dependence, which "is the core of rational risk management" (Geer, 2012).

Commodities markets

For energy producers, resellers, or consumers of sufficient magnitude, commodities markets are an indispensable component of risk management. Commodities markets emerged to help ensure more predictable cash flows for producers and consumers. There is some evidence that an early form of rice "futures" may have been traded in China more than 6000 years ago. There is also historic evidence from the Greeks' complete forward purchases of maize from Egypt more than 3000 years ago and seventeenth-century Japan futures trading on the Dojima Rice Exchange in Osaka (Wakita, 2001). A *futures contract* is the right to buy or sell a commodity or financial instrument at a fixed price on or before a specified date. Common modern commodities traded on US exchanges and regulated by the U.S. Commodity Futures Trading Commission include oil, natural gas, gold, silver, wheat, and corn.

Market history and fundamentals
Commodity futures differ substantially from stocks, bonds, or other financial instruments commonly used by individuals in that they

are derivative securities which are short-term maturity claims on real assets with substantial amounts of typical price volatility and seasonality (Gorton, 2004). As a result of these characteristics and their relative obscurity, commodities futures markets have always been viewed as confusing and even reckless by non-experts who mistakenly assume that derivatives themselves are responsible for increased prices of many necessary goods. The following declaration from the Japanese Shogunate in 1696 is reminiscent of more recent, sensationalized calls to ban *all* derivatives in the wake of the global financial crisis of 2008, despite their important role in stabilizing prices when used responsibly:

There are people who say they are just buying/selling rice, but instead set up a venue, invite many people, ask the participants to pay fees, set a due date, and speculate on prices in the market. As this is almost like gambling, we ordered them to stop this immediately. However, we have heard that these people are gathering again and frequently participating in this activity, which is totally outrageous behavior... (Sugie, 1985)

Informed by this perspective, Tokugawa Yoshimune, who became the Japanese Shogun in 1716, legalized rice futures trading between 1722 and 1725 in an attempt to increase commodities prices which had fallen as a result of good harvests and negatively impacted the nominal income of bannermen (Moss and Kintgen, 2010). The popular belief was that prices would be inflated by traders and that the markets encouraged hoarding (Moss and Kintgen, 2010). Yoshimune was wrong; prices fell after trading resumed.

American futures markets emerged after the Civil War as a result of efforts by the executives of household agricultural names like Archer Daniels Midland, General Mills, and Pillsbury. Because farmers and millers could sell wheat and other grains at a reasonable bushel price in advance, even before planting, they could reliably invest in their farms and businesses without exposure to the substantial price swings common in the actual global market. Just like the first rice bills in Japan, contract holders could resell the "paper" to a third party or hold it and complete the transaction, receiving the real asset in accordance with the prescribed timeline. When the Chicago Board of Trade (CBOT) was founded in 1848, it was focused on facilitating this kind of hedging between *bona fide* participants (Taylor, 1917). This meant that the counterparties in the contracts had a legitimate business use for the underlying products being exchanged and were not primarily seeking some form of purely financial arbitrage. The CBOT protected buyers and sellers by facilitating the initial transaction, inspecting payments, and ensuring that the ultimately delivered product met established quality standards from the contract.

This kind of hedging differs from speculation, where financial gain is the primary goal and is accomplished via long or short positions in specific commodities. The term *hedge* has been confused by *hedge funds*, which typically take a net long or short position but protect themselves from downside losses by simultaneously taking a smaller, but still opposite, position. Markets which are overly skewed towards speculators can suffer consequences, as the additional liquidity provided by additional participants can be eclipsed by increased volatility divorced from real-world business requirements. A recent exposé by the *New York Times* raised a number of questions about the role of financial players in the markets, but consumers and policy-makers should avoid snap judgments (Creswell and Gebeloff, 2014). Financial participants are a necessary component of ensuring sufficient liquidity for firms seeking to hedge their risk in both *physically* and *financially* settled electricity markets.

Electricity spot markets

In 1878, Chicago added an additional facility to facilitate "spot" transactions where wheat could be exchanged immediately for cash (Kotze, 2011). This same concept of a "real-time" market is used to facilitate financial transactions in energy delivery networks like natural gas pipelines and electric power. Spot markets and futures markets are highly correlated, but futures markets enjoy a *risk premium* which compensates the holder for being financially responsible for potential changes in the future spot price between the time of purchase and the execution date. This risk premium is clearly visible if both an inflation-adjusted and equally weighted index of spot commodities with equivalent collateralized futures contracts are plotted together.[2]

Since electricity is not storable and demand is strongly correlated with both seasonal and specific weather events, electricity spot market volatility is "unprecedented in other commodity markets,

[2] See e.g. the comparison from 1959 to 2004 in Gorton (2004).

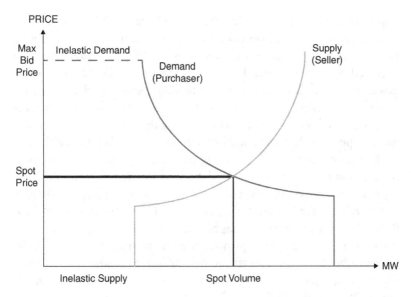

Figure 11.1 Principles for nodal price calculation for electric power systems

with annualized volatilities of 1,000% common" (Johnson and Barz, 2005). Investments in generation or transmission typically take years, or even decades, to plan and execute, while electric fields move near the speed of light. This coupling between the financial markets and the state of the electric power system is unique to electric power. Futures markets help bridge the gap.

For electricity markets, today's "spot market" is typically a day-ahead market offering hourly contracts, or blocks of hourly contracts, to cover electricity scheduled to be delivered the following day. Contracts are typically executed based on each node in the system, where locational marginal prices are varied by ISO/RTOs to maintain stability. A diagram of the basic principles for price calculation is shown in Figure 11.1.

Electricity markets use this spot market process and the resulting price to address the physical exchange of electricity but use derivatives markets to manage the financial risk for generators and retailers. Dramatic events like the bankruptcy of a Texas-based retail energy provider due to spot market price exposure during a 3-day ice storm in 2004, and the California electricity crisis of 2000–2001 where utilities had insufficient long-term supply contracts, underscore the importance of proper hedging (Deng and Oren, 2006). Electricity futures and options are traded on organized commodities exchanges, but substantial growth in over-the-counter (OTC) markets has also occurred in recent years. In North America, many of these OTC products are traded across the InterContinental Exchange (ICE). Bilateral contracts traded on platforms like ICE enable "market participants to tweak their positions during the day" and "can become quite important during periods of system stress" (Whitmore, 2008). While near-term physical balancing procedures effectively implement some version of the principles shown in Figure 11.1, longer-term financial and physical markets vary more substantially (Kaminski, 2012). Even in the US, bilateral markets for financial risk management differ deeply from RTO-managed markets and by region. Bilateral markets in the West help provide near-term price discovery, while their southeastern equivalents are primarily for long-term investment risk management within a traditional vertically integrated utility model (Whitmore, 2008).

European markets also differ from their North American counter-parts. The Scandinavian "Nord Pool" financial derivatives market has been in operation since 1995 and offers four purely financial contracts: base load futures, base load forwards, options, and contracts for difference (Botterud et al., 2002). The system is designed so that a financial contract holder, who can purchase supply or demand contracts as far as 8–12 months in advance, can submit an equivalent bid into the physical spot market administered by the Power Pool, RTO, or ISO during the week of delivery and can be completely hedged. This means that the financial exposure in the futures market is equivalent to the exposure in the spot market for contracts purchased at a fixed geographic location, for power of specific quality, delivered at the same time as the contract matures (Deng and Oren, 2006; Mercatus Energy Advisors, 2013).

Cyber risk management

The adoption of transactive approaches to facilitating interoperability and operational flexibility at all levels of the electric power system is a crucial component of long-term grid modernization. It is also central to managing operational risk in the short and long term, as illustrated by the brief overview of the futures markets. Transactive approaches are better suited to enabling more consistent and scalable forms of decision-making, especially since real-world electric power systems physically balance at rates exceeding human capacity to reason *or* act.

After identifying the critical services required to successfully operate an electric power system and the associated information flows which enable them, cybersecurity can be viewed in context with the other kinds of potential risks to electric power. Many cybersecurity reviews for business are myopically focused on technical challenges, with little regard for actions affecting value for the organization.

In practice, this extends to other systems, including bulk fuel supply and limitations stemming from waterways.[3] For business leaders and policy-makers, this is extremely challenging when attempting to estimate or manage risk, because "interdependencies in critical infrastructure have multiplied to the extent that it is difficult, if not impossible, to define defensive perimeters" (Clemente, 2013).

To manage risk stemming from our ongoing challenges with cyber*in*security, utilities, consumers, and regulators alike must heed the recommendation provided in defense policy specialist Richard Danzig's recent report. He prescribes a "four-part strategy of abnegation, use of out-of-band architectures, diversification and graceful degradation" to reduce large-scale cybersecurity risks to the US (Danzig, 2014). Energy decision-makers should acknowledge that cyber-enabled systems are now core to modern energy delivery networks, especially electric power, and attempt to estimate financial losses for known cyber and physical risks. The first step in applying a risk management strategy is the creation of a sufficiently representative model that can be used to help mitigate surprise. This is consistent with the general principle espoused herein, which is that achievable levels of security are maximized in the "absence of unmitigable surprise" (Geer, 2012).

11.2 Threats

When evaluating the security of cyber-physical systems such as those used in energy delivery, it is important to separate the notion of

---

[3] This is explored further in the Indian power outage case study on in Section 12.4 and discussion on the relationship between water and energy in Section 13.1.

vulnerabilities from threats and the broader notion of risks.[4] For our purposes, a *threat* can be described as a malevolent actor possessing specific intent and the capability to cause harm (Duggan et al., 2007; DHS, 2010). These threats must be categorized without regard for whether they are physical, cyber, or some combination. While it may be theoretically possible to achieve a particular effect via a cyber-action or a physical action, threats refer to actors with malintent who are likely to seek maximum impact at minimum personal or organizational costs—that is, we should not be surprised that the June 9, 2014, national power outage in Yemen caused by al-Qaida in the Arabian Peninsula (AQAP) was an act of physical sabotage and not, for example, a cyber-attack.[5] Given sufficient time and resources, *any* system or perimeter can be breached or destroyed by a motivated actor.

While many popular discussions of cyber threats are reduced to a list of potential representative actors, including script kiddies, hacktivists, non-state actors (e.g. terrorist groups or organized crime), insiders, and nation-states (Mateski et al., 2012), more detailed analysis is generally required to enable real risk management or improve future outcomes. Since real-world systems must allocate finite resources to address known vulnerabilities and protect key data, systems, and user credentials, some characterization of targeting and effort is required. In practice, this requires an ability to value the impact of specific components of the grid performing at a reduced level of effectiveness or being lost entirely.[6]

Duggan et al. (2007) developed an eight-level threat model using estimates of commitment and resourcing required to harm a system (Figure 11.2). Since most attackers are likely to expend the least amount of effort required to achieve their goals, identifying a range of potential attacks and common approaches to prevention or disruption can aid in the rigorous prioritization of potential mitigating actions (Mateski et al., 2012). Preventative or responsive measures must be catalogued and viewed against the cumulative set of threats to "avoid

---

[4] Risks are based on a measure of consequence and likelihood and can be physical, such as an earthquake, or virtual, such as a hacker.
[5] The attack occurred at Kilometre 78 on the transmission lines connecting the capital, Sanaa, with an AQAP stronghold in Marib—the majority of Yemen's 23 million residents were affected (Aljazeera, 2014).
[6] Some work has been done to value the impact of specific infrastructure losses within electric power systems—see e.g. Moslehi et al. (2006).

Threat Profile							
	Commitment			Resources			
					Knowledge		
Threat level	Intensity	Stealth	Time	Technical	Cyber	Kinetic	Access
1	H	H	Years to decades	Hundreds	H	H	H
2	H	H	Years to decades	Tens of tens	M	H	M
3	H	H	Months to years	Tens of tens	H	M	M
4	M	H	Weeks to months	Tens	H	M	M
5	H	M	Weeks to months	Tens	M	M	M
6	M	M	Weeks to months	Ones	M	M	L
7	M	M	Months to years	Tens	L	L	L
8	L	L	Days to weeks	Ones	L	L	L

Figure 11.2 Generic threat matrix

spending too much or too little on risk management" (Clemente, 2013).

According to a congressional report, utilities are under a constant state of attack with one utility reporting to be targeted by about 10,000 cyber attacks per month (Markey and Waxman, 2013). We focus on exploring specific cyber-threats, since physical threats to the diverse set of distributed utility assets are more straightforward and familiar. To enable a discussion of risk management, it is important to categorize specific cyber-physical system threats into an ontology. We adopt the following definitions in our discussion of cybersecurity and risk management throughout the remainder of the book:

- *Denials of service*: prevent legitimate traffic from being serviced, often as a result of overwhelming requests or demands on services by malicious agents (e.g. bot nets) acting on behalf of the attacker.
- *Data pilfering*: incidents where useful non-confidential but valuable business or government data is exfiltrated from protected networks capable of being utilized by an aggressor.
- *Breach of confidential information*: loss of customer data or sensitive network data which is controlled, subjecting the losing party to substantial legal or business risk beyond the loss of

advantage inherent in a *data pilfering* event. For example, the loss of sensitive customer consumption and billing data could be cause for lawsuits or fines since adversaries could potentially steal from clients and customers enabled by illicitly obtained personal information. Similarly, the loss of Department of Homeland Security–regulated Critical Energy Infrastructure Information with power grid topology and asset information could have additional consequences:

- *Theft*: use of data to directly steal power or other commodities (e.g. via manipulation of meter or billing information) or direct intellectual property theft.
- *Nefarious loss of control*: network intrusion or data manipulation resulting in physical control of critical electric power systems infrastructure with intent to injure; damage; or deprive electricity-dependent systems, facilities, regions, organizations, or people.

This business-focused threat ontology differs from a more technology-focused model commonly used by computer security–focused analysts. This is intentional because the resources devoted to preventing or responding to threats, whether cyber or physical, should be based on the economic value at risk. Some early efforts to estimate the value of prevention have begun.[7] Since absolute prevention is impossible, resources for response must be protected. In this matter, "cybersecurity is a lot like public health. You have to pay attention to the disease, its transmission rates, the vector of the transmission, and the acquired immunities."

For a more technically focused discussion of the particular effects within the cyber domain itself, as opposed to those more broadly affecting the cyber-physical system or business outcomes, we adopt the vocabulary of exploit, disrupt, deny, degrade, destroy, or deceive—which can be used in attack tree and graph modeling and cyber attack lifecycle analysis (Bodeau and Graubart, 2013).[8]

---

[7] See e.g. work in Koonce et al. (2006) to determine both the likelihood and extent of outages when system components fail to perform in accordance with their advertised design specifications.

[8] Cyber kill chain analysis was designed to aid organizations in leveraging empirical data about network intrusions and vulnerabilities to infer adversary intent with the goal of mitigating both the threat and vulnerability components of cybersecurity risks (Hutchins et al., 2011).

Exploitation and disclosure

Exploitations of networks are designed to gain access to information that the threat actor is otherwise not authorized to see. Although news about data breaches caused by hackers is increasingly making headlines, insider threats are another major source of risk from simple unauthorized disclosures. According to a 2011 security survey performed by McAfee, 42% of respondents said laid-off employees are the biggest threat to information security, and global intellectual property theft damage is estimated to be approximately $1 trillion (Ackerman et al., 2009). As costs have risen, new entrants are seeking to develop commercial solutions, applying heuristics and inferential statistics to reduce the risk. For example, in 2013, Guy Fillipelli of Red Owl Analytics partnered with In-Q-Tel, the CIA's affiliated venture capital arm, to commercialize a software solution that examines "organizational dynamics in support of governance, risk, and compliance" (Red Owl Analytics, 2013).

Companies are increasingly interested in reducing their risk of insider disclosures or actions as cases have become more numerous. One former Intel employee now faces 90 years in prison for data espionage by bringing proprietary information to competitor AMD (Jones, 2008). Another employee at the Federal National Mortgage Association, commonly known as Fannie Mae, uploaded malicious code designed to destroy data, including financial, securities, and mortgage information (U.S. Attorney's Office, 2010). Insider threat has been a long-standing risk to organizations, but in the post-Snowden era, it has garnered substantially increased attention and resourcing from business and government alike (Faturechi, 2014).[9]

Initial exploitation or disclosure of information that allows for others to access critical systems is often a key enabler of cyber attacks with more direct, tangible consequences. Examples include a control systems incident in 2000, when a supervisory control and data acquisition (SCADA) system attack on a sewage control system in Maroochy Shire in Queensland, Australia, released thousands of gallons of raw sewage. This likely remains among the most famous

---

[9] Edward Snowden was the NSA contractor who leaked large volumes of classified documents in 2013.

cases of insider threat (Slay and Miller, 2007). Other examples include a Sacramento man who pleaded guilty to attempting to shut down the California power grid (U.S. Attorney, 2007a), a hacker attempting to breach the Tehama Colusa Canal Authority computer system (U.S. Attorney, 2007b), and a case of an insider disabling pipeline leak detection systems off the California coast (Kravets, 2009).

Phishing attacks that cause employees to compromise personal information or credentials remain one of the most surefire techniques to initially compromise a network—67% of breaches in Verizon's 2014 Data Breach Investigations Report (Verizon Enterprise Solutions, 2014). Even the most sophisticated attacks (e.g. Stuxnet, as described on page 184) use an initial breach anywhere on a target network as a foothold to enable privilege escalation and lateral movement to reach their real target.

Deception

Deception is a powerful tool to influence outcomes. People and machines are not used to questioning the quality or veracity of the data they receive—especially on more critical systems where higher levels of protection are assumed. A standard military tactic, deception, is chronicled in Sun Tsu's *The Art of War* and immortalized with the Trojan Horse in Virgil's *Aeneid*. Less well-known and more recent examples include the phantom 10th Fleet of the US Navy, which was never put to sea and had no ships assigned to it despite being responsible for the destruction of enemy submarines (Maloney, 1991)—fittingly adopted by the Navy for its Fleet Cyber Command. The US General, George Patton, commanded an even more elaborate ruse—the First US Army Group designed to convince Nazi Germany that the Allies were massing forces for an attack on Pas de Calais instead of Normandy, where Operation Overlord would take place on June 6, 1944.[10]

Deception in the virtual world is relatively easy since it is harder to attribute specific actions to particular identities in cyberspace. Doing so requires an understanding of a combination of attributes, including the physical infrastructure of networks, logical layers, and

---

[10] See e.g. Bacon (1998) for more examples of major World War II deception operations.

personas in question. Almost all cyber attackers send traffic through (multiple) third-party systems in such a way that even knowing who attacked you is difficult. These "pivot points" or proxies separate all traffic received by the target network from the originator. Almost all cyberattackers use intermediaries for this reason. While technical difficulties in authentication and verification exist, networks face a bigger challenge: people. Even if it were possible to build a perfectly technically secure network, it would become insecure as soon as a human interacted with the system.

This phenomenon is why phishing attacks are of such concern to many companies. In a *phishing* attack, the attacker tricks the victim into handing over private information to gain access to credentials or information which might allow for retrieval of such information via social engineering. A close relative is the *spoof* attack, where a victim interacts with a fake website unwittingly. This interaction may facilitate data capture or allow for the website to implant malicious code on the user's device. These human-initiated actions are particularly challenging to guard against because they bypass most of the automated security protections designed to keep bad actors out and provide access deep into a protected network. With a foothold established, more advanced threats tend to focus on *privilege escalation* to gain more authority and move laterally through the target network (Colarik and Janczewski, 2007; Singer and Friedman, 2014).

Disruption

A *disruption* to a normal system, organization, or personal activity can impose costs resulting from the interruption of services. This could be a direct action, which would manipulate system performance such as network availability to prevent reliable communication. Small delays within time-sensitive elements of power systems can impede state estimation and energy management system performance and ultimately impact power quality and system stability.

Denial of service (DoS) attacks can disrupt or degrade services indirectly while also imposing direct costs on the network operator, who is faced with a choice between scaling up to serve fictitious malicious traffic or letting the legitimate traffic go unserved. In 2012 and 2013, large-scale distributed denial of service (DDoS) attacks on

US financial institutions resulted in significant costs. The malicious connections reached up to 30 gigabits per second and 18 million requests per minute, mostly "directed against insurance companies, brokerage firms, credit unions, asset management firms, small banks, Canadian and European financial institutions, and 30 of the top fifty US banks" (Akamai, 2013). These kinds of attacks have now been included in SEC filings, including recent 10-K filings from some of the largest financial institutions in the US (JPMorgan, 2013; Bank of America, 2013). DDoS attacks were the primary cause of the major 2007 Estonian disruptions first discussed in Chapter 9. These cyber attacks, often "botnet-based", are also commonly used to mask other concurrent intrusions into connected systems (CERT-EU, 2014; Alomari et al., 2012).

Degradation or destruction

The most extreme kinds of cyber attacks occur when an attacker gains control inside of a system—exemplified by the detailed discussion of Stuxnet in the Natanz uranium enrichment facility. Degradation of services can cause follow-on failures for systems that depend on specific levels of network or information-processing reliability, availability, latency, etc.... Stuxnet is far from the only example of cyber-physical system risk in industrial systems. The Idaho National Laboratory also researched how software can cause physical damage. In 2007, they filmed a demonstration project where they destroyed a more than $1 million diesel generator by sending it harmful software instructions to open and close a sequence of breakers, while a spinning generator was out of sync with the grid (Bernabeu and Katiraei, 2011). This control systems research effort was dubbed operation Aurora, and the gap in traditional protections was referred to as the "Aurora vulnerability". After 13 iterations, the 3.8-MVA diesel generator—operating at 60% of its rated power—experienced significant abnormal vibrations, and after 22 iterations it began to smoke (Bernabeu and Katiraei, 2011). Despite the fact that numerous events must correspond to enable destructive, Aurora-like attacks, consumers, energy companies, and governments are well advised to acknowledge that the electric power systems are likely to "be a target for cyber attack in any future conflict" (Lewis, 2010).

11.3 Towards operational risk management

When assessing the impact of specific threats or the exploitation of known vulnerabilities in critical infrastructure, it is advisable to begin with a significant effort dedicated to identifying transitive risks and common-mode failures in addition to a more generic baseline scenario-based risk assessment. Transitive risks are those risks which stem specifically from dependence, especially interdependence. A common-mode failure results from a single fault or fault set (NIST, 2014).

The Heartbleed Bug, a serious vulnerability in a popular cryptographic library used on the majority of servers, provides an excellent recent example of how the simultaneous exposure of many systems to a single fault can impose significant costs on real-world systems. Because of the rapid escalation of consequences from these faults, organizations should be especially attuned to their existence. Dedicated efforts to identify unique assets (e.g. GPS) and specific kinds of failures where compromised or non-performing assets have a propensity to become new attackers within the system are required. Both unique asset risk and cascading failure risk require specialized approaches to management, which differ from more localized risks.

If the core organizational functions for utilities (i.e. the provision of services to end users) are well understood, the economic impacts and liabilities associated with the failure to perform from any single event, or combination of events, can be approximated. Actuarial data exists for many types of service disruptions and can be used to determine appropriate levels of self-insurance versus external support. As this type of loss data has become more available and metrics have become more standardized over the past decade, capital markets and insurance markets have started to converge (Bauer and Kramer, 2007).

Understanding the direct and indirect financial loss information is important because it alone allows decision-makers to develop an understanding of expected, unexpected, and intolerable losses. This is traditionally depicted as a Value-at-Risk (VaR) curve—expressed as a probability density function—as shown in Figure 11.3. Estimation of expected, unexpected, and extreme losses is critical to developing contingent operational plans and securing adequate resourcing for response actions, if required.

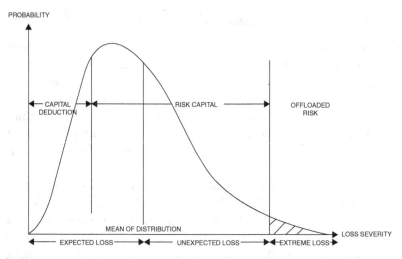

Figure 11.3 Loss severity versus probability

Business leaders and risk managers must agree on the amount of preventative (i.e. anticipatory) versus reactive (i.e. failure) capital to be deployed based on loss acceptability. For the utility industry, this means that utilities, consumers, and regulators have to determine reasonable assumptions and incentives to guide strategy in an uncertain future. In the US, this process involves substantial coordination between the Federal Energy Regulatory Commission and the state-level public utility commissions.[11]

Large-scale power disruptions are expensive for society—e.g. the August 14, 2003, blackout costs were estimated at well over $6 billion (Abraham et al., 2004)—and preventative measures can be justified in the rate base.[12] For example, vulnerable high-voltage transformers with long manufacturing times could be better stockpiled to reduce the likely time to recover from hurricanes, solar storms, or sabotage (Parfomak, 2014). This concept is illustrated in Figure 11.4—sufficient detail is required to demonstrate that preventative measures are in fact responsible investments, thereby enabling rate recovery mechanisms to be put in place. In practical terms, organizational strategy is expressed via resourcing levels.

[11] See e.g. Eto and Commare (2008) for a detailed analysis of the reliability information presented to public utility commissions in the US.
[12] A more detailed discussion of the blackout is provided in Chapter 12.

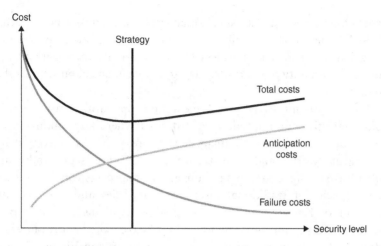

Figure 11.4 Preventative versus reactive costs

Developing and financing contingency operations

Developing and resourcing contingency operations to respond to critical incidents within electric power systems requires an understanding of the specific value destroyed and time required to restore normal operations in specific potential futures. This can be addressed using models to estimate the costs of particular disruptions. Considered events could be based on detailed scenario-based analysis like that already utilized for proactively positioning storm crews before major storms.[13] Analytical tools can also evaluate other, less probable hypothetical futures for which no ready explanation might readily exist. This could involve randomly simulating the impact of non-performing generators, line disconnections, etc. Regardless of the catalyst, costs mount quickly when major events impact infrastructure.

Financing response actions is a crucial but often neglected part of proactively responding to potential catastrophes. Organizations are frequently exposed to real and immediate costs without sufficient liquidity to regain normal operations and access to the long-term cash flows on which they rely in order to pay for the response actions.

---

[13] Mathematical tools like Monte Carlo simulations are commonly used to help evaluate a wide range of potential storm severities and paths to help craft statistically optimal advance locations for crews; see e.g. Cheng et al. (2014).

For most individuals and small organizations, insurance policies can help fill the gap by offering fixed amounts of financial protection at a known cost. This business transaction shifts responsibility for future uncertainty to the *insurer* at a tolerable financial burden for the *insured*.

For major infrastructure owners and operators, insurance is often effective for normal, localized risks where large numbers of assets are unlikely to be damaged simultaneously. As illustrated by Hurricane Katrina and Superstorm Sandy, the sheer scope and value of infrastructure that can be damaged in real catastrophes exceeds the capacity of traditional insurance companies and markets. As a result of Katrina, the fourth-largest personal insurer in Florida went bankrupt, and other major insurers have exited entire states or declined large numbers of policy renewals (Born and Viscusi, 2006). Even insurance companies typically pool risk into larger "books" which are sold into wholesale markets run by the larger re-insurers, who have access to substantial capital reserves capable of absorbing gargantuan losses.

To address the risk of catastrophic events, additional financial instruments called *catastrophe bonds* have been developed, which allow major infrastructure operators to put contingent contracts in place.[14] These contracts differ from insurance in that they are triggered by an agreed-upon event but remain independent of actual financial losses or costs. These insurance-linked securities are also typically issued to ensure the financial solvency of a primary insurer and in many ways are more similar to futures contracts than traditional insurance policies. Their growing role in protecting infrastructure from natural catastrophes is representative of the increasing convergence between insurance and capital markets (Johnson, 2012).

For example, the New York Metro Transit Authority teamed with GC Securities and Goldman Sachs to form MetroCat Re Ltd., which issued $200 million in catastrophe bonds after Sandy to guard against storm surge in an oversubscribed offering (Keohane, 2014). If the storm surge measurements taken by tidal gauges during named storms reach at least 8.5 feet in Area A (at Battery Park, Sandy Hook, and Rockaway Inlet) or 15.5 feet in Area B (East

---

[14] Contingent security concepts were first proposed by Nobel Laureate Kenneth Arrow in 1952 (Varian, 2001).

Creek and Kings Point), the securities are "triggered," meaning that, the MTA receives the money and investors keep only whatever interest payments have already been received. Otherwise, the MTA pays 4.5% interest (approximately $18 million per year) on the bonds, which were purchased by capital market investors over a 3-year term. The specified and one-half of storm surge in Area A corresponds to a "77-year storm" (1.30% annual probability), while the trigger level in Area B corresponds to a "108-year storm" (0.92% annual probability). Overall, investors have a 1.68% chance of loss occurring in any given year, with 45.0% of the estimated risk expected from events only in Area A, 22.6% expected in events only affecting Area B, and 32.4% from events in Area A and Area B.

Catastrophe bonds are effective at reducing exposure to high-consequence events at the "tail" of the risk profile or copula and converting them into manageable forms of known losses or unexpected losses (see Figure 11.3). However promising, insurance, reinsurance, and catastrophe bonds of various stripes will not replace the need for sovereigns to act as a lender of last resort for events of sufficient magnitude (Jaffee and Russell, 1996). Total catastrophe bond transactions in 2013 were estimated at $7.5 billion (AON-Benfield, 2014).

Unintended consequences of risk management

The complexity of modern financial systems can create unintended consequences. The financial crisis of 2008 was largely caused "by large financial firms taking on too much risk (leverage) in opaque trading environments" which was complicated by complex insurance products and counterparty credit failure in OTC trading of mortgage derivatives (Basu and Gavin, 2011). The increasing frequency and severity of the financial crisis is particularly striking given that industrialized economies have experienced increased levels of economic stability during the same period (Brock, 2002).

A similar example of unanticipated consequences that result from tightly coupled and complex systems is given in Richard Book-staber's description of portfolio insurance in *A Demon of Our Own Design*. Major firms with substantial portfolios of equities sought protection against losses by computing trading decisions based on the Black–Scholes options pricing model.[15] The method provided specified levels of protection for a known cost, effectively shifting the intolerable losses (shown at the right of Figure 11.3) into known costs of doing business. The dynamic hedging strategy, which had been brilliantly marketed as "insurance", assumed that the market would remain liquid and that volatility is manageable and properly estimated. By 1987, the technique had become a major driver of trading decisions, but illiquid markets and inability to adjust the hedges triggered a meltdown (Bookstaber, 2009). On October 19, 1987, the Dow Jones Industrial Average fell 22.61% in a single day as large portfolios attempted to simultaneously hedge against the same price fluctuations, akin to a giant playing in a swimming pool. Bookstaber's experience is a powerful example of the source of risk as dependence—"especially dependence on expectations of system state" (Geer, 2008).

The rush to commodities speculation in the early 2000s ultimately failed, just as Bookstaber et al.'s attempts to provide a "risk-free" way to hedge against equities had. Investors mistakenly believed that they could simultaneously move capital into commodity derivatives in search of better yields and hedge against equity risks (Basu and Gavin, 2011). Movement towards exchanges, which can provide more rigorous protections against fraud, aid in the estimation of the health of the entire system, and protect against the risks of counterparty credit failures can help. In many ways, this is a continuation of the same formalization process that culminated in the founding of the CBOT in 1848 to meet the patent needs of grain producers, millers and other market participants, primary business needs.

11.4 Chapter 11 Summary

- Estimating risk based on a combination of likelihood and consequence is core to success in the modern enterprise.
- Risk is a consequence of dependence. Security is a subset of reliability.

---

[15] The Black–Scholes model describes market behavior as a system of stochastic differential equations and is commonly used to price European-style options.

- Risk management is about improving the likelihood of better future outcomes. Costs which result from mitigable forms of surprise should be the primary focus of risk managers.
- Mitigating operational risks requires organizational leaders to consider cyber and physical threats. Contingency planning and mitigation efforts must consider direct effects of disruptions as well as indirect financial liabilities.
- Commodities futures markets are critical risk management tools that help farmers and energy companies manage cash flow and exposure to future price changes.
- Electricity spot markets can be highly volatile because of physical scarcity and other unique constraints on the electric power system. While the extreme volatility in the physically settled bulk electric power market bleeds over to purely financial contracts, electricity futures are a vital part of running any bulk electric power business.

12 | *Resilience as a core value*

Resilience is the capacity of individuals, communities, companies, and the government to withstand, respond to, recover from, and adapt to disruptive events.

Steve Flynn (2012)

Our society has steadily created more complex and co-dependent or interdependent critical infrastructure systems. We have become especially dependent on the availability of reliable electricity. As a result of this complexity and dependence, we have become subject to new risks that are often misunderstood or ignored entirely. Today's sudden fixation with big data—which really is a euphemism for the utilization of heuristics and inferential statistical models to aid in decision-making—is more dangerous as a result of our inability to view these dependencies and our increased levels of comfort with the increasingly stable services that technology has enabled, especially when compared to most of human history.[1] As the mean time between failures of all types increases, so does the perception of safety; thus, the allocation of capital to develop technologies or invest in infrastructure which prioritizes resilience over efficiency lags ever farther behind the actual level of risk.

Big data tools—which are often misunderstood as being capable of determining causality, despite their absolute limitation of determining degree of correlation as a black box—are further accelerating the optimization of our businesses, processes, investment strategies, etc. for an increasingly narrow *operational band*. Put differently, big data is only capable of utilizing the information made available to it to determine strategies that are highly correlated with a positive outcome based on its *training* data set. The training data is empirical: making decisions in this manner is akin to driving using only a rearview

---

[1] See e.g. Nate Silver's *The Signal and the Noise* (2012).

mirror. Empirical data, by definition, represents system states which have already been experienced, and extrapolation is often fraught with challenges.

New events, real-world occurrences which occur outside the band, and especially sustained system changes outside the expectations of the trained data set, are often inimical to such an approach.[2] For this very reason, there are numerous examples of algorithmic traders who claim to be able to make prodigious sums of money using proprietary strategies in commodities or equities markets—but with whom no respectable institutional investor would consider deploying capital until they successfully trade a live account and stop "backtesting" by running algorithms on training data.

When viewed from the scientific perspective of chaos theory and complex adaptive systems, the current trend is especially concerning. Complex business rules and regulations that underpin our energy systems were designed to protect the diverse set of vested parties and balance public and private equities, however these same foundational rules can also have unintended consequences. Our longtime quest to calibrate these systems to achieve an ever tighter, or more consistently optimal, outcome via rules and constraints can have an inverse effect on robustness or resilience in both natural and man-made systems (Miller and Page, 2007; Arrow et al., 1995). The growing set of rules which are tailored to a narrowing operational band can push large systems towards increasingly critical states where small changes can result in large-scale instability, such as the financial market collapse in 2008. A single-minded approach focused on efficiency can lead to *self-organized criticality*, where microlevel behaviors converge to an optimal state for a given scenario but collapse when an unanticipated action occurs (Bak, 1996). Said differently, purely maximizing the short-term efficiency of a system results in increased fragility to external shocks.

12.1 Exploring (inter)dependence

Today's infrastructure is an increasingly critical system of systems with unknown complexity. Approaches to its modernization require

---

[2] Advanced researchers are capable of utilizing approaches to mitigate these challenges and similar limitations associated with over-fitting training data, but the preponderance of commercial applications do not incorporate a rigorous assessment of when such techniques are apposite.

us to better understand the current state of affairs while simultaneously preparing to gain additional efficiencies. Rule-making in
such an environment is particularly challenging. The goal is to
determine a set of rules to guide interactions that help maintain
a state on "the edge of chaos" where productive output occurs
(Packard, 1988; Langton, 1990). These same rules, if not properly
implemented, can lead to a critical state of chaos or stasis (Miller
and Page, 2007). For society, reasoning about this new kind of
problem requires broader understanding and an acknowledgment of
the tension between maximizing immediate benefits and investing in
the future returns.

The magnitude of societal dependence on electric power is large
enough to surprise most people, making it a perfect vantage point
from which to explore resilience. Both the EU and Canada define 10
broad critical infrastructure categories, but the US officially uses 17
categories. Table 12.1 compares US critical infrastructure categories
with their Canadian and European counterparts. Seventeen of the 18
critical infrastructure systems, as defined by the US government and
on which we depend for key societal functions, are dependent upon
nearly continuous access to electricity. While more specific definitions
have since emerged, the Clinton administration's Policy on Critical
Infrastructure Protection from 1998 provides a concise definition:
"critical infrastructures are those physical and cyber-based systems
essential to the minimum operations of the economy and government"
(White House, 1998).

Our new societal challenge is to maintain reliable services despite
our widespread exposure to common-mode failures across these
interdependent critical infrastructure systems. We must also do
so while simultaneously managing rapid increases in infrastructure
variability and uncertainty as new technologies, services, and rapid
population growth stress each system, and the relationships between
them, in new ways.

12.2 Valuing diversity and future freedom of action

Resilience has become a fashionable term in the wake of major natural
and man-made disruptions over the past decade. Its importance,
illustrated by a series of dramatic and well-known events affecting
entire regions—including the Indonesian tsunami of 2004, Hurricane

Table 12.1 *US, Canada and the EU: critical infrastructure sectors*

US	Canada	EU
Agriculture and food	Food	Food
Commercial facilities	Government	Government
Government facilities	Government	Government
National monuments and icons	Government	Government
Defense industrial base	Manufacturing	Production
Nuclear reactors, materials, and waste	Safety	Safety
Emergency services	Safety	Safety
Chemical	Safety	Safety
Dams	Safety	Safety
Transportation systems	Transportation	Transportation
Postal and shipping	Transportation	Transportation
Banking and finance	Finance	Finance
Communications	Communications and IT	Communications and IT
Information technology	Communications and IT	Communications and IT
Public health and health care	Health care	Health care
Water	Water	Water
Energy	Energy and utilities	Energy

Katrina in 2005, and Superstorm Sandy in 2012—have caused most definitions to revolve around concepts of survivability.[3]

3 See e.g. the *Economist* (2004, 2005, and 2012, respectively).

Smaller, more subtle events that should emphasize less dramatic types of resilience are even more numerous but are commonly ignored. These events are those which expose us to the fragility of modern-day organizations, which can experience non-trivial consequences from any number of minor individual disruptions. Supply chain disruptions are a useful example. Boeing's 787 Dreamliner employs a global supply chain intended to decentralize manufacturing via a relatively small group of Tier-1 suppliers, better manage integration challenges, and drastically reduce overall development costs and time (Tang and Zimmerman, 2009; Marsh, 2009). Complications from the suppliers responsible for delivering millions of smaller parts to Boeing's Tier-1 suppliers and global logistics disruptions were major factors in the more than 3 years of delays in the first Dreamliners reaching customers (Kesmodel, 2011). Disruptions, large and small, are not unique to Boeing's 787 supply chain or production process; on July 4, 2014, the Burlington Northern train derailed in Montana, with multiple fuselages for Boeing's 737 and additional assemblies for 747 and 777 jets (Ostrower, 2014). Similar events can have large impacts on the availability of critical equipment, supplies, and fuels required during the building, operation, and maintenance of energy infrastructure.

When taken together, the definition of resilience that focuses on survivability is too limiting—it is temporally reductive. Resilience is particularly challenging because it requires balancing anticipation costs, for both disruptive and evolutionary change, with estimated response costs.[4] Organizations can consciously monitor their investments in resilience by monitoring the ratio of resources invested in future freedom of action versus the maximization of current benefits.[5] This broader definition of sociotechnical system resilience (i.e. organizational resilience) is best addressed by some of the same principles that are also applicable to technical challenges faced by modern-day energy systems, including cybersecurity and energy resource diversity.

---

[4] We use the same costs introduced in Chapter 11 on risk management. See e.g. Figure 11.3.

[5] For a detailed scientific discussion of future freedom of action maximization over finite time horizons to yield intelligent outcomes for a wide range of systems with a large number of potentially viable system states, see Wissner-Gross and Freer (2013).

Intentional heterogeneity in cyber-physical systems

Intentionally creating heterogeneity in cyber-physical systems is an important part of creating resilience for society. In today's monoculture-centric world, achieving this is challenging. Diversity costs more, takes more effort, and requires a conscious effort. Organizations seeking true diversity must commit to seeking out and rectifying emergent monocultures before they become too pervasive.

For example, Geer et al. (2003) authored a controversial paper asserting that the Windows monopoly was harmful to national security for developed countries—specifically that:

Computing is crucial to the infrastructure of advanced countries. Yet, as fast as the world's computing infrastructure is growing, security vulnerabilities within it are growing faster still. The security situation is deteriorating, and that deterioration compounds when nearly all computers in the hands of end users rely on a single operating system subject to the same vulnerabilities the world over.

Most of the world's computers run Microsoft's operating systems, thus most of the world's computers are vulnerable to the same viruses and worms at the same time. The only way to stop this is to avoid monoculture in computer operating systems, and for reasons just as reasonable and obvious as avoiding monoculture in farming.

While this perspective was apposite at the time (Geer and Aucsmith, 2003), it has since been eclipsed by other monocultures. Critical infrastructure's current "exposure to common mode risk is now small devices and the chips which run them" (Geer, 2014). Skeptics should take note that common-mode failures have already demonstrated that they are a significant threat to modern society—amplified by our increasingly absolute dependence on connectivity.

In 1990, AT&T experienced a common-mode failure that resulted in a 9-hour failure of their long-distance telecommunications network. This is a powerful illustration of why technical diversity is valuable (Elmer-Dewitt, 1990). A software bug that was common to backbone "4ESS" AT&T telephony switches caused a national telephone outage. AT&T's Director of Technology Development, Larry Seese, said that the company had updated the processors of all 114 "4ESS" switching systems in December 1989 (Neumann, 1990). On January 15, 1990, a switch in New York experienced a problem, causing the switch to reboot. In only a few brief seconds, network managers at AT&T's Network Operations Center in Bedminster, New Jersey, watched a massive 72-screen video array depicting their network turn into "a tangle of red lines as a rapidly spreading malfunction leapfrogged" across the network (Burke, 1995).

Rebooting a switch took 4–6 seconds and triggered a message to all connected switches notifying them of the fault via a message stating the non-availability of the downed switch—known as a call attempt message. Unfortunately, the mid-December update had changed the logic and caused the New York switch to continue sending additional failed call attempt messages. The additional messages caused the connected switches to believe that they too had problems and triggered the same reboot process that caused the first error. These messages, sent out via the common channel signaling system 7 (CCS7) links, continued to cause repeated failures across all of the 4ESS switches and caused the first major network outage in the company's then 114-year history (van der Linden, 1994; Neumann, 1990).[6] This failure mode only presented in deployed systems, when multiple switches interacted, and was dangerous because the entire system was simultaneously exposed because of the switching monoculture.

No *silver bullet energy source*

Monocultures in energy resource mixes can be as challenging to systems as monocultures in control systems, communications infrastructure, or other usually invisible technical infrastructures. This is especially true when discussing the problem of maintaining stable electric power systems—again resulting from the requirement of balancing supply and demand to maintain system stability. One timely example is the massive growth of particular types of distributed energy resources (DERs) in the United States—especially renewable DERs which rely on nature for power. A holistic, portfolio-based approach to managing energy resources is critical to efficiently providing electric services today,[7] just as load diversity was for Samuel Insull when he published his similar customer-focused monograph detailing the value of diverse customers in 1914.

---

[6] For additional information on the actual software code—a misplaced *break* statement in the C (language)—that caused the error, see *Expert C Programming* (van der Linden, 1994).

[7] See e.g. Hoste et al. (2009) and Taneja et al. (2012).

Solar power in Arizona has grown at breakneck speed, with approximately 3 million homes in Arizona equipped with photovoltaic panels to convert sunlight to electric current as of 2014 (Biello, 2014). According to the Solar Energy Industries Association, total solar electric capacity reached 1875 MW in 2013, with 701 MW installed in 2013 alone, helping Arizona edge closer towards its Renewable Energy Standard goals and placing it second to only California in total installed solar capacity in the US.

The changes in energy resource composition had unexpected consequences. Utilities sought additional fees from distributed solar owners to help mitigate the impact of solar-enabled reductions in electricity sales due. Utility executives and commercial-scale solar plant manufacturers such as James Hughes, CEO of First Solar Inc., argued that the Arizona net-metering rules resulted in an unequal distribution of "the overhead costs associated with the grid or other services provided by the utility, producing a very substantial 'cross-subsidy' funded by all other utility customers who must pay proportionately more in rates" (Hughes, 2013). Distributed solar companies and owners argued that increasing their fees would be a significant disincentive to renewable energy growth. Ultimately, the Arizona Corporation Commission decided to reduce solar subsidies and allowed a charge to solar producers of $0.70 per kilowatt of capacity per month to cover integration and infrastructure costs amounting to $7.00 per month for a typical residential solar owner.

Beyond the normal challenges associated with integrating a generation facility that is dependent upon its natural environment to control its power output—i.e. uncertainty and variability—there are two primary limitations for solar. The first is the problem described by Arizona utilities: most distributed solar-equipped facilities are not truly capable of running independent of the grid, especially for longer durations. Distributed solar is primarily displacing non-renewable energy requirements and can help ameliorate peak power limitations on shared infrastructure for generation, transmission, and distribution. However, distributed solar is not a "firm" capacity resource that utilities can count on for the purposes of ensuring that they have made the necessary operating plans and long-term investments such that a distributed solar customer (whose system was non-performant for any number of reasons) can seamlessly transition back onto the grid. For most distributed solar customers, this transition occurs each night as photovoltaics cease to produce enough energy to keep pace with consumption. System-level planning and analysis of short-, medium-, and long-term costs and benefits of distributed and utility-scale solar power are clearly required to address the public interest (see e.g. Baker et al., 2013).

The second limitation is more subtle but closely related—displacement of traditional generators by growing amounts of solar can increase the *ramping* needs of traditional generators. This has been termed the "duck curve" effect. The California Independent System Operator (CAISO) has been particularly vocal about this challenge. Three key time periods have been identified based on existing forecasts of solar and load profiles in 2020 (CAISO, 2014):[8]

1. A steep upward of approximately 8000 MW begins around 4:00 am as people prepare for the day.
2. A steep downward ramp begins between approximately 7:00 am as solar generation kicks in, slowing down but keeping net load requirements low until the sun descends beginning around 4:00 pm.
3. Finally, a steep and massive upward ramp of as much as 13,000 MW begins as solar generation ends and people return home.

These problematic ramping periods are depicted in Figure 12.1. The issue has been discussed extensively by CAISO in the public domain, and this graph depicts the problematic ramping periods for a hypothetical future scenario on the California transmission grid in 2020.[9] California's photovoltaic capacity is growing even faster than that of Arizona, with approximately 2700 MW added in 2013 alone (Kinsey, 2014). CAISO is on track to reach a total installed capacity of more than 20,000 MW of wind and solar and reduce mid-day load via distributed solar by much as 35% by 2020 (CAISO, 2014; Rothleder, 2013). This would be substantial but remains less (as a percentage of the resource mix) than Germany's 2013 mid-day load reductions of 40% from solar and wind (Burger, 2014).[10] The turmoil will be significant, as large amounts of CAISO thermal generators—more

---

[8] The CAISO analysis included wind as well, but wind impacts are significantly less significant.
[9] Data depicted in this chart obtained from Rothleder (2013).
[10] Germany produced approximately 5.3% (29.7 TWh) from solar and 8.4% (47.2 TWh) from wind turbines out of its total consumption of 560 TWh in 2013 (Burger, 2014).

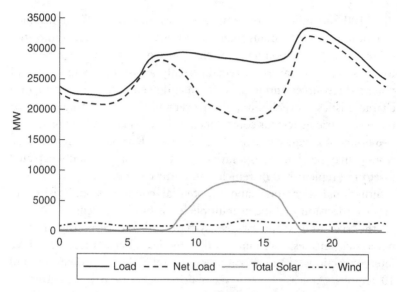

Figure 12.1 Future intra-day ramping challenges in CAISO

than 12,000 MW of the total installed capacity of 58,700 MW as of 2013—are scheduled to be re-powered or retired during the same time period (Rothleder, 2013).

More traditional sources of power have other limitations—e.g. the water used to cool thermoelectric and nuclear power plants is massive and commonly overlooked. For areas like Texas—which gets 33% of its power from coal, 10% from nuclear, and 48% from natural gas in 2012 and is predicted to more than double to reach a population of 55 million by 2050—balancing increasing water and electricity requirements may increase the attractiveness of other kinds of problems, such as the previously enumerated challenges with wind and solar (Faeth et al., 2014). A more detailed discussion of water scarcity challenges and their relation to energy is provided in Chapter 13. Growing requirements and constraints will mean that no single source of energy, style of generation or policy initiative can possibly meet societal requirements on its own.

12.3 Historic US blackouts

Synchronized electric power grids remain arguably the most complex system ever constructed. The Eastern Interconnection alone has more than 100,000 miles of transmission lines and 600 GW of installed generation capacity controlled by dozens of balancing authorities (Tate, 2011). Responding to changing operational requirements forces system operators to remain keenly aware of a wide range of potential resources and impacts, each endowed with unique temporal characteristics, reversibility, or irreversibility. For example, major thermoelectric generator commitment schedules typically take at least 4–6 hours of advance notice. Other assets such as pumped hydrostatic storage may be able to respond quickly but become a totally depleted asset until replenished or refueled at a future point.

Situational awareness about potential conditions and risks can improve decision-making at multiple levels but is challenging because of the size and scope of the system and the responsibilities of numerous entities. Within North America alone, there are three interconnections, 139 control areas, 18 reliability coordinators, and 10 regions. At least four of six major blackouts in North America since 1960 were caused in part by a lack of situational awareness resulting from these very real and present challenges.[11] Many of these outages are, in some way, related to the failure of the interconnected grid's reliability organizations to provide effective real-time diagnostic support capable of remaining synchronized with the combination of human and automated decision cycles for planning and operating the power system.

Nineteen sixty-five shatters the twentieth-century electric utopia

The cascading blackout of November 9, 1965, was a catalyst for the formation of the NERC in 1968 (Kirby et al., 1995). The event affected more than 30 million people and 20,000 MW—then the largest outage in history (FERC, 1965). The outage affected virtually all of New York; Connecticut; Massachusetts; Rhode Island; parts of Pennsylvania and New Jersey; and large swaths of Ontario, Canada.

The event was triggered by a faulty (backup) protective relay, a device intended to measure electricity flow and send control signals

---

[11] These four outages, which were significantly impacted by situational awareness and interregional coordination, are subsequently discussed in more detail herein: November 9, 1965; December 22, 1982; August 10, 1996; and August 14, 2003.

to circuit breakers at thresholds intended to prevent damage to infrastructure. The relay mistakenly opened one of five 230-kV lines transmitting power from the Sir Adam Beck Station No. 2 generation facility in Hamilton, Ontario, to greater Toronto. When the line disconnected, the remaining four lines were insufficient to carry current from the Beck plant and the 1600 MW already flowing towards Toronto from other generators in the south. The protective relays on the other four lines tripped successively in less than 2.5 seconds (FPC, 1965).

The electricity flowing into the Beck plant reversed course and immediately began overloading major transmission lines across upper New York state, tripping additional safety equipment and shutting down generators across the northeast in only a few minutes (Rosenthal, 1965). The outage is important as an illustration of why risks of cascading failure require special treatment—a small event can turn a single victim into an attacker, with amplified consequences. The initial outage occurred shortly after 5:16 pm and less than 15 minutes later, by 5:30 pm, 28 utilities and almost 30 million people were in the dark (FPC, 1965).

The significant impacts of the outage shattered the idyllic view that electricity pools—themselves a mid-century extension of Insull's pursuit of efficiency and profitability via load diversity—could be managed without improvements to coordination and reliability standards. Initial reactions to the outage were muted, as small disruptions were part of the common experience, but the scope and duration were new.[12] The *New York Times* was the only New York City paper that went to press during the blackout (NORC, 1966; Rosenthal, 1965).

That same evening, President Lyndon Johnson emphasized that "today's failure is a dramatic reminder of the importance of the uninterrupted flow of power to the health, safety, and well being of our citizens and the defense of our country" (FPC, 1965). He signed a memorandum directing the Chairman of the Federal Power Commission—the predecessor of the Federal Energy Regulatory Commission—to "immediately and carefully" investigate the failure "in order to prevent a recurrence" (FPC, 1965). Over the next 3 years,

---

[12] The National Opinion Research Center, which undertook a number of studies dealing with public reactions to national crises, interviewed 1300 people following the outage (NORC, 1966).

a series of major reports by the Federal Power Commission focused on the prevention (e.g. Prevention of Power Failures Vols. I–III [FPC, 1967]) of similar events (e.g. those detailed in NEPCC, 1966) and failed legislation (U.S. Electric Power Reliability Act of 1967) proposing a council on power coordination, and the Federal Power Commission recommendation of an industry-led council to improve interregional coordination led to the establishment of the National Electric Reliability Council (NERC) on June 1, 1968.[13]

Blackout trends

The challenges associated with coordinating across regions and dealing with the physical realities of managing large, interconnected electric power systems continued despite improvements led by the NERC and its members. Increased competition within the bulk electric power system and significant increases in interregional power transfers have contributed to the coordination requirements to maintain the stability of the grid—particularly rapid increases occurred during the mid-1970s after the oil embargo (Abraham et al., 2004). Some exemplary notable blackouts in the interconnected electric power system caused by operational or planning assumptions affected by interdependence between 1965 and 2003 are shown in Table 12.2.

Despite many assurances by utilities, regulators, and policy-makers, the frequency of large blackouts in the US is not decreasing and may be increasing (Hines et al., 2006, 2009). A common statistical test, known as the Kolmogorov–Smirnov test, compared outages between 1984 and 1995 that exceeded 400 MW or 100,000 customers with equivalent events from 1996 to 2006. It demonstrated that the blackout frequency of these substantial outages is not decreasing (Hines et al., 2009). The research even goes so far as to suggest that this and other properties of modern transmission systems are the result of self-organized criticality (Carreras et al., 2001, 2004).

When blackout frequencies from 1984 to 2005 are normalized to year-2000 customer numbers from the Energy Information Administration (EIA) and NERC records, it is clear that the frequency and

---

[13] The NERC later changed its name from the National Electric Reliability Council to the *North American* Electric Reliability Council in 1981 to better reflect Canadian participation (NERC, 2012b). In 2006, the organization changed its name to the North American Electric Reliability Corporation.

Table 12.2 *Exemplary major US bulk electric system outages*

Date	Affected area	Customers (thousands)	Capacity (GW)	Duration	Trigger	Cause
November 9, 1965	NY, CT, MA, RI, PA, NJ, Ontario	30,000	20+	13 hours	Faulty relay	Cascading failure
July 13, 1977	New York City City	9000	6	26 hours	Lightning	System islanded without sufficient generation
July 2, 1996	AZ, CA, CO, ID, MT, NE, NV, NM, OR, SD, TX, UT, WA, WY, British Columbia, Baja California Norte	2000	11.85	Minutes to hours	Tree flashover and relay	Cascading failure
August 10, 1996	AZ, CA, CO, ID, MT, NE, NM, OR, SD, TX, UT, WA, WY, Alberta, Baja California Norte	7500	28	9 hours	Line outages and system oscillations	Western Interconnect islanded in four sections

severity of outages are not improving (Hines et al., 2006). Of the 529 events recorded in the NERC Disturbance Analysis Working Group database between 1984 and 2000, 40% were from weather and fire, 31% resulted from equipment failures, 15% were from insufficient generation, 11% resulted from human error, and 16.3% were from sabotage (Hines et al., 2006).

The Northeast blackout of 2003

The more recent Northeast blackout of August 14, 2003, measured approximately 61,800 MW and affected 50 million people across large portions of the Midwest and northeastern US and Ontario, Canada. In 2004, the US–Canada Power System Outage Task Force estimated the total cost between $4 and $10 billion in the United States with billions more lost in Canada (Abraham et al., 2004). The disruption occurred when a combination of SCADA problems and lack of situational awareness enabled major failures to spread rapidly across the grid (Tate, 2011).

At 12:15 p.m. (EDT), the Midwest (in official title) Independent System Operator (MISO) received inaccurate data in the state estimator, which helps maintain grid stability without triggering any alerts to system operators. Just over an hour later, the Ohio-based First Energy Eastlake five-generation unit disconnected from the grid and shut down automatically (Abraham et al., 2004). That same afternoon, additional First Energy–owned 345-kV lines also disconnected as a result of overgrown trees causing contact faults in northern Ohio with disastrous effects on connected 138-kV lines. Control room operators did not grasp that the system was in trouble until 3:45 p.m. when a shift supervisor informed the manager that they were losing control of the local system—First Energy had no automatic load-shedding schemes and did not attempt to manually shed load (Abraham et al., 2004). From 3:46 p.m. to 4:06 p.m., conditions continued to degrade rapidly, but had 1500 MW of load in the Cleveland area been immediately shed, containing outage may have been possible.

At 4:06 p.m., the Sammis-Star 345-kV line triggered the major cascading failures across the Northeast portion of the continent due to the power swings and voltage fluctuations that triggered protective relay systems on other, functioning components of the grid (NERC,

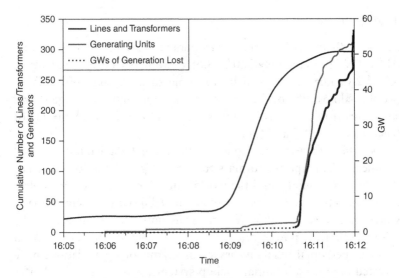

Figure 12.2 Timeline of 2003 Northeast blackout cascading failure

2004). The cascade started slowly but accelerated rapidly—taking less than 7 minutes in total—see Figure 12.2 from (Abraham et al., 2004).

The initial cause was a physical problem. The disturbance was exacerbated by software (e.g. the alarm systems and energy management systems) and the combined human–machine control system, which responded to each successive and escalating event. According to one report, "With the exception of the initial power equipment problems, the ongoing and cascading failures were primarily due to problems in providing the right information to the right individuals within the right time period. Also, the IT infrastructure failures were not due to any terrorist or internet hacker in the attack; the failures were caused by inadvertent events—mistakes, lack of key alarms, and poor design" (NIST, 2010a). The same increases in connectivity which had originally offered additional reliability for electric consumers were now being challenged by an electric power system grown so large and complex that a small disturbance in Ohio could have multi-billion-dollar impacts.[14]

---

[14] For a more detailed technical analysis, see the full NERC Technical Report (NERC, 2004).

12.4 India goes dark

The world's largest electric power outage occurred on July 30 and 31, 2012. Approximately 320 million customers were affected across 21 of 28 Indian states which are home to more than 600 million people—almost 9% of the world's population. India's outage shares some surprising similarities and striking differences with the major blackouts in the US and Canada.

Like the Northeast blackout of 2004, the Indian outage caused significant interruptions to mass transit and public services in Asia's third-largest economy and the world's fifth-largest consumer of electricity. Effects were felt most strongly in Delhi but were substantially worse on the second day of service disruptions. Compared to the North American power grid, the Indian grid is unreliable. As a result of perpetual reliability issues, many major organizations have invested in more substantial backup systems, which acted as a buffer and minimized impacts on the first day of the July 2012 outages. When the second outage struck, most backup systems were already stressed—backup fuel and batteries were largely depleted. Like US consumers in 1965, both rural and urban consumers in India were surprised by the duration and scope of the interruptions—the effects were unlike the repeated short-term and isolated outages typically experienced.

India's rapid growth since its independence from the British Empire on August 15, 1947, has created challenging conditions for electric power and energy in general. Electrified villages grew from 3,061 in 1950 to more than 593,000, while total generating capacity in the country rose from 1,362 MW in 1947 to more than 223,344 MW in March 2013 (Singh, 2013). However, substantial amounts of this capacity were developed privately and remain inaccessible to the grid. The lack of infrastructure, historically poor reliability, and lackluster power quality (voltage and frequency) provided a strong incentive for the largest consumers such as Tata Motors and Reliance Industries, to run independent energy operations without relying on India's growing national grid—an estimated 1.6 trillion rupees ($29 billion) has been spent on backup or isolated power generation capacity which remains unconnected to India's struggling grid (Singh and Katakey, 2012).

Two, nearly three, failures

India has five regional grids which are coordinated through the National Load Despatch Centre (NLDC), created as part of the 2003 Electricity Act and ordered into existence by the Ministry of Power in 2005. The NLDC has oversight of regional power grids and the transnational exchange of electric power and schedules all transmission. It is an intensely public system: roughly 50% of the power produced nationally transits publicly owned Power Grid Corporation of India Limited transmission lines. Because of the country's perpetual shortage of generation and transmission capacity, the NLDC rotates power between sections of the national grid. As the first outage was triggered, load curtailments did not occur in accordance with schedules, and substantial variations in voltage and frequency occurred.

Heavy power flow on the 400-kV Bina-Gwalior-Agra line exceeded 1,000 MW, well over its maximum surge impedance loading of 691 MW, while similar conditions throughout the Eastern Region caused lines rated to only 515 MW to reach upward of 550 MW (CEA, 2012). Almost identical conditions had only been narrowly avoided on July 29 and stemmed from the closure of the one of two circuits on the Bina-Gwalior-Agra line for an equipment upgrade on July 28. Only three units at Bardarpur TPS, supplying about 250 MW of power to Delhi, approximately 100 MW of generation at Narora supplying Simbhauli, and 100 MW supplying part of Rajasthan through Bhinmal avoided the calamity (CERC, 2012).[15]

The national peak demand during the event was 99,768 MW— 74,210 MW of which was inside the Northern-Eastern-Western (NEW) Grid, 35,669 MW in the Northern Region alone, and 3,384 MW to the city of Delhi. Power flowing into the Northern Region from the Eastern and Western Regions (2,585 MW and 2,862 MW, respectively) was simply not supportable with existing infrastructure (CEA, 2012). The magnitude of the capacity shortfall indicates that existing isolated power generation capacity which remains unconnected could have contributed substantially to avoiding the eventual collapse. After the event, power was restored from the

---

[15] The system was especially limited in its ability to receive power from neighboring areas as a result of two interregional lines—the 400-kV Gwalior-Agra (circuit 2) and 400-kV Zerda-Kankroli line—being shut down.

Eastern and Western Regions—flowing across the Vindhyachal A/C bypass to Singrauli at 4:53 a.m. and the 400-kV Gwalior-Agra line at 5:45 a.m. (CERC, 2012).

The second event happened less than 12 hours later at 1:00 p.m. on July 31 and was caused by a similar overload in the Northern Region, again on the Bina-Gwalior-Agra line. The second disturbance affected most of the NEW Grid except for the Western Region, approximately 100 MW of generation at Narora supplying Simbhauli, part of the Delhi system, and the CESC system. This outage compromised the Northern Region (NR), Eastern Region (ER), and Western Region (WR), but major business centers in Mumbai, Bangalore, and Hyderabad were largely unaffected since the Southern Region (SR) and North-East Region (NER) grids avoided collapse.

The process of recovering from the second disruption was mostly complete by 3:30 p.m.—initial efforts focused on facilitating thermal power plants required across the NEW Grid and restarting railroads, essential government facilities, and the Delhi metro. During the second outage, the NR imported a total of 3,925 MW: 1,825 MW from the ER and 2,100 MW from the WR (CERC, 2012).[16] The shifts in capacity were significant enough that the activation of governors is one of many factors which could have helped mitigate the outage conditions (CEA, 2012).

Understanding systemic failures

There are a number of systemic failures that contributed to India's very public failure to provide reliable electricity. The outages only reached between 83% and 86% of the peak load successfully handled by the grid in 2012: 115,446 MW in July, 118,066 MW in June, and 119,089 MW in May. While not the only cause of the outages, inadequacies in generation and transmission capacity will continue to plague India and other developing nations that are experiencing rapid growth in electricity demands. Decades of missed opportunities have contributed to private sector distrust of the electric system. During the eleventh Five Year Plan, from 2007 to 2012, India only intended

---

[16] The second outage resulted in six interregional lines being shut down as system operators attempted to maintain stability: 400-kV Balia-Patna, 400-kV Balia-Biharshariff, 400-kV Balia-Barh, 400-kV Zerda-Kankroli, 400-kV Ranchi-Sipat, and 200-kV Kota-Badod (CEA, 2012).

to install 78.7 GW of new capacity, but actual installations lagged behind the overall target by more than 30% (Enerdata, 2012).

The current energy portfolio also has substantial exposure to coal, which comprises nearly 67% of total nameplate capacity. The remainder consists of approximately 14% hydroelectric, 12% natural gas, 4% oil, 2% nuclear, and 1% renewable resources (Singh and Katakey, 2012). Coal India Limited, a state-controlled mining company, alone is responsible for more than 80% of coal production supplying the nation. Although India has the fifth-largest reserves of coal in the world, it remains perpetually unable to meet even its domestic demand, with imports expected to reach more than 200 million tons per annum at the end of the twelfth Five Year Plan in 2017 (PWC, 2012). India's unreliable electric grid is closely linked to its unreliable supply of bulk fuel—in 2012 around the time of the outages, many coal plants had stock for only 1–7 days. That same month, Coal India offered up to 70 million tons of coal stranded at mine entrances, also known as pitheads, to power producers that it remained unable to transport (Bose, 2012). Rising coal prices have contributed to breakneck inflation, and lack of supply has even caused baseload coal plants to idle regardless of grid conditions because of dwindling fuel. Unreliable gas supplies were also common in 2013; six gas-based power plants—representing a total nameplate capacity of 2102 MW—in Andhra Pradesh idled because of multiweek gas shortages from the Krishna-Godavari basin.[17]

India's power system also faces network losses that are more than twice the global average of 15% and the ramifications of poor policy decisions that ended metering of electricity for farmers in the 1970s (Smith, 2004).[18] The high losses are mostly due to large amounts

---

[17] The six idling plants were GMR's Vemagiri (370 MW) and barge-mounted (237 MW), GVK's Jegrupadu Extension (220 MW) and Gauthami (464 MW), Konaseema Gas Power (445 MW), and Lanco's Kondapalli Stage-II (366 MW) (PTI, 2013).

[18] India's transition to water-intensive crops in the 1970s created intense political pressure for relief from irrigation costs due to electricity (Smith, 2004; Singh, 2006)—saddling state-run utilities with billions of dollars in debt (Ahluwalia, 2002). Crisil, the Indian credit rating agency, estimates the total debt of the 28 state electric boards to be more than $30 billion despite only the Gujarat and West Bengal state electricity boards being profitable in recent years. Recently, nearly a dozen state governments, including Punjab, Tamil Nadu, West Bengal, and Rajasthan, initiated reforms to prevent public utilities from selling power

Table 12.3 *Indian electric power system demand and losses, 2003–2022*

Year	Bus energy (GWh)	Consumed energy (GWh)	Losses (%)	Capacity (GW)
2003–2004	471,639	362,799	30	111,444
2011–2012	968,659	755,847	21.9	152,746
2016–2017 (Forecast)	1,392,066	1,134,534	18.5	218,209
2021–2022 (Forecast)	1,914,508	1,608,187	16	298,253

Source: CEA, 2009

of fraud, non-payment, and theft enabled by a lack of metering infrastructure and enforcement. The non-metered use of electricity for agriculture creates short-term problems by increasing the uncertainty and variability of loads but also disrupts long-term planning because of the debt burden and unpredictable cost recovery.

Table 12.3 shows some of the historical energy consumption, capacity, and loss data and goals for India as published by the Central Electricity Authority (CEA, 2009).

Preventing another major outage will require a broad range of improvements to infrastructure, continued investment, and changes to operations. Disciplined operation of the power grid is required by all of the people and organizations responsible for the effective design, operation, and regulation of the system. In the short term, the need for disciplined enforcement by the Central Electricity Regulatory Commission and the aggressive implementation of load curtailments by the NLDC (and the regional RLDC equivalents) to maintain system health under emergency conditions is especially important.

12.5 Chapter 12 Summary

- Most of today's critical infrastructures, as defined by the US, Canadian, and EU governments, depends on reliable and nearly continuous access to electricity in order to function as designed.
- Disruptions to electric power systems can cost billions of dollars in damage and lost economic output because of growing dependence.

below cost. Total power sector indebtedness in 2014 is estimated above $77 billion by the World Bank (Pargal and Banerjee, 2014).

- Monocultures and system interdependency can create an environment conducive to systematic collapse if not addressed proactively. Increasing resource and technology diversity will help create resilience in the face of unanticipated future operating environments and system states.
- Vulnerabilities or failures can be different at the component and system levels, even for the same technology or process. Testing processes for real-world critical infrastructure systems require continuous testing and exercise of risk management practices to stay agile and adaptive enough to provide reliable services.
- Over-reliance on "big-data," or misunderstanding when heuristics and inferential statistics can aid with decision-making, can be dangerous. Complex, real-world systems consistently experience new actions and system states, requiring different and more sophisticated approaches to modeling and decision-making at scale.
- Cascading blackouts due to system interdependency occur throughout the world. They usually start as small, innocent faults caused by equipment malfunctions or operator errors that can quickly become national or international in nature.

13 | *Exploring energy security*

> To maintain energy security, one needs a supply system that provides a
> buffer against shocks. It needs large, flexible markets. And it's important
> to acknowledge the fact that the entire energy supply chain needs to be
> protected.
>
> Daniel Yergin (Spiegel, 2006)

In 1973, Arab states retaliated against US weapons shipments to Israel
with an oil embargo—slowing or stopping the 8.4 million barrels
per day (bpd) imported mostly from Middle Eastern nations. At the
time, President Richard Nixon thanked Americans for voluntarily
observing "gasless Sundays," announced that directed federal energy
consumption cuts had exceeded 20% by January 19, 1974, prepared
to ration gasoline and unveiled Project Independence. During the same
address to the nation, he stated: "We must never again be caught in a
foreign-made crisis where the United States is dependent on any other
country, friendly or unfriendly, for the energy we need to produce our
jobs, to heat our homes, to furnish our transportation for wherever
we want to go" (Nixon, 1974). Energy independence has been touted
ever since, the dates continuing to slip into the future and long passing
Nixon's goal of no foreign oil dependence by 1980. President Jimmy
Carter even went as far as to make energy independence the central
issue of his administration, calling for the "moral equivalent of war"
on energy (Carter, 1977).

Carter also established the Department of Energy in 1977, signing
the Department of Energy Organization Act on August 4. He even
made the contentious political decision to install solar panels on the
White House in 1977, an act next repeated by President Barack
Obama in 2010. The fixation over energy independence continued,
with Presidents Ronald Reagan, George H. W. Bush, Bill Clinton,
George W. Bush, and Barack Obama all failing to achieve Nixon's
goal despite sustained emphasis as a national priority. Each time,

political leaders promised independence and failed. What combination of factors could conspire to so reliably vex the US? How can we achieve reasonable energy security—and how might that differ from independence?

Energy security is again becoming a dominant and public geopolitical concern. In fact, energy and water—which in many ways can be thought of as processed energy—will be a dominant source of domestic and international conflict over the next 50 years. Tensions over energy supply will continue to ebb and flow as major supply changes are realized or fail to turn out as once hoped. Examples of this include the natural gas boom in the US and numerous delays to Brazilian commercialization efforts of major offshore oil reserves over the past decade, respectively.[1] Real energy security must be viewed as a function of reliable, economical access to energy and where risk is expressed as a function of societal dependence.

This chapter provides an introduction to the challenges of achieving energy security in the real world, despite the numerous and long-standing efforts to promote reliable, affordable, and sufficiently clean energy. To illustrate practical challenges with achieving energy security, we explore the relationship between water and energy, the Strategic Petroleum Reserve, the Fukushima Nuclear Disaster, the Tres Amigas Superstation, and geomagnetically-induced currents in electric power systems from solar storms or electromagnetic pulses.

13.1 Energy and water

In an increasingly populated world, water supplies, or disruptions to them, are becoming increasingly important to maintaining stability. This is not a problem unique to the developing world—it is equally applicable to major developed nations, including the US. Significant amounts of water are used to support the entire energy supply chain responsible for meeting the needs of typical consumers. There are four major areas of connection between energy and water availability and quality[2]:

1. **Energy extraction and production:** Large amounts of water are used for oil and gas well development and fracturing processes.

---

[1] See e.g. *Economist* (2013 or 2014a, respectively).
[2] See e.g. DOE (2006) for a more holistic exploration of water and energy interdependence.

Techniques like fracturing for natural gas, oil shale, and tight oil extraction are expected to significantly increase this form of consumption by 2030 (Carter, 2013). Once in production, large volumes of unusable water are produced.[3] Coal and uranium mining also remain water intensive. Coal mining in the US is estimated to use between 70 and 260 million gallons of water per day (DOE, 2006)—lubrication, dust suppression, fuel processing, and revegetation consume between 1 and 6 gal./MMBtu depending on the source of coal (Gleick, 1994).

2. **Electric power generation:** Thermoelectric power production is a major source of water consumption for cooling, maintenance, and environmental control technologies (e.g. carbon capture and scrubbing). Large amounts of hydroelectric power also have significant ramifications—e.g. for the millions of people displaced as a consequence of dam construction in countries including China, Pakistan, Sri Lanka, Egypt, Sudan, Bangladesh, Nigeria, Brazil, Turkey, and even the US since 1960 (Gleick, 1993b)—on their local environment while consuming significant quantities as a result of evaporation. US hydroelectric reservoirs produce more than 300 million MWh per year from 3,160 billion gal./day of flow through turbines (Solley et al., 1998) and average losses of 4500 gal./MWh (Gleick, 1994), resulting in losses totaling more than 3.8 billion gal./day (DOE, 2006).

3. **Refining and processing:** US petroleum refining operations exceed 800 million gallons of production per day while consuming between 1 and 2.5 gallons of water per gallon of output (Gleick, 1994). Additional water is dedicated to synthesis and steam reforming of synthetic fuels and hydrogen. Biofuels and ethanol production reached 1,359 million gallons in 2013 (EIA, 2014d), consuming an average of 11,062 gallons of water per MMBtu for corn ethanol and 45,004 gal./MMBtu during soy growth and processing (DOE, 2006).

4. **Energy transportation and storage:** Energy pipelines carrying natural gas or petroleum typically consume water for hydrostatic testing. Coal slurry pipelines use water for slurry transport, with

---

[3] Water produced as a byproduct of oil and gas production is commonly referred to as *impaired* because of contaminants or salt, which render it unusable for the most common societal uses of fresh water without some form of remediation.

the rates of water matching the volume of coal being consumed and rendering it unusable. Consumption for coal slurry pipelines is expected to increase from 50 million gal./day in 2005 to as much as 90 million gal./day in 2030 inside the US (Elcock, 2008). Oil and gas storage caverns such as the Strategic Petroleum Reserve (SPR) are typically created via slurry mining methods. Mining salt caverns, like those used for the SPR, requires approximately seven gallons of water for each gallon of storage capacity excavated (Kaufmann, 1960).

While these four areas cover the most significant connections between water and energy, they are not meant to be exhaustive. We focus on just two representative problems at the confluence of water and energy in this text: practical limitations on thermoelectric and nuclear power generation, which is water intensive, and the increasing scarcity of potable fresh water for human consumption and agriculture.

Physical water stress, scarcity, and absolute scarcity are considered to be when water supplies fall below 1,700, 1,000, and 500 cubic meters per person per annum, respectively (Falkenmark, 1989; WWAP, 2012). The UN currently estimates that approximately 700 million people in 43 countries suffer from water scarcity, and by 2025, 1.8 billion people are expected to live in regions with absolute water scarcity; as many as two out of three people may live under stressed conditions (WWAP, 2012). Two areas of particular importance are water used for energy extraction and transformation and the relationship between energy availability and reliable supplies of potable water which can be used for drinking and agriculture.

Over the past century, the global population has tripled, and water consumption has increased by roughly a factor of six. Additional forecast growth and changing weather and climate patterns could leave a gap between fresh water demand and supply from likely hydrologic cycles as large as 40% by 2030.

Water is processed energy

In August 2014, Toledo, Ohio's 400,000 citizens were told not to drink their tap water due to algae contamination of Lake Erie, which provides drinking water for more than 11 million Americans (Basulto, 2014). Three hundred thousand residents in and around Charleston,

Table 13.1 *Minimum water requirements for sustaining human life*

Purpose	Liters	Gallons
Drinking water	5	1.3
Sanitation services	20	5.3
Bathing	15	4.0
Food preparation	10	2.6

Source: Gleick, 1998

West Virginia, had a similar experience when a toxic spill of crude methylcyclohexane methanol (MCHM) and propylene glycol phenyl (PPH) ether escaped from a tank owned by Freedom Industries on January 9, 2014, and contaminated the Elk River, the drinking supply for the state's most populous region (Rogers, 2014).

The minimum basic water requirements required to sustain individual human life—not considering food production—in moderate climates for average activity levels are given in Table 13.1 (Gleick, 1998). Even at these very meager levels, amounting to 1,000 m^3 per day and categorized as the breakpoint between water stress and water scarcity, water requirements will grossly exceed available supplies. Many of the areas suffering from water scarcity also depend on water-intensive industrial and agricultural processes for economic advancement, making the challenges more formidable. For perspective, a more comprehensive estimate of US water consumption per capita, including agricultural and industrial uses, was 2,842 m^3 per year; the global average was 1,385 m^3 per year, with Indian and Chinese citizens at 1,089 and 1,071 m^3 per year, respectively (Hoekstra and Mekonnen, 2012).

While managing equities of numerous parties who depend on water for survival, income, or way of life is clearly required, it is difficult in practice. Long-standing efforts by organizations such as the International Law Commission have had limited success in applying general principles for resolving conflicts over shared water resources (McCaffrey, 1993; Gleick, 1998). Additionally, manufactured goods, natural resources, and agricultural products in a globalized economy represent complex and mostly hidden transfers of water between regions and countries outside of the natural hydrologic cycles (Hoekstra and Mekonnen, 2012).

One of the first known instances of states warring over water occured in seventh-century B.C.E. when Ashurbanipal of Assuria seized control of strategic water wells during a desert war against Arabia (Drower, 1954). The disputes over the Jordan River—which is shared by Jordan, Syria, Israel, and Lebanon—are a more approachable and recent example. A subtle component of the broader discord in the region, water conflict was a contributing factor to the 1967 war when several members of the Arab League attempted to divert the headwaters of the Jordan River away from Israel. Then-Premier Levi Eshkol noted that "water is a question of life for Israel" and that the country would "act to ensure the waters continue[d] to flow" (Gleick, 1993a). Even in 2014, the conflict between the extremist Islamic State (ISIS) was significantly influenced by the takeover of the unstable Mosul Dam because of its simultaneous ability to threaten the largest supply of water and electric power for northern Iraq, as well as the physical well-being of the cities of Mosul and Baghdad (Malas, 2014).[4]

In addition to improving wastewater reuse, increasing the availability of fresh water beyond what is normal for hydrologic cycles to provide can only be achieved via large-scale desalinization plants to process seawater and brackish water. Given its history, it is not surprising that at one point Israel was home to the largest desalination plant in the world at Ashkelon, producing more than 300,000 m^3 per day.[5] As of 2005, the total global capacity for desalinization plants proposed, in construction, or operating exceeded 55 million m^3 per day, but operational capacity was estimated at closer to 32 million m^3 per day (Gleick, 2014), rising to more than 40 million m^3 by 2008 and projected to exceed 100 million m^3 per day by 2016 (Schiermeier, 2008). Excluding agricultural and industrial uses, the infrastructure in 2005 could produce 11.68 billion m^3 per year—enough for basic

---

[4] ISIS targeted dams across Iraq—including the Fallujah Dam, Samarra Dam, Mosul Dam, and Haditha Dam—to gain control of water resources. The Mosul Dam, made especially vulnerable to breaches due to a design flaw, posed an additional threat because of its instability and large reservoir, capable of sending as much as a 65-foot wave towards the city of Mosul, located 30 miles downstream on the Tigris River (Malas, 2014).

[5] In 2014, the world's largest desalination plant was the 27 billion-riyal ($7.2 billion) Ras al-Kahir desalination plant with a first-phase production capacity of 1.025 million m^3 per day and 2,600 MW of electricity (Almashabi, 2014).

Table 13.2 *Global desalination plant inventory in 2005*

Plant capacity $(m^3 d^{-1})$	Production capacity $(m^3 d^{-1})$	% capacity	Number of plants
<100,000	29,577,101	53.5	10,487
100,000 to <450,000	19,497,398	35.2	99
>450,000	6,302,800	11	11
Total	55,377,299	n/a	10,597

Source: Gleick, 2014

sustenance for approximately 11,680,000 people. Table 13.2 shows a summary of the 2005 plant inventory from Gleick (2014).

While numerous desalination technologies exist, they remain energy intensive. Their tight coupling with energy means that there are no economies of scale to be gained, as typically seen of large-scale manufacturing today. Creating surface fresh water resources via treatment methods and moving water is simply energy intensive and costly at scale. The theoretical limit for efficiency, based on the laws of thermodynamics, is around 0.7 kWh/m^3 for 100% recovery and 1.06 kWh/m^3 for 50% recovery of seawater—practically, the process is optimally efficient at 1.56 kWh/m^3 when operating at the thermo-dynamic limit for a realistic operating pressure (Elimelech, 2012).

This represents a substantial improvement from 1970s technology, which consumed over 15 kWh/m^3 (Elimelech, 2012). Today, even the most expensive methods—e.g. boiling and condensing seawater as done by some of the thermal plants in the Gulf region—can produce potable water for less than $1 per m^3. Most current seawater reverse osmosis plants take about 2 kWh/m^3 and 3–4 kWh/m^3 for the entire operation of the facility (Schiermeier, 2008). To demonstrate the cost at scale, using an optimistic cost of approximately $0.25 per m^3—twice as cheap as the Israeli Ashkelon plant—with an energy requirement of 3 kWh/m^3 to address the domestic shortfalls forecast for 1.8 billion people expected to live in absolute water scarcity by 2025, it would cost on the order of $113 billion per year and require delivered energy exceeding 1350 TWh.[6] This still excludes agricultural

---

[6] Assuming average water resource availability for this group of 750 m^3 per annum, leaving a shortfall of 250 m^3 per annum for each person.

and industrial uses, as well as the capital and marginal operating costs of transport and storage.

Clearly, more than desalination is required to address the shortfall. But business leaders and policy-makers alike need to recognize that energy and water shortages seed discord. As of 2013, 17 desalination plants were proposed in California to increase the reliability of water in the face of another prolonged drought, but forecasted inflation adjusted electricity price increases of more than 25% from 2008 to 2020 will increase costs of desalination between 9% and 15% over the same period (Cooley and Heberger, 2013). This coupling makes planning and risk management more complex. Meeting our growing requirements while simultaneously adapting to changing climate conditions will undoubtedly be energy intensive, capital intensive, and take substantial amounts of time.

Cooling power plants

Physical water scarcity is closely related to energy extraction and transformation efforts. Electricity production is the single largest source of water withdrawal around the globe, surpassing even agricultural uses. In 2005, as much as 41% of fresh water withdrawn in the US was for thermoelectric and nuclear cooling (Kenny et al., 2009), with particularly large amounts of water consumed by baseload coal and nuclear power plants (Faeth et al., 2014). The distinction between *withdrawals*, where water is returned, and *consumption*, where water is lost, is important to understand. In 2008, surface water resources accounted for 94% of thermoelectric water withdrawals and 86% of consumption in the US and were substantially higher east of the Mississippi River—buoyed by higher levels of coal-fired plants and the legacy of once-through cooling systems (Averyt et al., 2011).

Growing demand for water and electricity integrated energy and water policy planning. Non-thermal sources of electricity generation are likely to prove increasingly important where communities face growing water scarcity. Other approaches, such as switching from wet cooling to dry cooling—dry-cooled systems use air instead of water to cool the steam exiting a turbine but still require substantial volumes of water for cleaning and maintenance—may be possible in some areas, but at an increased monetary cost and lower efficiency. Average annual losses in performance for the switch in the US are 6.8% for

Table 13.3 *US thermoelectric generator cooling type frequency and generation percentage in 2008*

Cooling type	Percentage of facilities	Percentage of generation
Once-through	43	30
Recirculating	56	45
Dry	1	2

Source: Union of Concerned Scientists, 2012

nuclear plants, 1.7% for CHP plants, and 6.9% for other fossil-fueled facilities, including coal and natural gas plants (EPA, 2009b). The U.S. Union of Concerned Scientists (UCS) published data for 5000 thermoelectric generators detailing the cooling type and percentage of generation, as shown in Table 13.3 (UCS, 2012).

Minimizing environmental impact can also be challenging. Low-carbon-emitting technologies such as pulverized coal with carbon capture systems boast many of the highest water consumption factors when geothermal and hydroelectric facilities are excluded. Table 13.4 shows typical approximations of water requirements for modern electric generation facilities (data sourced from Macknick et al., 2011; Kenny et al., 2009; Meridian Corp., 1989; Clark et al., 2011).

The substantial differences in water withdrawals and consumption for specific facilities makes policy and planning challenging. Addressing environmental externalities and minimizing the risk that water-related issues will prevent a plant from producing electricity for the grid is a specialized process, tailored for the unique requirements of a particular watershed. When planning assumptions fail due to low water levels (e.g. from drought) or increases in water temperature (i.e. during hot summer months), facilities may be forced to reduce output or become completely unable to contribute to the grid—increasing prices and the risk of capacity shortfalls and instability.

Shutdowns of French nuclear facilities due to critical shortage of water for cooling contributed to power shortages during a major heat wave between August 4 and 18, 2003, ignominiously capping the hottest recorded summer in Europe since 1500 (Poumadere et al., 2005). Large numbers of elderly people were injured or killed. The Indian Point Energy Center, a nuclear power facility which produces approximately 25% of the electricity consumed in New York City

Table 13.4 *Median and maximum water withdrawal and consumption for selected electric generation facilities*

Fuel	Cooling type	Median (gal./MWh)		Maximum (gal./MWh)	
		Withdrawal	Consumption	Withdrawal	Consumption
Nuclear	Tower (recirculating)	1,101	672	2,600	845
	Once-through	44,350	269	60,000	400
Oil	Tower (generic)	500	480	1,000	600
	Once-through	20,000	100	50,000	300
Natural gas	Tower combined cycle (CC)	253	198	283	300
	Tower steam	1,203	826	1,460	1,170
	Tower CC w/CCS	496	378	506	378
	Once-through CC	11,380	100	20,000	100
	Once-through steam	35,000	240	60,000	291
Coal	Tower (generic)	1,005	687	1,200	1,100
	Once-through (generic)	36,350	250	50,000	317
	Tower (subcritical)	531	471	678	664
	Tower (supercritical)	609	493	669	594

Table 13.4 (*continued*)

Fuel	Cooling type	Median		Maximum	
		Withdrawal	Consumption	Withdrawal	Consumption
		(gal./MWh)		(gal./MWh)	
Coal w/CCS	Tower (subcritical)	1,277	942	1,329	942
	Tower (supercritical)	1,123	846	1,148	846
Geothermal	Tower (EGS)		4,784		5,147
	Dry (EGS)	1,000	850	2,000	1,778
	Hybrid (EGS)	2,000	1,406	2,764	1,999
Solar PV	n/a	0	26	0	33
Wind	n/a	0	0.8	0	1.1

Source: Macknick et al., 2011; Kenny et al., 2009; Meridian Corp., 1989; Clark et al., 2011

and the lower Hudson Valley, is in a similar predicament because of the 2.5 billion gallons of Hudson River water used daily for cooling (Avila, 2014). In 2014 the New York Department of Environmental Conservation proposed annual closures of the plant between May 10 and August 10—in 42-, 62-, or 92-day increments—unless the plant owner installs a closed-cycle cooling system, a move billed as too expensive by Entergy Corporation. Environmental pressure from water requirements, spikes in tritium (a radioactive isotope of hydrogen with a half-life of 12.32 years) in groundwater near the plant (Revkin, 2014), and the Nuclear Regulatory Commission announcement naming Indian Point as one of ten sites considered most in need for a re-evaluation of earthquake vulnerability (Wald, 2014) could significantly change the costs and incentives for the single largest electricity-producing facility serving the largest city in the US.

13.2 Strategic Petroleum Reserve (SPR)

The new millenium has demonstrated the uncertainty in global energy markets, with noted experts like Daniel Yergin remarking that "we are living in a new age of energy supply anxiety" (Spiegel, 2006). Each successive wave of instability in the Middle East triggers another call for price stabilization efforts via the release of strategic reserves. On June 23, 2011, the International Energy Agency (IEA) announced a plan deemed "Libya collective action" that ran until September 25 in an attempt to offset the estimated 132 million barrels of oil that had failed to reach the market and contributed to a 20% rise in spot markets over six months (Clayton, 2012).

The IEA-coordinated releases of nearly 60 million barrels produced a small but immediate reduction in prices that was soon outpaced by changes in the strength of the dollar versus the euro and the improved economic outlook (Clayton, 2012). This was similar to Nixon's signature energy policy authorization of the $10 billion Alaskan pipeline in November 1973—which was significant and helpful but, ultimately overshadowed by overall economic growth that was far greater than the increase in capacity. The sheer volume of petroleum consumed in the US, an average of 18,887 thousand barrels per day (bpd) in 2013 (EIA, 2014e), renders insurance stockpiles like the SPR to little more than potentially helpful tools to ensure adequate supply for national defense and disaster response purposes—too small to be of significant import to global price changes.

Table 13.5 *Strategic petroleum reserve facilities and capacities*

Facility	Location	Storage (mb)	Drawdown (mbpd)
Current			
Bryan Mound	Freeport, Texas	254	1.5
Big Hill	Winnie, Texas	160	1.1
West Hackberry	Lake Charles, Louisiana	227	1.3
Bayou Choctaw (current)	Baton Rouge, Louisiana	76	0.55
Future			
Big Hill	Winnie, Texas	250	1.5
Bayou Choctaw	Baton Rouge, Louisiana	109	0.6
New Facility	Richton, Mississippi	160	1
Retired			
Weeks Island	Iberia Parish, Louisiana	72	n/a

The SPR was created by President Gerald Ford on December 22, 1975, under the Energy Policy and Conservation Act of 1975 as the US component of the IEA in OECD countries' International Energy Program (IEP) (Eaton and Eckstein, 1984). It was in response to the Organization of Petroleum Exporting Countries (OPEC) oil embargo against the US and the Netherlands which marked the energy crisis of the 1970s and caused the price of crude oil to rise from $3 to $30 a barrel between 1973 and 1980. Participants in the IEP pledged to establish reserves equaling 60 days of consumption, a number later increased to 90 days in 1980 (Eaton and Eckstein, 1984). By March 1982, the US reserves had grown in spurts to reach 250 million barrels; they would go on to reach 700 million barrels on August 17, 2005 (Eaton and Eckstein, 1984; Bamberger, 2009). The oil is stored in five giant caverns operated by the Department of Energy, which are excavated from natural salt domes via water injection and brine removal. Current, retired, and proposed facility locations and capacities are given in Table 13.5.

Today, some 40 million bpd cross the oceans on tankers, and that number could rise as high as 67 million bpd by 2020 (Spiegel, 2006). In a globally connected energy market, pipelines and shipping lanes must remain secured to facilitate smooth transfers of energy resources between regions. The US alone averaged 6.2 million bpd in net imports during 2013, but the SPR can only support a maximum withdrawal rate of 4.4 million bpd—a shortfall of nearly 2 million bpd if the unlikely scenario of absolute import cessation, including from Canada, were to occur (EIA, 2014e; Bamberger, 2009). The Libya sale, which drew 30.64 million barrels from the SPR, left the US with 80 days of "import protection" in 2012 (Andrews and Pirog, 2012).

Modern, energy-intensive societies can only be insulated from their ravenous demand for energy for short periods. Policy-makers and business leaders need to plan for fuel disruptions and not rely on mythical "insurance" reserves to prevent them from being exposed to large-scale price shifts in global energy markets, whether they be coal, natural gas or oil.

13.3 Fukushima

The Fukushima nuclear disaster illustrates the impact of ongoing climate and weather shifts on energy infrastructure and the consequence of a major source of electric generation capacity being unavailable to support national energy requirements.[7] On March 11, 2011, a magnitude 9.0 earthquake struck northeastern Japan and created a tsunami which destroyed the six-reactor Fukushima Daiichi nuclear power generation facility and numerous transmission and distribution assets in Japan's northern grid, killed some 20,000 people along with the tsunami, and led to the imposition of rolling blackouts across large swaths of the country. Four nuclear facilities were directly affected, including Fukushima Daiichi, Fukushima Daini, Onagawa, and Tokai Daini (Japan Cabinet Office, 2011).

Fukushima has been widely described as the worst nuclear disaster since Chernobyl in 1986, eclipsing the 1979 Three Mile Island incident

---

[7] Many large thermoelectric generation facilities are increasingly at risk, because they were historically located near coasts to aid with cooling, facilitate lower-cost transport of goods for construction, and ensure that bulk fuel could be reliably supplied with low marginal costs and minimal dependence on any singular entity for transport.

in the US. At the time of the accident, 36 nuclear reactors were actively providing 33.5 GW of power, but by May 2011, only 17 of the 50 reactors—excluding the Monju and damaged Fukushima Daiichi plants—were in operation. The number of operating reactors soon dwindled to zero in support of maintenance, regulatory, and safety inspection requirements (Wilson, 2014). Nuclear power is poised to restart in late 2014, after comprehensive safety inspections, licensing reviews, and periodic shutdowns, including periods of nuclear-free Japanese power in 2014 and 2012 (Matsuyama, 2013).

The failure at Fukushima Daiichi could have been mitigated or prevented by higher plant elevation, sea wall elevation, and location and watertight protection of the 13 crucial emergency diesel generators—12 of which failed, causing the backup cooling systems for the reactor cores and spent fuel pools to fail (Lipscy et al., 2013). Lax oversight by compromised regulators—Japan's Ministry of Economy, Trade and Industry (METI) was responsible for expanding nuclear power generation *and* regulation until September 19, 2012, when the Nuclear Regulation Authority was established—and consistent failures to heed repeated warnings about explosions and radiation release vulnerabilities in the General Electric reactors used at Fukushima Daiichi from the U.S. Atomic Energy Commission since the 1970s helped create the condition for the nuclear component of the disaster (Kaufmann and Penciakova, 2011; Wilson, 2014). The combination of poor governance and the composite catastrophe exposed major gaps in Japan's energy security and disaster response. Prime Minister Naoto Kan, who initially established a joint crisis response headquarters with Tokyo Electric Power Company (TEPCO), would eventually resign in August 2011 (Funabashi and Kitazawa, 2012; Harlan, 2011). The Japanese evacuation zone was defined as the area 20 to 30 km from the nuclear plant, while US officials set the zone at 80 km (Danzig and Hosford, 2012).

The highly centralized Japanese electricity industry is dominated by 10 privately owned and vertically integrated utilities which grew by interconnecting with Tokyo Electric Light or Osaka Electric Lamp, the second Japanese electric utility. Osaka Electric Lamp, in the west, initially purchased General Electric equipment on the US-based 60-Hz standard, and the legacy investments led to a bifurcated power system. Addressing the shortage of power in northern Japan and the greater Tokyo area was made virtually impossible by the lack of adequate *interconnection facilities* to convert power from the Western grid, despite available capacity. At the time of the tsunami, only three interconnections, with a capacity of approximately 1,000 MW, linked the separate systems. TEPCO's immediate shortfall was more than 10 times the capacity of the interconnections. Even with 2011's record snowfall, which boosted hydroelectric energy resources, TEPCO and other utilities were forced to rapidly retrofit and reactivate mothballed facilities, cancel scheduled maintenance, and arrange for the purchase and delivery of previously unneeded fossil fuels. Japan was also forced to immediately implement a mix of utility-imposed blackouts on roughly 10 million people, a number that soon fell to 3 million, along with a campaign of "self-restraint" involving rationing and coordination of energy consumption.

Before Fukushima, Japan was not reliant on any particular source of energy for more than 30% of its electricity requirements, but in 2012 nuclear fell to 2% while liquid natural gas and oil rose to serve 48% and 16%, respectively (EIA, 2014b). Japan has also reduced environmental regulations for fossil fuel plants to encourage utility investment, especially in coal. But even when rushed, the construction phase of large natural gas, oil, or coal facilities takes 2–4 years, making new construction an important part of an overall strategy but far from sufficient on its own. Japan experienced significant growth of fossil fuel–based generation in 2011 and 2012 to compensate for lost nuclear generation capacity, despite rationing and societal restraint (EIA, 2014b). Electricity prices rose along with fuel imports, METI approved tariff increases ranging from 7% to 11% for six of the ten major utilities since July 2012 alone—since 2010, households averaged a 19.4% increase in rates with 28.4% for industrial consumers—and Prime Minister Shinzo Abe has announced plans to seek legislative approval for unbundling generation, transmission, and distribution (EIA, 2014b; Aldrich and Platte, 2014).

Despite these changes and anti-nuclear sentiment in the Japanese public, METI's 4th Basic Energy Plan, approved in April 2014, continues to emphasize nuclear power as an "important power source that supports the stability of the energy supply and demand structure" (METI, 2014). Masakazu Toyoda, Chairman of the Institute of Energy Economics, Japan, said "we don't have to have 50%, but 20 or 25% might be necessary. [...] I think that we cannot survive without nuclear" (Strickland, 2013). Toyoda's statement is unsurprising, as Japan's energy requirements far outstrip its natural resources; importing uranium, even during conflict, is much more feasible than maintaining global sea lanes to insulate against disruptions to a society that is overly reliant on coal or LNG imports. That said, 23 reactors will reach their 40-year operating license limit by 2025, meaning that many of the older existing reactors may not restart due to significant costs associated with meeting new regulatory requirements and lack of time available for cost recovery (Aldrich and Platte, 2014). Investors are skeptical of the total numbers given these conditions. In July 2014, JP Morgan cut its forecast from 42 to 31 reactors restarting by 2019, and financial adviser Raymond James Ltd. forecast that as much as two thirds of the pre-accident fleet may never resume operations (Sharples and Inajima, 2014).

Due to sustained reductions in power quality and availability on the grid, Japan is witnessing significant growth in co-generation, combined heat and power, and CHP facilities. This matches the US experience with Superstorm Sandy in the New York metropolitan area and the Indian experience with robust private power generation owned by companies like Tata and Reliant Industries. Sustained lack of reliable power has motivated Japanese businesses. In June 2014, three major business lobbies—the Japan Business Federation (Keidanren), the Japan Chamber of Commerce and Industry, and the Japan Association of Corporate Executives (Keizai Doyukai)—sought early restart of the nuclear reactors, arguing that the nation's top priority for energy policy should be a quick return to inexpensive and stable supplies of electricity.

A misunderstanding of risk to interdependent systems, a lack of acknowledgment "that catastrophes often do not come in neat categories" and that "systems built to respond to narrowly defined risks tend to fall into disrepair when those risks do not materialize," and a lack of investment in societal resilience—especially within the electric power system—caused significant harm to the Japanese economy and people (Danzig and Hosford, 2012). Japan's increased need for additional liquid natural gas, diesel, and crude oil will continue to have a global impact on energy markets and has materially affected global regulation and perceptions of nuclear power. The Japanese experience should serve as a warning to other nations that suffer from a similar lack of interconnection capacity to rapidly and cost-effectively move large amounts of electricity in order to maintain critical services and minimize economic disruptions.

13.4 Integration of the North American power grid

The Tres Amigas project is an unprecedented effort to connect North America's three major asynchronous electric grids in a single 22-square-mile "superstation" located in Clovis, New Mexico. The project is important because it will increase the grid's ability to integrate large amounts of renewables and reduce the risk of capacity shortages stemming from drought conditions, lack of wind, regulatory closures (e.g. nuclear power following Fukushima in Japan), or fuel shortages. The project will allow the North American grid to proactively manage the direction and level of power flowing between the Western Interconnect, the Eastern Interconnect, and Texas, which runs its own separate power system—also known as the Electric Reliability Council of Texas (ERCOT).

The project was initiated by Phil Harris, a former CEO of PJM Interconnection—the largest regional transmission organization in the United States responsible for interregional power coordination across 13 states and the District of Columbia—and is currently scheduled to complete financing and break ground in the fall of 2014 after a series of delays. Tres Amigas will own and operate an electricity clearinghouse and operate as a merchant transmission company, meaning that it is responsible for all construction, operation, and maintenance costs at the superstation. The FERC also granted Tres Amigas rights to sell its transmission services at market rates via auction or negotiated contracts (FERC, 2010).

The FERC has approved interconnection agreements between Tres Amigas and the Public Service Co. of New Mexico in the Western Interconnection and Southwestern Public Service in the Southern Power Pool (SPP) in 2013 (FERC, 2012, 2013a). The project intends to grow along existing railroad and highway rights of way where possible, initially connecting the Four Corners power market in northwest New Mexico to the SPP across the superstation, using a 73-mile 345 kV line to provide up to 750 MW of bidirectional capacity at a cost of $429 million (Boswell, 2014). Further expansion would involve roughly 360 miles of transmission line development throughout southern New Mexico to enable solar and wind development in the region (Platts, 2013). The overall effort is extremely capital intensive and is being financed via industrial revenue bonds, with second and third phases estimated at $436 million and $793 million, respectively (Williamson, 2014; Boswell, 2014). All told, the superstation would enable 4,750 MW of interchange capacity, which could grow to as much as 30 GW over the long term.

Initial interconnection agreements have been signed between all parties, excluding ERCOT, which is determined to ensure that the Texas grid remains independent of FERC oversight. When asked about Tres Amigas and Southern Cross, another $2 billion potential interconnection project seeking to connect ERCOT with utilities in Mississippi and Alabama, the chairwoman of the Texas Public Utility Commission, Donna Nelson, emphasized that "there's no way we would support any of those if we didn't have commitment from FERC that it didn't threaten our jurisdiction" (Galbraith, 2012). ERCOT's current FERC exceptions cover five operational DC ties to the other grids, totaling 1,106 MW of capacity.

Tres Amigas is also integrating a number of new technologies together into a first-of-its-kind commercial deployment in North America—similar technology has been employed in China's high-voltage direct current (HVDC) grid and parts of the planned European Extended Grid Plan.[8] The project plans to demonstrate the first HVDC subterranean superconducting lines in the US, voltage source converters to connect AC power in each interconnect with the DC superstation, and on-site energy storage to shape and smooth power transfers between the independent systems. Perhaps the most unique characteristic of the project is that there are no guarantees by the government and no fixed-cost recovery from ratepayers. Harris notes that his project is "unique in that it totally stands on its own merits like a toll bridge. If this electrical toll bridge has value to you in either moving energy or in the ancillary support that our equipment can provide, then you'll pay for it" (Kessler, 2014).

13.5 Electromagnetism and the power grid

Another energy–security issue is the impact of strong electromagnetic fields—whether from space weather or man-made—on electric power systems. Space weather is often used as a synonym for solar activity consisting of solar flares (large explosions of magnetic energy) or

---

[8] The European extended grid plan would include parts of the existing network with additional HVDC lines up to 40,000 km in length—extending from Iceland to Israel.

coronal mass ejections (the violent release of plasma). Despite the frequency of solar events and a number of recent advances in solar forecasting by heliophysicists, the field remains in its infancy.[9] With the cost of a large-scale geomagnetic storm potentially reaching over $1 trillion (NAS, 2008) and comparatively small, recent, and still large-scale blackouts like August 14, 2003, costing between $7 and $14 billion (Hilt, 2005), improving performance is an imperative. Waiting for another catastrophe—like the 1965 blackout that drove the formation of the North American Electric Reliability Council in 1968 (Kirby et al., 1995)—to serve as a catalyst is a dangerous and ill-advised strategy.

Geomagnetically induced currents

When the energy released in solar events strikes the Earth's atmosphere and magnetic shield, the Earth's magnetic field shifts. These shifts in the field induce currents in the long, high-voltage wires used primarily in the bulk electric power system. The resulting *geomagnetically induced currents* are a function of power system, earth, and storm parameters but are generally worse in east-west transmission lines than those running north-south (Barnes et al., 1991).[10] While threats to the grid are particularly severe, and the focus of this report, space weather events can also damage satellites and avionics and disrupt radio communications etc. (Morrow, 2011; NAS, 2008; Cannon, 2013; Odenwald and Green, 2008).

Induced currents can reach significant magnitudes, melting copper windings in transformers; melting lines until they sag and short on trees or the ground; and causing damage to large, expensive infrastructure components. Most existing response models involve utilities

---

[9] NASA has three major related projects: the Solar Terrestrial Relations Observatory, the Solar Dynamics Observatory, and the Advanced Composition Explorer. The Advanced Composition Explorer remains the best early warning system for Earth, as it orbits between the Earth and the Sun at L1 and intercepts coronal mass ejections, radiation bursts, and solar wind. Unfortunately, the Advanced Composition Explorer, launched in 1997, is already past its intended retirement date but is equipped with enough fuel to operate through 2025 (Margolies and von Rosenvinge, 1998; Baker and Lanzerotti, 2008).

[10] The earth surface potential induced by a disturbance generally has a steeper east-west gradient, but the gradient direction is dependent on the orientation of the earth to the disturbance interacting with the planet's magnetic field (Barnes et al., 1991).

pooling resources to reduce costs while maintaining preparedness within a region. Examples include sharing trained crews, lateral agreements to share replacement transformers, etc. Major space weather events or physical attacks can cause simultaneous failures that render such processes/resource pools inadequate and ineffective. Unscheduled flow between regions—also known as loop flow—is also strongly correlated with solar activity and occurs on both transmission and distribution lines, challenging grid operators.

Historical events

On March 13, 1989, one such geomagnetic disturbance caused by a solar flare triggered cascading failures in the North American bulk electric power system with estimated damages reaching into the hundreds of millions of dollars. The outage lasted 9 hours—until 83% restoration—and was power-system-wide in the Hydro-Quebec power system (Barnes et al., 1991). A larger storm occurred in 1921, from May 10 to 12, when telegraph and telephone companies saw service interruptions and voltage variations from the geomagnetically induced currents that were both dramatic and volatile (*New York Times*, 1921b). Permanent damage to telegraph cables, fuses, etc. occurred (*New York Times*, 1921a), and aurora were widely observed throughout the United States, United Kingdom, Samoa, and other uncharacteristically low latitudes (Campbell, 1921; Silverman and Cliver, 2001; Angenheister and Westland, 1921).

The largest solar flare ever recorded, known as the Carrington event, named for the astronomer who recorded it, occurred on September 1, 1859. The large coronal mass ejection was violent, nearly doubling the amount of sunlight produced at the flare location, which was the first recorded space weather event as a result of its visibility in white light (Loomis, 1859; Stewart, 1861). Magnetic storms are measured based on disturbance storm time (Dst) minimum estimates, which are measured in nanoTeslas (nT). Today's best estimates place the Carrington event at roughly -900 nT ($+50$, -150) from original estimates of $-1,760$ nT. This magnitude is closer to the storm of 1921, leaving it approximately 50% more powerful than the 1989 storm (Cliver and Dietrich, 2013).[11] As a comparison, the Carrington event is approximately 18 times the magnitude of

[11] The previous nT estimates likely included an ionospheric contribution leading to early overestimation of magnitude (Cliver and Dietrich, 2013).

ordinary geomagnetic storms that produce the Northern Lights at northerly latitudes. Focusing on induced currents, as opposed to direct observations and measurements of the storm intensities unsurprisingly provides yet another estimate of 1921 to 1989 storm magnitudes. An AT&T telecom cable in northern Illinois failed on August 4, 1972, when it experienced a 7 V/km geoelectric field with a storm of 800 nT (Kappenman, 2010). Similar observations for the storm of 1921 indicated a geoelectric field strength of 20 V/km, indicating that similarly oriented conductors experienced effects as great as ten times the impact of the 1989 North American power failure that impacted Quebec (Kappenman, 2010).

As a result, a Carrington-like event is thought to be a worst-case scenario for space weather and is often used as a benchmark for evaluating the impacts from electric grid disruption to an industrialized nation like the US (Cliver and Svalgaard, 2004). Power system collapse in both the Pacific Northwest and northeastern United States is considered a likely outcome from a severe geomagnetic storm centered at 50° northern latitude, impacting a total estimated population exceeding 130 million. High-voltage transformer destruction is likely to be more costly and damaging than the large-scale blackout in such a scenario. There are more than 300 large 345-kV, 500-kV, 765-kV transformers at risk of permanent damage or loss—all devices which normally average 15 months' time from order to delivery and are often made outside the US (Jappenman, 2010; Morrow, 2011). The lack of incentives for utilities to maintain bench stock could be quickly addressed by regulators by authorizing rate recovery to mitigate the likelihood of prolonged outages from such events.

While the specific forecasts for a future Carrington event are unclear, space weather will become increasingly important to our digital world. Some degree of likelihood can be gleaned from the 2013 Royal Academy of Engineering Report on Extreme Space Weather (Cannon, 2013):

One contemporary view is that a Carrington-level event will occur within a period of 250 years with a confidence of ~95% and within a period of 50 years with a confidence of ~50%, but these figures should be interpreted with considerable care.

Other research generally agrees, placing the probability of another Carrington event occurring within the next decade at between 4%

and 6% (Kataoka, 2013). The National Research Council workshop in 2008 identified an estimated cost of a Carrington-scale "severe geomagnetic storm scenario" at $1 trillion to $2 trillion in the first year, with a recovery time ranging from four to ten years (NAS, 2008). While many of the specifics regarding the impact of such an event on modern society remain contested, it is clear that today's grids—as a result of accumulated design decisions, increased amounts of high-voltage transmission, and existing operational processes and procedures—are more susceptible to geomagnetically induced currents than ever before (Kappenman, 2010). Evaluation of contingency procedures, temporal complications to mitigating processes, network resilience, and safety margins must be assessed, ultimately weighting the cost and complexity against potential risks to large-scale systems and society (IRG, 2010). The physical susceptibility of the electric power system to space weather disturbances requires the development of mechanisms to adequately price risk, ensure responsible mitigation, and understand downstream dependencies of other critical infrastructure systems which rely on the constant operation of the electric power system at scale.

Electromagnetic pulses

Electromagnetic pulse (EMP) weapons can also create solar storm–like events that threaten electric power systems and a wide range of modern telecommunications and electronics. Reports by the congressional EMP Commission in 2004 and 2008 both warned that terrorists or state actors may calculate that an EMP attack will cause the most significant disruption if used or leverage if threatened (Foster et al., 2004, 2008). By 2014, a 2009 congressional Commission on the Strategic Posture of the United States[12] and studies by the National Academy of Sciences, the Department of Energy, the Federal Energy Regulatory Commission, and the National Intelligence Council all urged action, but little has occurred. The most outspoken efforts to raise awareness and encourage action have come from a few outspoken individuals, including former CIA Director Jim Woolsey, Dr. Peter Vincent Pry, and billionaire hedge fund manager Paul Singer, who told his clients that EMP is "the most significant threat" to the US

---

[12] The Commission was co-chaired by former Secretaries of Defense William Perry and James Schlesinger (Perry et al., 2009).

and its allies (Woolsey and Pry, 2014). Other policy-makers like Jim Lewis, director of the Strategic Technologies Program at the Center for Strategic and International Studies in Washington, remain skeptical, arguing that "the wild hysteria that's greeted EMP attacks lately is wildly overstated" (Mulrine, 2014).

There are three primary levels of EMP risks, not all of which are of practical concern. The primary threat from an EMP weapon is a high-altitude nuclear blast creating a broad-band, high-amplitude electromagnetic pulse (HEMP). The pulse is caused by gamma rays released in the blast interacting with the atmosphere and would be most damaging to the US if released at an altitude between 250 and 300 miles in the center of the country. This type of high-altitude nuclear explosion produces a form of time-changing magnetic behavior, often referred to as a magnetohydrodynamic electromagnetic pulse (MHD-EMP) which induces quasi-DC currents in long conductors that are similar to solar geomagnetically induced currents (Barnes et al., 1991). Low-altitude nuclear detonations can be used to achieve similar effects on hardened military or communications targets via a source region electromagnetic pulse (SREMP) within a radius of approximately 3–8 km with even higher late-time voltage levels. Finally, high-power microwaves (HPMs) can leverage batteries or chemical explosions to damage electronics in a much smaller target area without nuclear weapons technology.

Skeptics who doubt whether or not specific threat actors could obtain or would choose to utilize such devices aside, there is little doubt that the physical laws governing electricity will damage key infrastructure if a HEMP detonation occurred. During the 2008 Commission to Assess the Threat to the United States from an EMP Attack, every SCADA system component or DCS component tested failed when exposed to the simulated EMP environment (Foster et al., 2008). The equipment cannot survive the fast-rising electromagnetic pulse, which is similar to lightning but generally considered more dangerous because of cumulative damage occurring in the three basic successive EMP pulses and the faster voltage rise that occurs in the first EMP pulse.

The first pulse is caused by a free-field energy pulse which simultaneously—within approximately one billionth of a second—affects an entire region, with large induced transient voltages that damage printed circuit boards, wires and conductors common to modern electronics.[13] The second pulse is typically referred to as the middle time component and is most similar to lightning but lower in surge amplitude. It would not normally affect many systems, but the first pulse can disable protective systems that would detect and protect against this type of "surge." The third and final pulse is also called a late-time EMP and is generally experienced by long conductors like power lines, copper telephone lines, and submarine cables which are most susceptible to electric or magnetic induction and, because of being commonly buried, resistive coupling. Induced currents which run in the same direction as the conductor's length are *electrically induced*, while current flows in closed conducting loops triggered by a generated magnetic field perpendicular to the plane of the loop are *magnetically induced*. Currents which are introduced to the ground, water, or other conducting materials and shift to better conductors nearby (e.g. subterranean power or telecommunication lines) are caused by *resistive coupling*.

Those who prefer empirical data from real-world systems can take heart, or despair, because the U.S. Starfish Prime test on July 9, 1962, as well as the seven Soviet nuclear tests conducted above an altitude of 12 km between September 1961 and November 1962 (as part of the K Project in Kazakhstan), confirm the devastating effects of HEMP. The U.S. Starfish prime test occurred 898 miles away from Honolulu but still caused major electrical problems in Hawaii when effects were larger than initially anticipated. The Soviet tests, the most significant of which was Test 184 on October 22, 1962, were even more concerning because it appeared that automatic relays designed to prevent line and transformer damage were not successful despite successfully tripping the circuits. Test 184 destroyed electronics, shut down long-distance telephone lines, burned out insulators on above-ground power lines, fused underground power lines, and burned out (even burned down) generating plants (Foster et al., 2008).

---

[13] The free-field energy is triggered by gamma rays, which are high-energy photons that interact with atmospheric molecules and result in decaying Compton electrons which then emit photons in turn.

13.6 Chapter 13 Summary

- Energy security is best viewed as a function of reliable and economical access to energy. Societal dependence on a particular form of energy should be carefully considered when attempting to weigh specific risks.
- Water and energy security issues are increasingly inseparable. Vast quantities of water are vital to the energy supply chain, and large amounts of energy are required to clean and transport water suitable for supporting growing human populations.
- Temporarily storing fuels can provide some short-term relief to supply disruptions, but existing efforts, such as the Strategic Petroleum Reserve, are to generate sustained price effects in an increasingly integrated global energy market.
- Increasingly robust energy and power transport systems, such as the Tres Amigas project in the US, will help improve societal resilience to particular energy resource disruptions such as the temporary shutdown of all Japanese nuclear plants in the aftermath of Fukushima.
- Integrating large-scale electric power systems is central to developing more economically efficient, planetary-scale energy delivery networks. However, such efforts must be accompanied by appropriately resourced risk mitigation efforts to guard against intentional attacks and naturally occurring threats to these increasingly critical infrastructures.

14 | *Energy-as-a-service*

Companies [...] are at a crossroads—they need to reengineer business
models that worked well for more than 50 years if they want both to
survive and retain their position in the market.

White et al. (2013)

In 2013, nearly 100,000 Texans elected to join TXU Energy Right
Time Pricing, agreeing to pay more during the week for "anytime"
electricity usage in exchange for free power between 10 p.m. and
6 a.m., via TXU Energy Free Nights, or 12 a.m. Saturday to 11:59
p.m. Sunday, via TXU Energy Free Weekends, all year (Little, 2014).
With 1.5 million customers and more than 40% of the market share
for retail electricity sales in Texas, TXU's offering was significant,
both in terms of size and as the first "free" electricity incentive plan
in Texas. These plans steeply incentivize consumers to concentrate
their energy during the non-peak hours in the program to reduce
strain on the Texas grid, but there have been concerns over how
the plans were marketed to consumers because of an 18-month
contract agreement, steep cancellation fees, and general confusion
about electricity purchases (Lieber, 2013).

The convergence of energy, information, and financial flows is
creating new opportunities for economic growth and improvements to
how we relate to the energy we consume. The programs in Texas are
just the beginning of diverse energy services made possible by modern
technologies. Beneath these new kinds of services offered to businesses
and individual consumers, decision-making or support tools rely on
software platforms "for myriad tasks that previously were performed
in traditional, non-tech ways" (Evans et al., 2008).[1] New business

---

[1] Evans, Hagiu, and Schmalensee's *Invisible Engines* provides a detailed exploration of how software platforms have accelerated technological innovation
over the past quarter-century.

models seeking differentiation by selling energy in creatively priced offerings to incentivize non-peak consumption, pool-buying power of many consumers with complementary load profiles like Samuel Insull in 1914 (see page 168), or charging customers based on energy outcomes rather than the commodity will increasingly emerge.

This chapter explicitly discusses how we currently buy electricity; energy services versus products; improved metrics to improve energy visibility; and energy efficiency as a resource for individuals, businesses, and nations.

14.1 Buying electricity today

When we plug a computer or a TV into a wall socket, what are we buying? Are we buying the *electrons* that travel from the generator to our homes? Or are we buying *access* to those electrons whenever and wherever we might want? Do we care about electricity at least cost with defined scheduling, about 24-hour access to electricity with no scheduling, or just about the devices and services which are reliant upon it? Are we buying energy as a commodity or a comfortable environment as a service?

A typical laptop power supply offers a surprising number of clues. Most of the symbols, such as UL and CE, signify compliance with different countries' safety and performance requirements.[2] Some symbols represent disposal and recycling instructions. First and foremost, we are buying *safety* that the power supply will not electrocute the user. Figure 14.1 shows a specifications from typical laptop voltage transformer and power conditioner.

The variables, outlined in Table 14.1, represent the electrical characteristics of the adapter. Typical values are also given for exemplary regions. The input column depicts the power quality required from the grid in order for the adapter to work. The adapter converts the AC signal (represented by the symbol $\sim$) to a DC signal (represented by the symbol $=$). Since DC does not have a frequency, there is no frequency (or Hz number) associated with the output.

---

2 Underwriters Laboratories (UL) was established in 1894 and provides safety and performance certification testing for listed products and recognized components. The CE mark has been mandatory for certain products sold within the European Economic Area since 1985, standing for Conformit Europenne.

```
┌─────────────────────────────────────┐
│  AC ADAPTER                          │
│  INPUT: 100–240V~, 50–60Hz 2.0A      │
│  OUTPUT: 19V⎓ 3.42A                  │
└─────────────────────────────────────┘
```

Figure 14.1 Specifications from a laptop adapter

Table 14.1 *Variables from an AC power adapter*

Input	Output	Examples	Notes
100–240 V ~	19 V ⎓	Japan: 100 V, North America: 110 V, United Kingdom: 230 V, Kuwait: 240 V	~ indicates AC and ⎓ indicates DC.
50–60 Hz	n/a	Japan: 50 and 60 Hz, North America: 60 Hz, Germany: 50 Hz	50 Hz is most common globally; 50–60 Hz were selected because of lighting technology requirements in Europe and the US during initial electrification.
2.0 A	3.42 A		Fewer amps are required with the higher-voltage AC input than the DC output.

Based on these specifications, this adapter can safely be used on any standard power system in the world.

What is missing from the label is, perhaps, more important than the specifications captured in Table 14.1. Nowhere on this adapter is there an *energy* requirement or kWh specification. This device requires uninterrupted access to electricity within the specified performance parameters in order to function. Many of the devices we purchase fall into this category—we have become accustomed to always-on electricity services where the only action required is flipping the switch. Programs like those offered by TXU may change this anytime access relationship with electric power over time, but in the near term it is clear that we expect access to power anytime, anywhere we please. Clarity about purchasing electricity versus electricity access *plus* the commodity is central to a rational discussion about energy. Confusing these two closely coupled, but distinct, problems leads to problems in individual decision-making and contributes to unhelpful comparisons and hyperbole common to energy discussions.

14.2 Energy products versus energy services

We buy energy products in many forms—chemical fuels and electric energy—and we experience benefits in the form of services—comfort and mobility. Business people are well versed in the difference between products and services and the spectrum in between. Products can generally be measured and counted. They are tangible and answer the question of what one makes. Services are generally the result of applied skills or expertise. They are less concrete, frequently intangible, and answer the question of what one does. In energy, fuels are the products that have energy. Infrastructures provide the service of on-demand access to perform work.

Most conversations in energy are derived from this structure. Many revolve around the product of what we buy—e.g. coal electricity versus renewable electricity, petroleum vehicles, versus battery-electric vehicles, or the environmental externalities associated with oil sands versus those experienced with uranium extraction for nuclear power of similar magnitude. Explorations of specific enjoyed benefits and alternatives and debate over the best metrics for decision-making are much rarer. Urban transportation dilemmas are representative. One can choose to travel by foot, bicycle, taxi, car, bus, subway, or train in most major metropolitan areas. Policy-makers can incentivize specific choices (e.g. walking and subway travel) by keeping fares low and keeping alternatives priced high.

New York City's proposed but failed 2007 congestion pricing initiative would have done just that by charging between $8 and $21 for cars and trucks entering parts of Manhattan during weekdays—generating an estimated $380 million per year for a new city transportation authority and funding a $31 billion regional transportation improvement effort (Schaller, 2010; Smith, 2014). Less radical approaches like toll reform have been proposed along the same lines since the 1980s (Smith, 2014). If the societal goal is to minimize average commuting times and maximize *urban mobility*, public transportation efforts like the one proposed by Mayor Bloomberg triumph. But if other metrics, such as a particular representation of the cost and time for a typical *driver* to reach part of Manhattan from one of the other four boroughs comprising New York city, is selected, then the limited available capital resources may be devoted to improving roadways, impact on regional congestion aside.

Society's unrelenting thirst for energy is not based on our desire to extract coal or mine uranium. Our demand for energy is driven by the energy services which make our lives more safe, comfortable, productive, and enjoyable. We buy energy for warmth on a cold day, the speed of modern travel, fueling our computing-centric infrastructure, and wonderment at our favorite laser-equipped rock show. But the services we consume depend on underlying energy products.

The energy balance diagram first introduced in Chapter 2 (Figure 2.1) depicted the relationship of energy products in an energy balance. It is updated in Figure 14.2, which represents the energy services that businesses provide for each component. Recall that *energy in* must equal the sum of useful energy out and wasted energy out. *Energy in* is only another name for the *energy product*; it represents the fuels we use and consume. *Useful energy out* is only another name for *energy services*.

Useful energy is *why* we consume energy. It represents the work done and tasks accomplished on our behalf. The energy transformation is the appliance, equipment or device, or third party that transforms energy products into energy services. In some cases, this is simple; the countertop electric or gas-powered heating element differs little from a woodfire stove common in yesteryear. In other cases, like Software-as-a-Service (SaaS) companies, there may be hundreds or thousands of interrelated supply chains supplying

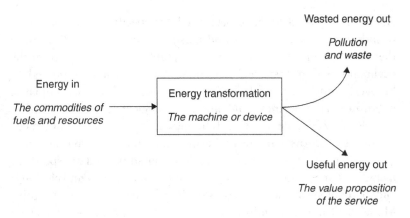

Figure 14.2 The balance of energy for services and products

physical components or providing services, each with an energy footprint, to service deceptively simple applications we access on computers or mobile devices.

Focusing exclusively on fuel choices will neither reduce current consumption nor address future growth in energy requirements. Replacing coal with renewable power plants where feasible, installing energy-efficient LED lightbulbs, and driving fuel-efficient cars may be great things to do, but they are insufficient to address the issues we face today. Fuels represent only one half of the energy balance. The more important challenge, which has received substantially less attention, is addressing the value proposition of why we buy energy to begin with.

Historically, measuring the kinds of energy outcomes that we use to describe energy-related services has been infeasible. Even where metrics could be readily defined, the cost of measurement devices, data storage, and transport has far outstripped the economic value of gathering and analyzing such data on a continual basis. With the advent of the smart grid, more intelligent buildings and pervasive sensor technologies—best embodied by smartphones which have impressive onboard data capture, storage, and transmission capabilities—data is becoming cheaper and more available. Our improved ability to quantify the quality of light, air, and comfort inside major energy-consuming sectors like homes and offices is a powerful tool. For example, low-cost sensors within common spaces and personal workspaces can be used to adjust environmental conditions to maximize benefit to occupants. Historically, standards like

ASHRAE Standard 55 have been used "to specify the combinations of indoor space environment and personal factors that will produce thermal environmental conditions acceptable to 80% or more of the occupants within a space," but these static documents proved to be overly restrictive and lacked responsiveness to regional climates, organizational priorities, and non-physiological factors in comfort (ASHRAE, 2004; de Dear and Brager, 2002). Cheaper and more ubiquitous sensors can move beyond the standard environmental factors (temperature, thermal radiation, humidity, and air speed) to include active measurements of personal factors (activity and clothing) and psychological factors—collectively referred to as the Adaptive Model of Thermal Comfort and Preference—that impact perceived comfort levels to simultaneously reduce energy consumption and improve occupant well-being during design and operation (Brager and de Dear, 2001; de Dear and Brager, 2002).

Some improvements can be made with little visible change to current facilities, business models, or operational practices. Others, like incorporating data feeds from occupants who are equipped with personal sensor devices capable of detecting activity levels or environmental conditions, will take more effort and investment. The availability of low-cost, high-quality data will allow better, outcome-focused energy metrics to be used in decision-making. As data sets of sufficient size and quality are generalized over time, association and correlation will form the basis for energy service models more akin to content delivery on modern telecommunications networks. Just as some consumers choose to purchase internet services and separately procure all content via providers like Amazon, Netflix, or Hulu, future energy service delivery models will enable unbundling of energy services. Connections to the grid may be provided by a local utility where a natural monopoly still exists given the capital-intensive nature of the grid, but this "wires" business need not be integrated with the energy retailer who buys "firm" power in the bulk market and resells it to numerous smaller customers on a more flexible basis.

Opportunities during the transition

The transition away from vertically integrated utilities to competitive bulk markets run by regional transmission authorities, to competitive retail services, and towards a transactive and open-access electric power system is one of the most exciting and significant economic opportunities of the coming century. Energy's central role in society means that modernization within the sector will have downstream effects on virtually every aspect of life. The pace of change will accelerate as individuals, businesses, and governments learn that energy costs can now be effectively measured and attributed to specific business outcomes. This attribution is central to improving planning and capital allocation to increase returns on investment.

Evolutionary changes like the TXU program or similar programs promising lower off-peak electricity rates are now more commonly finding their way to the US. Upgraded metering infrastructure allows time-specific pricing or other, more sophisticated rate structures to be implemented. Potential cost savings are significant, especially for consumers who would prefer to save money. In Duluth, Minnesota, a new pilot project created pricing incentives for off-peak hours, which include weekend days and from 10 p.m. to 8 a.m. on weekends. The program, run by Minnesota Power, is the first in the state (Myers, 2014).

More exciting programs will leverage this kind of dynamic pricing to accomplish even more. While some consumer groups are concerned that new rates are too complex, it is important to realize that human actors may or may not be involved in the day-to-day decision-making processes. More sophisticated and low-cost building control systems and energy management tools have the capacity to help schedule typical kinds of energy-consuming tasks in the home to account for changes in current or forecast energy prices today. However sophisticated such tools are, people will remain especially important when new situations arise or in addressing faults or anomalies which are not easily incorporated into software products which automate more typical, menial tasks.

With some forethought, we can offer higher-quality services that consume less energy via effective long-term planning before purchasing energy-consuming devices. Consumers are not typically incentivized to be concerned with energy consumption when purchasing major appliances, but differences add up over time. Finding a low-cost and old refrigerator on Craigslist may seem like a bargain at first, but over the long haul, energy costs can eclipse the capital investment required for many home appliances. A comparison of representative energy requirements and average national electricity

Table 14.2 *Comparison of refrigerator and freezer energy standards,*
1978–2014

Year	Energy use (kWh/year)	Estimated bill ($/year)
	Bottom-mount	
1978	2,742	315
1990	1,432	165
2014	550	65
	Side by Side	
1978	2,951	340
1990	1,755	200
2014	624	70

Source: ASAP, 2011

costs for the U.S. Department of Energy Refrigerator and Freezer
standards is shown in Table 14.2 (ASAP, 2011).

Similar purchase dilemmas are faced by most small businesses,
non-profits, and government organizations. A more holistic approach
focused on estimating the total cost of ownership, is required to make
more consistent, economically optimal decisions. When possible, this
starts long before purchasing devices like appliances. A home or
business with large windows that faces the sun will reduce lighting,
and often heating, requirements. By maximizing the availability and
reliability of energy services and minimizing waste during design,
construction, and operation, we can significantly reduce the need for
fuel. In addition to reducing negative externalities, this just makes
good economic sense.

14.3 Making energy visible

Managing energy for end users requires a fundamental change to the
metrics for consumption: redefining energy management by linking
costs to actual energy outcomes, not commodity consumption. Most
current utility efforts remain commodity focused and use demand-side
management to describe utility control over end-user devices to meet
the needs of the larger system. As the movement towards energy

services business models is enabled by new technologies, consumers will see increasingly sophisticated offerings with significantly reduced requirements for energy or engineering expertise. This transition to focusing on energy outcomes will help mask growing complexity in energy service delivery systems, despite significant ongoing changes. When evaluating potential energy services business models or services, this complexity, and the associated risks, must be properly considered.

Additionally, service providers and consumers alike must recognize the importance of contextual information to the establishment and long-term viability of energy services businesses. Discounting context will doom established companies or new entrants because of the substantial exposure to so many physical, behavioral, regulatory, and economic factors. Sufficiently tailored services will be better postured to avoid unpleasant mishaps in such a dynamic financial–physical–information system. Increasingly bespoke services are also likely to emerge to harness special requirements of particular kinds of customers. This section explores this concept for regional services capable of generating revenue by meeting system needs and by offering bespoke services for an exemplary, significant energy consumer: data centers.

Increased complexity below the surface

While consumers will be able to more easily relate to and control their energy-related expenditures, increased data will allow service providers to take more complex actions behind the scenes on their behalf. For example, service providers that help bridge the retail market with the "firm" wholesale power market can also be responsible for the physical and financial risk of non-delivery or production of electric power. Physically settled commodities contracts in the wholesale market do not let energy providers "opt out" for any reason. Any differences between the amount of energy scheduled for delivery or consumption on the grid and what occurs is charged to the company. Since typical retail consumers are too small to purchase power in the wholesale markets, and few would wish to give up the freedom of the "always-on" grid, more specialized approaches are required. For example, end consumers can maintain their independence while still being rewarded for reliably drawing power from, or injecting power into, the grid.

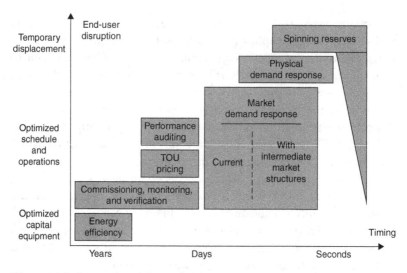

Figure 14.3 Categories of demand-side management

There are generally three major components to predicting energy usage for typical consumers and distributed generators: behavior, physical requirements, and the demand-side management model. The demand-side management model itself has four component models which affect overall energy requirements, capacity profiles, and the degree of control over a demand-side asset. These include energy efficiency measures, time-of-use characteristics/elasticity, market-based demand response, physical demand response, and spinning reserve, as shown in Figure 14.3.[3]

While predicted energy requirements for most retail customers are still generally calculated as net load today, these categories translate well to a more developed open-access model for the grid.[4] With the right composite metrics, it will be possible to keep track of how reliably a typical consumer or distributed generator adheres to future schedules and to value their participation accordingly. Firm power is worth more; uncertain energy contributions are worth less. One approach is to keep track of consumer reliability via a behavioral score

[3] This figure was influenced by the categories and depiction given in Palensky and Dietrich (2011) and Joo and Ilic (2011, 2012).

[4] Recall from Section 10.2 that "net load" treats distributed and non-controllable generation or storage assets (e.g. wind and solar energy) as negative load.

similar to a credit score for financial transactions. While admittedly imperfect, this type of composite metric might be used to help classify the quality of more complex physical and financial power contracts to hedge risk between counterparties participating in the electric power system. Intermediaries acting as service providers can utilize density functions for each considered time period to develop more accurate forecasts of future physical performance (Crabtree et al., 2010a, b, c; 2011).

Context matters

Without adequate data, offering fixed-price or long-term energy delivery contracts to customers is impossible. The massive amounts of volatility due to physical scarcity and balancing requirements in the wholesale markets, seasonal variance in requirements, underlying volatility in bulk fuels, environmental limits on externalities associated with energy extraction and transformation, etc. are too difficult to overcome. Current approaches to decision-making and the supply-focused industry model are simply too inflexible.

To underscore the importance of context, a quick foray into energy consumption during holidays or sporting events is useful. Similar, if less dramatic, changes in energy consumption are also observable on weekdays versus weekends—hence the TXU weekend rate program. One powerful example is the effect of soccer on electricity demand. Demand data from the UK National Grid for the July 4, 1990, World Cup semi-final between England and West Germany shows a 1,600 MW increase in consumption at halftime and full time, foreshadowing a record-breaking 2,800 MW surge following the penalty shoot-out to end the game (Bukhsh, 2012). According to the National Grid company, the 1990 World Cup event remained the top "TV pickup" surge only reaching 550 MW during England's 2014 World Cup opener, less than 20% of the 1990 penalty shoot-out surge (National Grid, 2014; Beech, 2014).

Load forecasting, and therefore power purchasing efforts, can be improved substantially when made aware of these kinds of major exogenous events or stimuli to the target system. Since these kinds of deviations in energy utilization require a great deal of contextual knowledge, it is likely that many novel energy-as-a-service (EaaS) models will be quite specialized for the communities they serve.

Harnessing invisible revenue streams

Regional services are likely to emerge in order to capitalize on their relative advantage in contextual knowledge and community relations. They are also likely to offer pricing models and manage risk very differently based on the unique characteristics of energy usage and infrastructure in their region. For example, Pacific Northwest services will need to incentivize customers to consume more power from April through June, when large amounts of runoff transit from the Columbia River system combine with wind output and light seasonal loads to create recurring capacity surpluses. Those same customers are likely to see disincentives during normal winter morning peak load conditions in the region. Programs in other locales may be conceptually similar but will feature completely different incentive structures.

Shaping the composite load and distributed generation is important to economically serve energy-as-a-service customers and to generate additional revenue for service providers themselves. Energy companies, like most of the data-centric internet companies, will seek to develop additional revenue streams that will allow them to offer price-competitive services to clients while increasing their own profitability. Recall that services which automate processes on behalf of customers can more intelligently schedule loads to occur during times of increased net economic benefit. California programs are likely to encourage shaping and storage by end users to mitigate the "duck curve" (shown in Figure 12.1) and monetize their efforts within the CAISO-managed physical power markets. Coordinated heating and cooling between large numbers of small residential or commercial customers can also be used to cost-effectively manage increasing system variability and uncertainty resulting from renewable energy portfolio increases such as wind without the harmful cycling costs associated with volatile ramping for commercial thermoelectric power plants (Kumar et al., 2012; Jordan and Venkataraman, 2012; Lew et al., 2012). It is much easier and cheaper to modify many thermostat settings in networked homes than it is to rapidly change the output of a coal plant. This service could be effectively sold to a variety of bulk power system participants to aid in maintaining system stability and hedging against system volatility in a variety of markets, including California and the Northwest.

Other examples of invisible revenue streams might include large companies with immense purchasing power acting as power resellers themselves, leveraging third-party services to provide white-label benefits back to their employees. For large organizations like IBM, which had more than 430,000 employees at the end of 2013, the return could be significant. As with other commonly provided employee benefits, organizations that can increase effective discretionary income for employees while reducing overall compensation costs stand to gain a competitive advantage. Large entities may someday coordinate electricity purchases with the collective purchasing power of business operations and employees in a future regulatory environment on a smart grid.

More specialized energy services

Some specialized consumers, such as data centers, may require more advanced load models when conducting site analysis or detailed economic modeling for their operations. Much more simplistic load curves and analytics may be suitable for general consumers, but because data center operating costs are tightly coupled with energy costs due to power-hungry processors and major cooling requirements, bespoke models are justified.

For the most sophisticated, large data center operators, rigorous cyber-physical system models will be developed for use in energy trading and even routing. Since typical cloud-based data storage services replicate data in geographically distributed data centers, the decision to route a request to California, Virginia, or some other location is likely to be increasingly tied to specific energy and bandwidth costs to service the request. Cyber-physical system models are required because of the interaction between the physical environment and the computing units which cannot be adequately captured in pure event-based time-stepped simulation. The more detailed model can consider both transient and steady-state behavior and has been validated on operational deployments (Banerjee et al., 2013).

The power quality sensitivities and requirements for data centers make them particularly good candidates for advanced energy management analysis prior to, and post, construction. Fundamentally, this kind of detailed energy analysis is based upon a workload model and metric structure describing arrivals, intensity, and type which

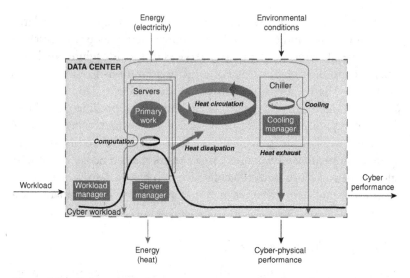

Figure 14.4 Cyber-physical energy system representation of a data center

specify computational process energy requirements and heat outputs. This is then passed as an input into a heat circulation model where a cooling process considers the heat exhaust requirements along with actual external environmental parameters to determine total energy output (Banerjee et al., 2013). The model components and exogenous inputs are shown in Figure 14.4.[5]

Energy efficiency

Energy efficiency is commonly touted as a significant and readily available energy resource, but there are significant impediments to attaining this energy resource commercially, including a highly fractured market for clients and energy-consuming devices.[6] Techniques such as disaggregation of specific end uses from existing meters using non-intrusive load monitoring can be helpful in identifying candidate equipment for efficiency efforts at minimal cost (Berges et al., 2008). Consumers, utilities, and policy-makers should be keenly aware that

[5] This figure is simplified and extended from a cyber-physical energy system simulation model diagram in Banerjee et al. (2013).
[6] See e.g. Behrens and Glover (2012) and Bouton et al. (2010) for a more detailed discussion of impediments to attaining energy efficiency resources at scale.

energy efficiency improvements do not necessarily equate to reduced environmental impacts or dependence on fossil fuels. Discussion of efficiency is not effective at constraining energy services growth *per se* but can be used to be much more cost- and resource-effective to meet current uses of energy. The move towards energy services will be instrumental in enabling effective deployments of private capital into increasing energy efficiency.

Energy efficiency is not equivalent to reducing consumption, a common myth which is well refuted, but efficiency can be a major opportunity for economic growth.[7] Benefits include reducing negative environmental externalities required to maintain current standards of living and liberating additional future capital flows for more productive uses. That being said, policy-makers and business leaders should be keenly aware that our thirst for energy is only likely to increase over the long term, irrespective of efficiency. As energy costs decrease, new technologies, services, and industries will emerge—taking advantage of marginal efficiency gains or becoming viable for the first time as a result of rapidly democratizing advanced technologies and low-cost sources of power. The fact that liberating capital from being spent on energy and reducing marginal costs of consumption will increase overall demand should not be construed as a reason to *avoid* investing in energy efficiency.[8] Rather, our quest for cleaner and sustainable energy sources must continue in parallel with efforts to improve the efficacy of the energy handled by man-made extraction, transformation, delivery, and consumption systems.

Some of the popular confusion about energy efficiency stems from its episodic role in politics during periods of increased geopolitical instability, when fossil fuel delivery systems in our source regions are threatened with interruptions. Since the most significant of these, the 1973 oil embargo, the US has made substantial improvements to energy productivity. In 1970, the country generated $60 billion in GDP per quad, but this number more than doubled to $135 billion

---

[7] See e.g. Herring (2006), Polimeni et al. (2007), Smil (2010a) and Jevons (1865) for advanced reading on the Jevons paradox and its sometimes misguided application to energy efficiency.

[8] See e.g. "The Efficiency Dilemma" in the *The New Yorker*, which drew responses from Amory Lovins of the Rocky Mountain Institute, Michael Levi of the Council on Foreign Relations, and many others (Owen, 2010).

Table 14.3 *Percentage of actual and forecasted US industrial delivered energy consumption by source, 2010–2040*

Source	2010	2040
Coal	1.6	1.6
Renewables	2.3	3.6
Electricity	3.3	3.9
Liquids	8.4	8.7
Natural gas	8.0	10.4
Total	23.6	28.2

Source: EIA, 2013c

in GDP per quad by 2011 while energy consumption increased from 67.8 quads to 97.3 quads over the same period (EIA, 2012b).[9]

While some of the gains in GDP relative to energy consumption reflect the increased outsourcing of certain high-energy processes to other countries—the numbers do not include energy imports in the form of manufactured goods or products such as steel or aluminum which take large quantities of energy to manufacture—much of these gains stem from actual efficiency improvements which are most easily viewed in three major categories:

1. *Industrial:* The industrial sector is generally considered to include manufacturing (e.g., food, paper, chemicals, refining, iron and steel, non-ferrous metals, non-metallic minerals) and non-manufacturing (e.g. agriculture, mining, and construction). Global industrial energy requirements are formidable, consuming 200 quads in 2010 and forecasted to reach 307 quads in 2040 (EIA, 2013c). In the US, the industrial-sector-delivered energy consumption accounts for roughly a quarter of the national total. Table 14.3 shows the percentage of industrial-delivered energy consumption by source (EIA, 2013c).

   In 2012, President Barack Obama signed Executive Order 13624, announcing a goal of adding more than 30 GW of industrial combined heat and power (CHP) and efficiency projects by 2020

[9] Total energy consumption in the US was 63.495 quads in 1970 and 78.096 quads in 2011 (EIA, 2012b).

via additional incentives (Administration of Barack Obama, 2012). Additional US federal government–led efforts to increase efficiency include research programs and incentives enacted via the Energy Independence and Security Act of 2007 and the Energy Policy Act of 2005.

2. *Transportation:* Transportation includes energy utilized to move people or goods on land (e.g. roads, rail, and pipelines), water, and air. Energy efficiency efforts in transportation tend to focus on petroleum and liquid fuels, such as fleet fuel efficiency standards within developed nations. Transportation sector energy consumption of petroleum and liquid fuels is expected to increase by 36 quads by 2040, accounting for the majority of a forecast overall delivered energy consumption increase from 101.0 to 139.5 quads between 2010 and 2040 (EIA, 2013c).

Evaluating transportation efficiency improvements is difficult because of numerous changes to land use and economic activity levels which demonstrate high degrees of correlation to actual energy consumption. For example, US 12-month total moving vehicle miles peaked at 3,038,889 miles in November 2007, up from 1,125,491 miles in January 1971, when records begin (U.S. Department of Transportation Federal Highway Administration, 2014). Miles rose steadily from 1971 to 2007, but since the peak, 12-month moving vehicle miles have remained under 3,000,000 miles during the anemic economic recovery.

3. *Buildings:* Energy consumption within the built environment has outpaced transportation and industrial consumption growth in developed nations like the US since the 1950s, when these three sectors were of similar magnitude. Much of this is attributable to the massive growth in household appliances and other energy-consuming modern conveniences, such as air conditioning. A similar transformation is now occurring on a global scale, as electricity access is gradually extended to more than 1.3 billion people for the first time and as many as 2.6 billion more people transition from biomass for everyday tasks like cooking (EIA, 2013c; ExxonMobil, 2014).[10] A 2010 study by McKinsey and

---

[10] A major initiative to extend electricity access to sub-Saharan Africa known as "Power Africa" was unveiled by the White House in 2013 (Administration of Barack Obama, 2013) in an attempt to address major impediments to

Company estimated that the US economy alone could glean more than $1.2 trillion in benefits at a cost of $520 billion within stationary energy-consuming facilities (Bouton et al., 2010). Attaining these potential benefits is entirely another matter. Lack of capital, behavioral risks, structural purchase and ownership issues, and complex transactions and operational relationships have prevented providers from reaching substantial scale.

The remainder of this chapter primarily focuses on energy efficiency and its relationship to EaaS within the built environment and industrial users with significant reliance on electricity and natural gas.

Energy efficiency and EaaS synergies

In an EaaS paradigm, energy efficiency services are but one of many avenues for end users to control their total utility spending while maximizing the benefits they derive from the energy they consume. Most consumers, especially small and medium businesses, are price-takers when it comes to energy. With limited time, expertise, and tools to manage energy as an operational resource, it can be difficult to make the right financial and energy management decisions, especially when a highly fragmented set of competing services is independently presented. As new service models are introduced, the number of options to consider will only increase further. Overall, the increased competition will benefit consumers, but managers will face even more uncertainty and conflicting advice.

Active demand-side management approaches, like those pioneered by Boston-based EnerNOC and improved upon by Philadelphia-based Viridity Energy, have outpaced energy efficiency services in recent years, but the two fields are converging. Many of the early benefits promised by the "smart grid" and "internet of things" revolve around concepts of smarter buildings for living and working. Real progress in managing more intelligent buildings will come only with improved capital allocation—for both long-term investments and short-term operations. In the immediate future, commercial and industrial users will remain the most cost-effective and promising sectors for energy

economic growth and the numerous health, literacy, and other issues stemming from energy poverty (Jolly, 2010; Wines, 2007).

efficiency investment.[11] As more expensive techniques which are typically used on facilities exceeding 500,000 square feet become commercially viable for small and mid-sized facilities, even those below 150,000 square feet, residential prospects will increase.

Long-term investments in energy efficiency are stymied by purchase complexity and long timelines. A homeowner who plans to own a home for less than 10 years is unlikely to be enthused about making investments on a 30-year timeline, especially when asset appraisals are unlikely to account for substantial investments in energy-saving equipment or designs and mortgage underwriters in the US have been openly hostile towards Property Assessed Clean Energy (PACE) loans which place a lien on property to secure energy efficiency improvements (The *Economist*, 2014b). Commercial projects face similar challenges because appraisals and lease comparisons often fail to consider the operational cost of a facility. Even if improvements to a facility would benefit owners and tenants for substantial amounts of time, any economic benefit is normally considered only within the expected period of ownership or tenancy. Until energy efficiency investments can be more effectively transitioned between occupants over the expected service life of the equipment, investment will disproportionately lag.

Long-term energy management strategies must consider a combination of reliability, price variability, and volatility. Because of the long-term exposure to these factors and actual performance risk—the question of whether or not promised savings will actually materialize—on projects, most end users choose not to go it alone. Most commercial and industrial energy efficiency projects use specialists that can assume the performance risk for the savings from which capital equipment upgrade finance costs are paid. Energy efficiency service agreements or power purchase agreement–like structures are typically used as contractual mechanisms during the life of the project.[12]

Energy service companies (ESCOs) have been a dominant force in energy efficiency projects for some time. But even as specialists, these

---

[11] While residential efficiency is also substantial, accounting for roughly 28% of US electricity consumption, it is simply too fragmented and fraught with hidden educational, regulatory, and transaction complexity costs.

[12] Power purchase agreements have also been used to help manage relationships in the burgeoning distributed solar energy industry.

firms have had substantial challenges with non-performing efficiency projects. ESCOs typically hold the debt used to pay for the actual capital upgrades and are paid out of reduced costs experienced by the client. ESCOs typically take the preponderance of the "savings" from an estimated "would-be" cost and share part of the difference with the client. But project non-performance has been a recurring problem, and "savings" models often cause issues over time when hypothetical cost estimates seem less reasonable. As a result, some of the largest mechanical services contractors in North America have largely avoided ESCO-like guaranteed or shared-savings contracts. The risks to long-term customer relationships, project performance risks, and credit risks have simply outweighed the potential benefits.

Integrated operational and capital budgeting

New approaches that take advantage of more data and more closely link operational budgeting with long-term capital budgets show more promise. Energy efficiency projects are most easily financed, via whatever source of capital is most appropriate, when cross-cutting analysis is linked to energy outcomes and not reductions in commodity consumption. The process of developing appropriate business metrics, identifying inefficiencies based on those metrics, determining financial incentives, and making actual improvements is surprisingly nascent given our lengthy experience with consuming energy in homes and businesses, but determining the right metrics and goals for optimization is inherently reliant on the availability of data—data which is only now becoming available. Once identified and actioned, strategies require consistent and continual tending which is adequately responsive to changes in energy markets.

Over the longterm, more services that manage cash flow volatility and uncertainty for end users of energy and translate energy consumption information into relevant operational metrics will emerge. With the right processes, intentionally designed approaches to privacy, and performance requirements, substantial energy efficiency resources can be harvested. The sheer size of the market and capital-intensive nature of these projects will ensure steep competition.

Traditional utilities, many of which have had Public Utility Commission–mandated robust energy efficiency programs since the 1970s oil crisis, will be competing with new entrants like ESCOs that will increasingly be vying for direct client relationships with end users. Saving instead of producing more energy does still remain less expensive than developing new capacity in most of North America. But energy efficiency projects are facing increased resistance from communities and are clashing over many program specifics. In 2014, Xcel Energy testified before the Colorado Public Utilities Commission that its state-mandated demand-side management program was expected to deliver reductions of 2913 GWh at a direct cost of $962 million between 2015 and 2020 (Jaffe, 2014). At the hearings, regulators were faced with advocates for expanding the program, who cited an estimated return of $600 million for ratepayers, and the Colorado Office of Consumer Council that represents ratepayers, which argued that an estimated $1.3 billion in rate impacts was not realistic.

As capital equipment budgets and operational utility budgets converge, utilities will face increasing risk of disintermediation by current partners. Major large-scale electricity retailers like Hess, new entrants like Choose Energy (a Kleiner-Perkins venture-backed company offering competitive utility comparison services to retail customers in select deregulated markets like Texas), and traditional engineering consulting firms will each seek to expand service offerings to maintain or gain market dominance.

In the near future, capturing efficiency is not likely to overpower the combined effects of reduced energy costs, human capacity to develop new energy-consuming tools and services, or income growth. Fortunately, energy efficiency need not do so to be economically feasible, technically possible, and environmentally responsible.

14.4 Chapter 14 Summary

- Consumers are more interested in buying energy services (i.e. comfort, mobility, and convenience) than buying energy products (i.e. kWh, gallons, and therms). This is similar to normal differences in marketing product benefits versus features for traditional manufactured goods.
- The proliferation of sensors, analytics, and automation tools is enabling the quantification of tangible energy benefits in addition to energy consumption.

- Business opportunities will arise as better information enables a more efficient utilization of resources to deliver superior services that are narrowly tailored for particular applications.
- Energy efficiency is one example of an industry that can greatly benefit from an EaaS approach to increase the number of commercially viable projects and deliver societal benefits.
- Gaining substantial efficiencies within existing building stock remains challenging because of the efficiency performance and credit risks, diverse stakeholders, multiple business models, balkanized regulatory market, substantial local and regional pricing variance, and the lack of sufficient long-term incentives.
- Efficiency opportunities within the industrial base, vehicle stock, and other areas with faster technology refresh rates, such as information technology, will continue to see broader participation thanks to increased visibility and simpler financial incentives.
- The convergence of energy, information, and financial flows is a critical trend with profound implications for a number of established industries.

PART V

Societal advancement

15 | *Bringing it all together*

Traveling in time and space from the Big Bang to the Sun and through the ages of the Earth we note that energy conversion is the prime mover in the evolution of life and the universe.

Reiner Kümmel (2011)

Energy exploration, extraction, transport, transformation, and utilization are all driven by human demand for energy outcomes. Unfortunately, most conversations deal with energy products and resources. The development of specific resources and the emergence of energy-related technologies center around the desire to liberate human labor for higher-value tasks. In many ways, the control of energy resources is synonymous with wealth. For early agricultural societies, technologies of animal-driven mills or water pumps were transformative. In Latin, the word for money or wealth, *pecunia*, originally referred to cattle property (Kümmel, 2011). This is unsurprising, given that land owners had the resources to support the domesticated animals which served as the primary means of transforming solar energy into forms more useful to human societies of the era. Moderately more advanced societies gained tremendous benefit from the first wind- and water-based prime movers which directly harnessed nature without the overhead costs associated with animal engines. The same territory-centric model of harnessing diffuse energy from harvesting natural resources and agricultural products is an oft-ignored but foundational component of the Westphalian state. Most of human history is dominated by *wealth* as a derivative product of land ownership and *value* as a flow (strongly correlated with, or even coupled to, the amount of expended human labor) stemming from commodities markets (Hall and Klitgaard, 2011).

The mechanization of society at scale, beginning with the Industrial Revolution, decoupled the accumulation of wealth from ownership of vast natural resources or lands. This shift is largely responsible for enabling unprecedented economic growth and improvements to amenities enjoyed by average people. Thanks to a combination of electricity, liquid fuel–based transportation systems, and modern machinery (with huge embodied energy requirements in the constituent components), humanity has quite literally reshaped the Earth—largely within the past 150 years. This drive to create useful forms by doing work to gather and derive value from chaotic and disparate resources is central to the evolutionary experience and the emergence of economic systems (Avery, 2012; Weber et al., 1989; Georgescu-Roegen, 1971; Chen, 2005).

The productive power of energy is central to the development of human civilization, and surplus energy has been viewed by a number of pre-eminent scientists and scholars as the key factor underlying many of the most significant advances in human civilization (Kümmel, 2011; Hall and Klitgaard, 2011). But any evaluation of useful, let alone surplus, energy first requires an accounting of *available* versus *unavailable* energy (Georgescu-Roegen, 1986). The Second Law of thermodynamics states that all systems, without being impacted by any external agency, move towards disutility (Barbier, 1989). This is applicable to energy and to matter more broadly, as "illustrated by rust, by wear and tear of motors or automobile tires" (Georgescu-Roegen, 1986). The ability to control energy to increase useful work expands the capacity of individuals, groups, or humanity to lower entropy in the environment—moving scarce resources to more easily accessible states. Living beings modify their environment and take action to combat naturally increasing system entropy. Life seeks utility. A living organism quite literally "feeds upon [...] negative entropy" (i.e. low-entropy energy and matter), which is utilized for sustenance before degrading into high entropy (i.e. waste) (Schrödinger, 1944; Georgescu-Roegen, 1986). Energy systems for human civilization emerged, and have since evolved, to fulfill this need.

Because of its singular focus on meeting demonstrated demand signals, the energy industry remains one of the few places where the oft-misused design cliché, "form ever follows function" clearly remains apposite.[1] Energy delivery networks emerged relatively slowly

---

[1] "Form ever follows function" was coined by American architect Louis Sullivan (Rawsthorn, 2009).

to meet society's growing thirst for energy. Each successive age of human development has required major increases in energy products and services. Thus far, this book has explored cases of relative energy scarcity or abundance and often unexplored connections between energy delivery systems, dependent infrastructures, and the physical world. This final chapter is intended to place the numerous vignettes and concepts introduced throughout the text in better global, temporal, and even cosmological context.

Human societies have long been enabled, hindered, and organized around energy resources. Primary energy sources—and, later, refined energy products and services—have evolved to meet societal demands. The energy intensity of society has changed dramatically as man has gained the ability to do useful work. When plotted over time, as shown in Figure 15.1, the growth in energy utilization with increasingly sophisticated human societies is phenomenal.[2] When viewed in real terms, it is even more fantastic—at the dawn of agriculture, approximately 5 million paleolithic *Homo sapiens* consumed on the order of 1.785 million BTU, or 292 *billion times less* than in 2013 (IEA, 2013; Haub, 2002). The growth in energy utilization is reflective of mankind's ability to harness energy to do ever more work on its behalf. The gradual mechanization of society formed the basis for increased intellectual development and scientific advancement, accounting for the fact that the major increases in energy consumption occurred in parallel with substantial increases in how efficiently energy was used during the same period (Smil, 2003).

15.1 Ages of energy

Man's transformation of energy into ever more desirable forms is truly ancient. Early transformations of energy by primitive man were limited to natural biological energy transfers from daily food intake, estimated between 2000 and 3000 kilocalories per day. With the addition of fire, wood fuel increased the total energy use by early man by about 2.5 times (Ngo and Natowitz, 2009). With the move towards stationary agricultural societies, energy consumption roughly doubled again. While early use of animals by agricultural societies

---

[2] Data for this figure sourced from Goldemberg (2012), Cook (1971), and Ngo and Natowitz (2009).

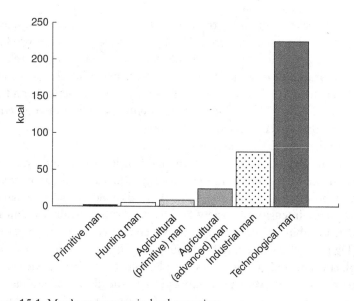

Figure 15.1 Man's energy use in kcal over time

and the gradual progression towards wind- and water-powered aids is commonly used to chart man's energy progress, an evaluation of fuel sources and the evolution of smelting for tool production is also instructive. Even the ancient ages of mankind—Stone, Bronze, and Iron—can be traced based on the available fuel sources and the technologies used to control their transformation into useful energy. Each age is characterized by increased human capacity for more intensive use of energy and growing energy surplus (Chen, 2005).

Harnessing energy for tool-making

Wood was the fuel of choice during the Stone Age, also known as the Paleolithic Age. Many paleoanthropologists agree that while sporadic evidence exists for primitive man's control over fire in the Lower Paleolithic (ca. 2,500,000–300,000 B.C.E.), "unequivocal evidence" for the use of fire dates to the beginning of the Middle Paleolithic, ca. 200,000–300,000 B.C.E. (Karkanas et al., 2007). Early fires could be used for cooking, heating, and illumination, but temperatures and stability remained insufficient for metal work other than tin and lead, which have low melting points.

The earliest evidence for crucible smelting of copper dates to the mid- to late sixth millenium B.C.E. in the Balkans and southeastern Iran ca. 5200–4500 B.C.E. (Thornton et al., 2010). These early crucible smelters were likely the foundation for furnace smelters, which could achieve higher temperatures. At 500 °C, malachite, a green rock rich in copper, disintegrates into copper oxide; however, at 700 °C, carbon monoxide steals additional oxygen from the copper oxide and becomes carbon dioxide, leaving copper behind (Keys, 2010). Separating out pure copper to serve as the base metal for bronze alloys (which also required much rarer tin) required even higher temperatures—roughly 1100 °C. The furnace techniques proliferated, and by around 3000 B.C.E., metallurgists were rapidly producing bronze tools throughout Eurasia. Ultimately, shortages of tin were likely partially responsible for the transition away from the Bronze Age and the start of the Iron Age beginning around 1100 B.C.E. (Wilford, 1994).

Iron was extraordinarily plentiful when compared to tin, but the early *bloomery* process (also known as the direct process) for reducing ore to iron incorporated very little carbon, making it ductile and too malleable for direct use (Joosten, 2004). Some of the earliest evidence of iron-smelting operations is at Tell Hammeh in Jordan ca. 900 B.C.E., but evidence of even earlier iron production and working in the region also exists (Veldhuijzen and Rehren, 2007). More detailed explorations of early iron production are given in e.g. Tylecote et al. (1971), Serning (1979), and Craddock (1995), but the overall process ultimately required temperatures of 1176 °C to separate the raw iron, or *bloom*, from the waste product known as *gangue* material. Early bowl furnaces gave way to domed furnaces, slag-pit furnaces, and slag-tapping furnaces, each with hundreds of years of service. Ultimately, the blast furnace (also known as the indirect process) produced liquid iron with higher carbon contents known as cast iron, but it would not be developed until the twelfth century C.E. (Lidberg, 2012).

Early processes relied on charcoal for smelting, but the invention of coal-derived coke, also known as degassed coal, enabled the use of coal in metallurgy for the first time. Despite being relatively plentiful and structurally superior to charcoal, without prior processing of coal, the sulphur released when burned would make iron too brittle for use. Coke is made by carbonizing coal—known within technical circles as pyrolysis—and is used in smelting to fulfill the critical role of supporting ore and limestone charges, allowing reducing gases to ascend while molten slag and metal descend within the furnace and acting as the reducing agent for the oxidation process (Valia, 1994). Today's society remains dependent on this same Iron Age process to produce steel; more than 520 million tons of coke are used annually to smelt a billion tons of iron (Smil, 2009). A number of futurists and environmental groups have argued that solar power output versus cost has been doubling every 2 years and that we are less than 20 years away from having all our energy needs met by inexpensive renewables (Wadha, 2014). Even putting aside the significant changes required to modernize and integrate diverse renewables in today's electric power system, this normative future is not possible until specialized processes like the use of coke in smelting have established commercially viable alternatives. This dependence provides strong evidence that coal will remain critically important to the global economy for the foreseeable future, even if its use as fuel for thermoelectric generation declines substantially.

The first prime movers and consuming embodied energy

If history offers any lessons about innovation with respect to energy, it is that harnessing new sources of energy usually takes a long time. New ways of capturing energy often emerge from the fringe and are slow to catch on because of the relative levels of risk when compared to competing, proven technologies and the overall capital requirements. The integration of the wheel and the sail to form a windmill or the development of the vertical waterwheel as the two dominant prime movers, until the emergence of steam engines and turbines, is illustrative. Lewis Mumford, in *Technics and Civilization*, went as far as to separate technologies developed since 1000 C.E. into eotechnic, paleotechnic, and neotechnic epochs based on the development of water power, coal power, and electric power, respectively (Mumford, 1934). Dutch historian R. J. Forbes called prime movers the "keystone of technology" and suggested five eras: human powered, animal powered, water powered, steam powered, and an emerging age of nuclear power. Nikola Tesla viewed the acceleration of human movement—enabled by man's possession of "appropriating more or less energy from other substance, and turning it to motive energy"—as the "chief task of the man of science."

Motive power means work. To increase the force accelerating human movement means, therefore, to perform more work. (Tesla, 1900)

The wheel emerged as a technology in the Eurasian Steppes around 3500 B.C.E. but quickly spread across Eurasia and the Middle East (Anthony, 2010). Interestingly, the development of the wheel and axle concept at scale was likely held back by a dearth of quality tools for carpentry, because cast copper chisels and gouges only became common in the Near East around 4000 B.C.E. (Wolchover, 2012). Sails were likely invented in Egypt between 3300 and 3400 B.C.E. to take advantage of northerly and northwesterly winds blowing off the Mediterranean to propel boats against the current (Hassan, 1997). Roughly 3000 years passed before these two technologies would be combined to harness power from the natural world.

The vertical waterwheel was the first prime mover which liberated human and animal labor from milling and increased productive capacity in other industries. Horizontal waterwheels for milling appear to have been developed earlier, but the first definitive description of a vertical wheel occurs around 25 B.C.E. when Vitruvius describes a vertical undershot watermill in detail (Reynolds, 2002; Moritz, 1956):

Wheels are used in rivers . . . round the outside, paddles are fixed, and these, when they are acted on by the current of the river move on and cause the wheel to turn . . . Mill wheels are turned on the same principle, except that at one end of the axel a toothed drum is fixed . . . . The teeth of the drum which is on the axel, by driving the teeth of the horizontal drum, cause the grindstones to revolve.

Despite their advantages, water mills appear to have been slow in their diffusion; very limited amounts of surviving evidence suggest that it took until the fourth century before water mills began to play important economic and social roles (Reynolds, 2002).

Windmills had a similarly slow initial adoption and origination from horizontal windmills across Persia, Tibet, and China (Hills, 1996). The first known historical reference to windmills occurs in *Pneumatics* from circa 1 C.E. and corresponds with water-powered grinding wheels that also existed at that time (Shepherd, 1994; Woodcroft, 1851). The development of windmills, especially towers, was important because of their ability to do work without being co-located with running water. Horizontal windmills were used by the

Persians for irrigating crops and grinding corn in stories as early as 644
C.E., but the more important vertical designs which heavily leveraged
attributes from then-contemporary water mills only emerged in the
twelfth century (Hills, 1996).

Windmills opened up new agricultural regions to cost-effective
cereal production; legislation was even passed in 1784 to encourage corn production in Ireland via the construction of additional
windmills (Rhodes, 1962). Windmills were desirable because of their
flexible siting options, but they still required substantial capital
investments and produced less useful power than their water-driven
counterparts. Despite limitations as an intermittent power source and
the demonstration of James Watt's rotative steam engines to drive
millstones in London in April 1786, the construction of large tower
mills—tower mills eventually came to dominate the earlier wooden
post mills common to Europe—continued into the late nineteenth
century (Hills, 1996; Mosse, 1967).

The Industrial Revolution and consuming kinetic energy

The Industrial Revolution allowed for energy to provide useful work
previously obtainable only from man, beast, wind, or water. The
ability to move fossil fuels, such as coal or petroleum, to arbitrary
locations and then transform the fuel into useful work enabled previously unimaginable growth in productivity. Coke-smelting propelled
coal into mainstream status, but it took time to spread from Britain,
as coke-smelting only became profitable in France or Germany in the
mid-nineteenth century (Fremdling, 2000). Outside of the remarkable
exploitation of coal in England and Wales beginning in the Roman era,
coal had remained a largely obscure resource (Smith, 1997). Abraham
Darby I is generally credited with the invention of coke-smelting,
but his real innovation was in merging sand casting methods learned
from the Netherlands with coke-smelting methods recreated from an
experienced smelter named Shadrach Fox who had already proven
that coke iron could be used for castings (Allen, 2006).

Up until this transition, most forms of non-kinetic energy were
consumed to make tools, cook food, or heat common spaces.
While the embodied energy needed to create metallic tools is very
high, employment generally required little additional natural resource
wealth. Until industrialization, wind and water could only power a

very small number of machines at specialized and static locations—human and animal power was still required to employ the majority of man's creations, including those made for war. This meant that very little fuel was required to actually use the completed tools but that early societies did not gain significant advantages in labor like those experienced in later energy revolutions. It also accounts for the more rapid industrialization of energy-rich Britain—which was the world's dominant coal producer and most powerful nation in 1860—and the US as compared to land-rich (and therefore photosynthesis and human labor–rich) but fossil fuel energy–poor France (Jevons, 1865; Hall and Klitgaard, 2011). The combination of energy resources, raw materials necessary for industry, and derivatives of Watt's steam engine meant that Britain could expand beyond its previous limits as an agricultural society (Chen, 2005).

While increased accessibility to coal reserves undoubtedly aided in a substantial way—coal's price to consumers in real terms fell by more than 40% over the course of the Industrial Revolution—its rise in consumption is linked more to increases in demand than to any technological innovations in mining, which could have expanded the industry long before actual growth occurred (Clark and Jacks, 2007). With the introduction of steam, mobile power's transformative effects were immediately apparent. Motive power was now available to accomplish real work as never before—meeting Tesla's definition of "accelerating human movement" (Tesla, 1900). Despite the enormous potential of steam to usher in a new era in waterborne transport, sailing ships remained the preferred vessels for interocean voyages with substantial economic advantages for bulk cargo (i.e. all but mail and passengers) into the twentieth century. As supply chains for coal, port size, propulsion efficiency (i.e. better engine designs *and* the replacement of the paddle wheel by the screw), and ship size all increased, steamships gradually gained prominence. Steam technology only assumed a primary role in transport following the opening of the Panama Canal in 1914. Steam's prominence roughly corresponded with the apogee of sailing technology, best exemplified by the German-built, steel-hulled *Preussen* whose five masts, 430-foot length, and 47 sails drawn from 60,000 square feet of canvas could propel to between 9 and 11 knots (Davis et al., 1997).

Another oft-forgotten component of the Industrial Revolution, even from an energy perspective, is the series of revolutions in the scientific understanding of kinematics: the mathematical understanding of motion. Before the time of James Watt and his steam engine, the limited number of historic prime movers had created few incentives for the development of mechanisms to translate various types of movement into others.

Even though the Newcomen steam engine was developed in 1712, the development of Watt's steam engines—which allowed steam to expand and condense in separate containers—was not developed and proven until 1768, when Watt first sought patent protection (Boldrin and Levine, 2008). The development of numerous linkages and ingenious devices was required to make the newfound power sources more useful. As late as 1781, Watt himself was skeptical of the now-ubiquitous flywheel design to regulate rotary power output from expanding pistons, and other experts, such as noted steam engineer John Smeaton, still advocated that steam-powered grain mills ought to rely on pumps to drive a traditional waterwheel for "producing smooth circular motion" (Ferguson, 1962). Watt's development of a kinematic mechanism allowing the reciprocating translational motion of a piston into the rotational motion of a working beam in 1784, commonly referred to as "Watt's parallel motion," was much more important to the Industrial Revolution than his actual steam engine (Müller-Sievers, 2012). As human understanding of kinematics grew, it combined with additional improvements to steam power which steadily accelerated following the expiration of Watt's first patent. Between 1810 and 1835, fuel efficiency of UK-based steam engines is estimated to have increased by a factor of five, while total horsepower was added at a rate of more than 4000 per year (Boldrin and Levine, 2008).

15.2 Putting the present in context

Despite the tremendous pace of technological innovation and change in our modern society, when it comes to energy or the adoption of new prime movers, things are remarkably similar to historical energy transitions. For example, ocean thermal energy conversion (OTEC) can leverage large temperature differentials between deep ocean waters and those at the surface. Makai Ocean Engineering is one of a number of companies pursuing this form of baseload renewable generation. Current OTEC plants commonly rely on the warm water and a heat exchanger to vaporize a low-boiling-point liquid, driving a turbine, and relying on cool water from depth to condense the liquid and complete the cycle (Courtland, 2014). Given the number of other approaches to electricity generation currently competing for capital and the fact that the first commercial-scale plant—a 16-MW nameplate capacity facility with 11 MW of expected delivered electricity which is owned by New Energy for Martinique and Overseas—is expected to be operational in 2018, displacements to the 94% of baseload thermal generation are unlikely to occur anytime soon (IRENA, 2014).

Understanding when to make the leap to new prime movers or fuel sources requires estimates of likely costs and benefits. Because energy sources and costs have historically been managed as largely non-competitive and non-differentiating cost centers, capital costs for modernization are large, and significant structural impediments to change are prevalent, most companies have few incentives to bear the risk of new energy-related technologies until presented with overwhelming evidence. These factors explain why technological discontinuities (innovations with dramatic implications for work factor versus performance) exist deep within the energy sector but are both less common and slower to proliferate than in other sectors of the economy (Anderson and Tushman, 1990; Foster, 1986). Historical energy transformations for fuels and prime movers provide ample evidence that incremental innovations within infrastructure systems are generally more dominant and that often a wide range of mutually enabling technological, regulatory, and even geopolitical developments are often required to shift prime movers or fuel sources. As energy sources have become more diverse and society has become increasingly electrified, competitive forces within the energy sector are growing. As government regulation within the energy sector is further reduced, technological change will increase commensurate with the increased competitive pressure on firms that are no longer protected monopolies; the elimination of telephone monopolies provides a useful corollary.

It is frequently said that if Alexander Graham Bell were alive today, he would not be able to recognize the modern telephony and communications networks, but that Thomas Edison or Samuel Insull would likely recognize much of the equipment and many operational processes and tools used in the electric grid. The energy industry faces significant uncertainty from new primary fuel sources, changes in consumption needs and behaviors, and environmental regulation which is further complicated by substantial regulatory alignment. In many ways, the cumulative impact of the accelerating ongoing changes is similar to the transformation of the telecommunications industry in the US within the last 40 years, but the analogies between the breakup of the Bell System telephone monopoly on January 1, 1984, and the ongoing changes to the industry must not be taken too far. The convergence of energy, information, and finance will enable transformational change, but the sheer breadth and existing capital investments involved in energy exploration, extraction, transport, transformation, and delivery networks will prevent rapid shifts to new prime movers or fuel sources where legacy investments and workforces exist. The most rapid developments are likely to be in the energy services businesses where new technologies and business models create significant value and have fewer vestigial structures or investments to overcome. The utility industry is currently best positioned for widespread disruption.

During a 1991 interview, Bell Atlantic's CEO Raymond Smith offered insights into the challenges of transforming a business and culture accustomed to operating as a protected monopoly to a competitive environment, noting that "the ways we were accustomed to operating impeded our ability to achieve our goals" and that many of the problems were "a consequence of our heritage" (Kanter, 1991).[3] Regulators, utilities, and customers should heed Smith's lessons. Utility transformation will require cultural change to provide substantial benefits. Smith focused Bell Atlantic on five strategies to increase growth and deliver results: improving efficiency, marketing, new products and services, entirely new businesses outside of existing service territories and the US, and *incentive regulation* to allow the regulated core business to benefit from internal initiatives "while protecting the telephone rate-payers" (Kanter, 1991).

In the preface to the published interview, Rosabeth Moss Kanter, the Harvard Business School professor who interviewed Smith, emphasized that Bell's impressive performance following the breakup stemmed from its response to both competitive forces, "increasing communications speed and blurring the distinction between information technologies and communication technologies" (Kanter, 1991).

---

[3] Bell Atlantic was one of seven regional telecommunications holding companies formed in the US during the forced AT&T breakup of 1984.

Today, this statement is equally applicable to today's utilities, which face blurring distinctions between energy, information, and financial products and services which are all required to meet personal or organizational energy management challenges.

Towards information as energy

Despite some of the more fantastic scenarios acknowledged herein, the grid, and other energy networks, have only started to evolve, and many more advances are sure to change the way we interact with energy—especially electricity. The most transformative of these are likely to stem from current work on *transactive energy* which can usher in an era of open-access and market-based participation in energy network optimization and utilization (GWAC, 2013). The accelerating merger of financial, information, and energy infrastructures is creating a new field of interdisciplinary science and business out of necessity. Success in today's increasingly integrated energy system requires explicit consideration of these three distinct but interrelated asset flows, especially within the electric power system:

- *Economic benefit* is typically measured as a function of financial utility in local currency, based on the value derived from energy resources injected or consumed from the system (e.g. the electric grid).
- *Information and information capacity* are derived from raw data via intentional processing where useful output and connectivity are measured in bytes and bytes per second, respectively. The question of whether or not information can be gained about any observable system is increasingly one of energy availability in suitable forms.
- *Energy and capacity* are governed by the laws of physics—including conservation of energy and increasing entropy within isolated systems—and refer to the energy available to humans for useful work, generally measured in joules and joules per second by scientists or kilowatt-hours and kilowatts for typical consumers.

The communication and extraction of useful information is governed by information-theoretic limits—recall Claude Shannon's work—which have profound practical implications for the implementation of "smart" energy systems. Designing, building, and operating intelligent energy infrastructure, such as the smart grid, requires careful consideration of technical and economic constraints on the storage, communication, analysis, and transformation of data into useful information. Overall system capacity, efficiency, and performance will be improved in an increasingly interoperable and interdependent system during normal operating conditions; however, interoperability and dependence also bring additional complications and introduce new vulnerabilities and risks. Significant challenges remain to be addressed to manage new relationships, data, and stakeholders.

Examples of this convergence are easy to come by. Given the immense energy costs associated with modern data centers and telecommunications operations, Google, Microsoft, and other internet darlings have already experimented with energy initiatives. In 2014, Amazon even unveiled plans to heat a skyscraper using a heat recovery chiller that will extract useful energy from the heat waste generated by an adjacent data center in a new Seattle complex (Stiles, 2014). Utilities, led by larger investor-owned utilities, are also exploring data-enabled services and alternate rate programs. Telecommunications firms are offering products and services geared towards the connected home, promising improved security, convenience, and efficiency. As promising approaches are identified, consolidation and expansion will help reshape the relationship between consumers and today's leading retail energy providers.

Financial incentives, regulations, and monitoring strategies will also require a holistic reevaluation to mitigate financial fraud and control against other negative externalities as the role of government shifts to enable new markets to develop. These changes offer several interesting possibilities for the future where utilities, financial services providers (e.g. banks), and communications or data-centric companies from the internet revolution participate in all of these traditionally independent services.

15.3 Earth's energy future

Energy will be an increasingly important factor in the course of economic and human development. Understanding the scope of mankind's readily available energy resources, the development of energy networks at a planetary scale provides a unique perspective from which to consider business, environmental, and even geopolitical issues of the day. The physical realities of the Earth as an energy system inform the realm of the possible for human action, because the availability of fuels and the infrastructure shape feasible approaches to fulfilling man's incorrigible thirst for energy.

As energy delivery networks for fossil fuels have become increasingly global, such as the development of robust global shipping and pipeline networks for oil, market fundamentals have become increasingly important. Robust *midstream* natural gas infrastructure, which links producers and consumers primarily via pipelines, is similarly central to overall commodity spot market price and price volatility. Heavy regulation in the US followed the Natural Gas Act of 1938 and continued until the FERC was given the authority to liberalize interstate natural gas markets following the passage of the Natural Gas Policy Act in 1978 (Juris, 1998). Open access to pipeline transportation and unbundling of sales from pipeline transportation (to varying degrees by state) complemented emerging wholesale markets and further accelerated system efficiencies. Integration of natural gas pipelines within Eurasia and North America corresponded with the movement towards wholesale markets and will continue to show increasing integration via liquid natural gas (LNG) transport and the likely eventual trading of environmental externalities (e.g. carbon) (Kaminski, 2012).

The Henry Hub, the first and most important in North America, opened in 1988 and now serves as a widely used benchmark for US natural gas prices. The hub helps balance energy supply delivery contracts across multiple pipeline participants—via operational balance agreements similar to Area Control Error concepts in the electric grid—to help shippers avoid individual penalties for imbalances in physically settled commodities contracts across the 13 interconnected pipelines which have a cumulative capacity of 1.8 billion cubic feet per day (Juris, 1998). The development of liquid natural gas terminals and shipping networks promises a global natural gas market in the very near future, spurred by the US shale gas boom and Russia's Vladimir Putin's repetitive disruptions to European gas supplies, in an extension of how the Henry Hub and wholesale markets have created more efficient continental delivery networks. The rush to increase the mobility of natural gas spawned a major set of LNG carriers, with 119 ordered between 2011 and 2014, expanding on the end-of-year 2010 fleet of 360 ships whose combined capacity was 53 million m^3 (Kaminski, 2012; Saul and Vukmanovic, 2014).

The growth in competitive supply chains and viable alternate fuels appears to have, at least partially, reduced the impact of geopolitical unrest on energy prices. Available supply (including resource availability, extractability, and transportability) has kept pace with demand. Fundamentals have kept oil prices low—sluggish economic growth and a glut of supply—despite widespread violence in the Middle East, unrest in the Ukraine, and frequent North African security disruptions. Energy and oil expert Daniel Yergin noted that "if ever there were a geopolitical world that should be driving oil prices higher, it would seem to be right now [...] But what it tells you is how powerful the fundamentals of the market are, and right now the fundamentals are winning out" (Johnson, 2014). While the duration of this relative calm in oil markets despite ongoing energy security concerns is unclear, even the U.S. Energy Information Administration has reduced expected price volatility in its future oil price paths from the past two decades, which averaged around 30%. The 2014 International Energy Outlook's revised oil price cases reflect adjustments in energy demand growth, OPEC investment and production decisions, non-OPEC supply, and the supply of other liquid fuels (EIA, 2014a).

A cosmic perspective

Evaluating mankind's energy on a more cosmic scale has long been an aspect of futurists and science fiction, but it is useful for better placing the discussion on energy in a cosmic perspective. Over the long term, society must be configured such that its overall demand for energy, on which its productive industries and life-sustaining technology are dependent, do not exceed the resources available for consumption. To this end, scientists have developed the concept of an energy budget for Earth as a system.

Surface-based efforts began in the early 1900s but were rapidly accelerated by satellites—especially with the launch of the Nimbus-7 satellite in 1978, the Earth Radiation Budget Experiment in the 1980s, and the Clouds and the Earth's Radiant Energy System which started in the late 1990s—which ultimately helped produce widely used energy budget diagrams (Li et al., 1997; Hun et al., 1986; Kiehl and Trenberth, 1997; Chambers and Bethea, 2013). Satellite-obtained data sets also enabled major improvements to global heat flow modeling, including the central role of oceans in transporting heat

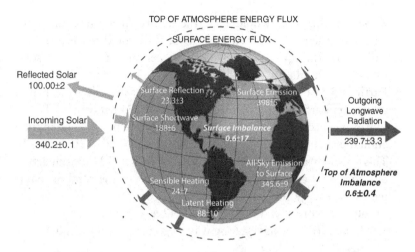

Figure 15.2 Global annual mean energy budget of Earth from 2000 to 2010

around the globe (Zhang et al., 2014; Stein, 1995). The energy budget attempts to calculate energy entering the Earth system, primarily from the Sun, as well as the energy which is absorbed or radiated back out into space. This concept is useful when considering the amount of energy that human civilization could theoretically extract without substantially changing the energy balance in such a manner that climactic shifts would disrupt ways of life and already tenuous geopolitical stability. While some of the data underlying current estimates of the energy budget have very large uncertainties, current scientific work estimates that more energy is being retained than radiated into space, meaning that some warming is occurring as this energy imbalance is stored in the Earth system as heat (Trenberth et al., 2009; Hansen et al., 2011). Figure 15.2 depicts the total "flux," also known as energy transfer, in and out of the Earth system at the top of the atmosphere (TOA) and between the TOA and the surface; values are given in Wm^{-2} and data is sourced from Stephens et al. (2012).

The best-known speculative scale for measuring the energy use of an entire civilization is known as the Kardashev scale. So named for Soviet astronomer Nikolai Kardashev, it was published in 1964 with three tiers of civilizations, differentiated by the amount of usable energy at their disposal[4]:

---

[4] Others later proposed extending this to five tiers, but only the first three are discussed herein.

- **Type I** civilizations would have the ability to harness all of the power on their own planet. For Earth, this would mean using the entirety of the approximately 341 W per m^2 of incoming solar radiation (Trenberth et al., 2009). Since the surface area of the Earth is 5.101×10^8 km^2, the total energy available to a Type I society on Earth is roughly 1.74×10^{17} W or 174 petawatts.[5] To put this in perspective, the International Energy Agency estimated total annual primary energy supply at 520.4 quads in 2013 (IEA, 2013). This would yield a Kardashev rating of only 0.72, meaning that our society is consuming only about 0.16% of the total available planetary energy budget.[6] Even though worldwide primary energy supply has more than doubled from 242.4 to 520.4 quads between 1973 and 2013, we have a long way to go before approaching Type I status (IEA, 2013).
- **Type II** civilizations would be able to utilize all of the power from a single star. The Sun, as a representative case, outputs on the order of 3.848×10^{26} W—this is more than 2.2 billion times the amount of energy reaching the Earth system from the Sun.
- **Type III** civilizations would be capable of using the energy of an entire galaxy, which vary substantially in size and composition. Kardashev chose the Milky Way Galaxy as a benchmark, estimating power output at 4×10^{37} W or 100 billion times the energy output of the Sun.

Back to Earth

Returning from the cosmic thought experiment, an equally fantastic transformation has occurred with little fanfare or acknowledgment—the emergence of planetary-scale energy delivery networks. The integration of multiple global energy delivery networks is accelerating with the movement towards market-based systems for electricity and natural gas. Electricity, because it balances in real time, can be viewed as a continuous energy delivery network. Natural gas pipelines are also continuous, slower-balancing systems but are connected across long distances via the increasingly substantial fleet

[5] Kardashev originally defined Type I civilization as 4×10^{12} W which he deemed a "technological level close to the level presently attained on earth" in 1964 (Kardashev, 1964).
[6] This uses Carl Sagan's Kardashev rating scale, which is base-10 logarithmic (2000).

of LNG carriers. The mix of continuous and discrete energy delivery modes for physical fuels creates a variety of unique constraints and market behaviors which must be accounted for by energy producers, transport/distribution entities, retailers, and consumers.

The development of integrated global energy delivery systems is not only reflective of increased maturity within energy sectors from organizational, regulatory, and market perspectives but also increasing standardization of energy-consuming devices and management technologies around the world. Piecemeal development of networks occurred naturally because of business and regulatory heterogeneity, but consolidation and standardization have become increasingly necessary to maximize economic benefits. As we seek additional efficiencies in operations and capital via integration, new information and communications technologies like DYMONDS will appear alongside additional physical superprojects such as Tres Amigas. Yet, from a systems perspective, these significant individual contributions remain merely the means to an end: minimizing disruptions across delivery systems to ensure that we can reliably attain the desired energy outcomes or services.

15.4 The enabler of civilizations

Surplus energy is the central engine for economic growth and the advancement of civilization. The human journey from hunting and gathering to our current era of widespread mechanization and the nascent emergence of information man is most effectively viewed from this perspective. The emergence of artisans, the development of new technologies, even advances in language and cultural expression, are strongly correlated with episodic advances in energy harvesting, transformation, and utilization. With the adoption of new prime movers and the liberation of human time, we have steadily gained agency as individuals and as a species.

The depiction of individual energy utilization in Figure 15.1 is incredible but less dramatic than an analysis of man's tools for harnessing energy to do useful work. The degree of change from our early precursors to modern society is even more pronounced when energy and information transformation capacity are viewed in concert. The acceleration of humankind's computational capabilities is even more dramatic than their energy counterparts. At its most basic, this phenomenon is known as Moore's Law, which states that computing capability doubles every 2 years. While this has held to date, physical laws still apply, and performance gains are increasingly difficult and costly. The heat produced from the resistance to electron flow within CPUs inside today's silicon-based chips is immense and approaching theoretical limits; this is driving a transition towards distributed computing and away from additional enhancements to processing speed within an individual instruction queue. The profound improvement in human tools for accessing useful energy and extracting useful information from data is shown in Table 15.1 (Cook, 1976; Ifrah, 2001).[7] The values listed represent the output from widely available technologies during each time period.

Useful energy and information are the key to the future. Useful energy is highly ordered (i.e. has a low entropy), distinguishing it from unusable energy which is of no practical importance to the conduct of useful work. Information which is germane to individual or organizational decision-making is also distinguished from data by low entropy. The work required to create useful objects, first demonstrated at scale during the Industrial Revolution, and the work required to create useful information is immense. While our internal biological energy requirements (also known as *endosomatic* energy needs) remain around 2400 kcal per day, our lifestyles and society (also known as *exosomatic* energy needs) are hugely intensive (Lotka, 1956). The amount of low-entropy matter and information which is used in the current economy dwarfs that used in our species' past. Life extracts negative entropy from its surrounding environment. As a civilization, our tools for harnessing energy and data to do useful work are only beginning to emerge and shape our possible futures.

15.5 Still driven by demand

The availability and cost to acquire useful energy have fundamentally shaped the formation of human civilization. Energy outcomes—the useful work which has liberated human labor to solve problems and develop culture—require the consumption of numerous energy products, each with specific roles in our emergent economic system.

---

[7] Computing perfomance is commonly measured using floating-point operations per second (FLOPS). A computer operating at a rate of 1 giga-FLOPS (GFLOPS) can conduct 1 billion operations (e.g. multiplication) over rational numbers.

Table 15.1 *Evolutionary power and information requirements and outputs of human machinery*

Era	Energy machine	Power produced (hp)	Information machine	Information processed (FLOPS)
Agrarian	Man pushing a lever	0.05		
Agrarian	Ox pulling a load	0.5	Abacus	0.1
	Waterwheels	0.5–5		
1600	Versailles water works	75		
1712	Newcomen steam engine	5.5		
1768	Watt's steam engine	40		
~1850	Marine steam engine	1,000	Babbage's Analytical Engine (proposed)	1
1900	Marine steam engine	8,000		
~1940	Steam turbine	300,000		
1961			IBM 1620	56.5
1970s	Coal or nuclear power plant	1,500,000		
1984			Cray X-MP/48	800,000,000
2013			Intel Pentium G550-based Commodity Server	4.848×10^{12}

The relationships between the energy outcomes desired by society and the energy products consumed to attain them are deeply intertwined, surprisingly subtle, and central to good decision-making. We consume products ranging from fossil fuels to solar energy in order to attain comfort (i.e. heat or cooling), mobility, or light or transform our physical environment. Increasingly, we expend huge amounts of energy to amass data and extract useful, low-entropy information to take action, make decisions, or shape our world.

Today's energy systems did not spring from the mind of an eccentric lone genius or emerge from a single transformative technology. Real progress in energy products and energy outcomes has been spurred by quiet revolutions in kinematics, processes, or business which made new technical approaches or fuel sources economically attractive or competitive. Access to raw energy products is but one component of meeting our societal demands, which control our institutional, economic, and political approaches to discrete issues. The rapid process of electrifying industrial powers in the 1900s is paradigmatic. Society sought rapid growth to universally accessible, low-cost, and comparatively reliable access to power. The physical, financial, and regulatory infrastructure which exists today is reflective of the historical requirements and the available technology of the period.

As our society has grown more energy intensive and our world has grown more populous, a new set of emerging demands have become apparent. Economic and reliable access to energy remains controlling, but the timescales being considered matter. Over the long term, economic and reliable access to energy means making targeted investments in resilience, security, and environmental considerations. Climate shifts, changes in weather patterns, susceptibility to natural disasters, and risks from targeted human disruptions—from cyberspace or in the physical world—must be considered and priced. These demands will dictate the evolution of energy for the next century. For society, the primary concern is to manage the risk of delivery.

Managing risk in complex systems with large numbers of interactions between players and stakeholders is challenging. Energy systems, especially electric power systems, are quintessential complex systems and are full of interdependencies which challenge our ability to observe the system at all, let alone to manage the inherent risks. The myriad of technologies and systems introduced in this book are representative of today's available tools to address these challenges and improve decision-making. Advancement requires additional transparency, democratization of information about our infrastructure to better enable all members of society to make better economic and risk decisions, and a willingness to encourage diverse solutions to our changing problems.

Meeting our pressing energy and water requirements in a world with growing energy needs and a rapidly burgeoning population requires a consistent and concerted effort to chart out and implement a number of energy transformations. Consumers, energy companies, and policy makers are well advised to avoid the fallacy of singular solutions to energy challenges. Scarce resources—whether scarce in terms of supply or because of the lack of acceptable substitutes in key processes (e.g. coke for steel smelting)—should be protected for use in their most critical function where possible. Outside of these limited cases, energy systems need to seek increasing interoperability between raw fuels and societal transmission and transformation capabilities. Increasingly interoperable sources of energy will facilitate increased system resilience and serve as a near-term hedge against geopolitical instability.

Improvements to education and awareness of energy issues are needed due to energy's central role in most other critical infrastructure systems and its increasing inseparability from the sustenance of human life. Given that world population stabilization appears to be less likely in the near term and increasing global population estimates now range between 9.6 and 12.3 billion in 2100, there is real urgency (Gerland et al., 2014). The cross-cutting nature of effective energy policy and operational strategy requires societal participation, not just engineering effort. Approaches to energy often fail to provide adequate consideration for larger, long-term societal impacts and values. Because of the immense capital requirements and long lifespan of many energy assets, these decisions can become controlling if not managed carefully. Fortunately, the same actions required to improve geopolitical stability, hedge against long-term climactic shifts, and most drastically improve the lives of the world's poor are completely aligned. Creating the conditions to increase access, affordability, and reliability of energy around the world remains of singular importance, today and for the future.

References

Abraham, S., Efford, J.R., Keen, L., McLellan, A., Manley, J., Vollman, K., Diaz, N., Ridge, T., Wood, P., Glotfelty, J., et al. (2004: April) *Final report on the August 14, 2003 blackout in the United States and Canada*. Tech. rept. U.S.-Canada Power System Outage Task Force.

ABB group (n.d.) 'Xiangjiabe-Shanghai: connecting remote generation'. Available HTTP: <https://library.e.abb.com/public/57af6cb9ca0204-ffc1257dcf004d7495/POW0056%20Rev%202.pdf>. Accessed March 2015.

ABS Group (2001) 'Seattle (Nisqually), Washington earthquake of February 28, 2001 (Magnitude 6.8): An EQE Briefing'.

Abur, A. and Exposito, A.G. (2004) *Power system state estimation, theory and implementation*. London: CRC Press.

Acar, E., Agrawal, B., Conn, A., Ditlow, G., Feldmann, P., Finkler, U., Gaucher, B., Gupta, A., Heng, F.L., Kalagnanam, J., et al. (2011: February) *Framework for large-scale modeling and simulation of electricity systems for planning, monitoring, and secure operations of next generation electricity grids*. Special report in response to Request for Information: Computation Needs for the Next-Generation Electric Grid. IBM Research.

Ackerman, N., Anderson, R., Arora, A., de Barros, A.P., Blum, R.O., Carter, L., Edwards, L., Farnsley, G., Gercke, M., Kannan, K., et al. (2009) *Unsecured economies: Protecting vital information. The first global study highlighting the vulnerability of the world's intellectual property and sensitive information*. Tech. rept. McAfee, Inc.

ACLU (2014: February) *Metadata: Piecing together a privacy solution*. Tech. rept. American Civil Liberties Union of California.

Adams, E.D. (1890) Telegram: 'The international Niagara commission letter of invitation to the cataract construction company' June 25, 1890. Online. Available HTTP: <https://archive.org/stream/niagarapowerhist00adam/niagarapowerhist00adam_djvu.txt>.

Adams, E.D. (1927) *Niagara power: History of the Niagara Falls Power Company*, Vol. 1. Niagara Falls, NY: Niagara Falls Power Co.

Administration of Barack Obama (2012: August) Executive Order 13624, *Accelerating Investment in Industrial Energy Efficiency*. U.S. Government Printing Office.

Administration of Barack Obama (2013) Fact Sheet: *Power Africa*. Office of the Press Secretary, The White House, June 30.

Agrawal, D., Left, H. and Menon, V. (1996) 'Efficiency and efficacy of incandescent lamps', *American Journal of Physics*, 64(5): 649–654.

Ahluwalia, M.S. (2002) 'Economic reforms in India since 1991: Has gradualism worked?' *The Journal of Economic Perspectives*, 16(3): 67–88.

Aiken, J. and Aiken, R. (1962) *Power: the gift of Niagara*. Buffalo and Erie County Historical Society, Buffalo.

Akamai Technologies, Inc. (2013: July). Financial services firms defend against distributed denial of service attacks and data breaches while maintaining high performance and availability. White Paper. Cambridge, MA: Akamai Technologies.

Aldrich, D. and Platte, J. (2014) After the Fukushima meltdown, Japan's nuclear restart is stalled. *Washington Post*, August 15.

Aljazeera (2014) Power-line attack plunges Yemen into darkness. *Aljazeera*, June 11.

Allen, R. (2006) *The British Industrial Revolution in global perspective: How commerce created the Industrial Revolution and modern economic growth*. Oxford: Department of Economics, Nuffield College, Oxford University.

Allyn-Feuer, A. (2009) 'Pay-as-you-drive insurance, privacy, and government mandates', *Ars Technica*, July 17.

Almashabi, D. (2014) Saudis start production at world's biggest desalination plant. *Bloomberg*, April 23.

Alomari, E., Gupta, B., Manickam, S., Karuppayah, S. and Alfaris, R. (2012) 'Botnet-based distributed denial of service (DDoS) attacks on web servers: Classification and art', *International Journal of Computer Applications*, 49(7): 24–32.

Amin, M. and Wollenberg, B.F. (2005) 'Toward a smart grid: Power delivery for the 21st century', *IEEE Power and Energy Magazine*, 3(5): 34.

Anderson, P. and Tushman, M. (1990) 'Technological discontinuities and dominant designs: A cyclical model of technological change', *Administrative Science Quarterly*, 35: 604–633.

Anderson, R. (2002) 'Security in open versus closed systems—The dance of Boltzmann, Coase and Moore'. Paper presented at Open Source Software: Economics, Law and Policy Conference, Toulouse, France, June 20–21, 2002.

Andrews, A. and Pirog, R. (2012) *The strategic petroleum reserve—Authorization, operation, and drawdown policy.* Tech. rept. Congressional Research Service.

Andrews, E. (2011: December) Barbara Hepworth: 500k 'Two Forms' sculpture stolen by metal thieves. *The Telegraph.* Online. Available HTTP: <http://www.telegraph.co.uk/news/uknews/crime/8968237/Bar bara-Hepworth-500k-Two-Forms-sculpture-stolen-by-metal-thieves. html>. Accessed August 2014.

Angenheister, G. and Westland, C.J. (1921) 'The magnetic storm of May 13–14, 1921: observations at Samoa Observatory', *New Zealand Journal of Science and Technology,* 4: 201–202.

Anon. (1890) 'Overhead telegraph wires', *The Telegraphic Journal and Electrical Review,* 27(658): 4.

Anthony, D. (2010) *The horse, the wheel, and language: How Bronze-Age riders from the Eurasian Steppes shaped the modern world.* Princeton, NJ: Princeton University Press.

AON-Benfield. (2014: July) *Insurance-linked securities: Fourth quarter 2013 update.* Tech. rept. Chicago, IL: AON-Benfield.

APPA (American Public Power Association) (2013) 2014–14 *annual directory and statistical report.* American Power Association, Arlington, VA.

APS (American Physical Society) (2009) This month in physics history: April 25, 1954: Bell Labs demonstrates the first practical silicon solar cell. *American Physical Society* 18(4): 1954. Online. Available HTTP: <http://www.aps.org/publications/apsnews/200904/physicshistory. cfm>.

Arrow, K., Bolin, B., Costanza, R., Dasgupta, P., Folke, C., Holling, C.S., Jansson, B.O., Levin, S., Maler, K.G., Perrings, C. and Pimentel, D. (1995) 'Economic growth, carrying capacity, and the environment', *Science,* 268(April), 520–521.

ASAP (Appliance Standards Awareness Project) (2011: August) Refrigerator and freezer standards Q&A. Online. Available HTTP: <http://www.appliance-standards.org/product/refrigerators-and-freezers>.

ASHRAE (2004) *Thermal environmental conditions for human occupancy 2004.* Tech. rept. ASHRAE.

ASME (1977: April). *State line generating unit no. 1* (1929) Tech. rept. The American Society of Mechanical Engineers and the Commonwealth Edison Company of Indiana, Inc.

Avery, J. (2012) *Information theory and evolution.* Singapore: World Scientific Publishing Company.

Averyt, K., Fisher, J., Huber-Lee, A., Lewis, A., Macknick, J., Madden, N., Rogers, J. and Tellinghuisen, S. (2011: November) *Freshwater use by U.S. power plants: Electricity's thirst for a precious resource. A report of the Energy and Water in a Warming World Initiative.* Union of Concerned Scientists.

Bacon, D. (1998: December) 'Second World War Deception: Wright Flyer paper no. 5'. Maxwell Air Force Base, Montgomery, AL: Air Command and Staff College Air University.

Bacon, R. Bhattacharya, S. and Kojima, M. (2010) *Expenditure of low-income households on energy; evidence from Africa and Asia.* The World Bank.

Baetz, J. (2011) "Germany decides to abandon nuclear power by 2022". Associated Press.

Bak, P. (1996) *How nature works.* New York: Springer-Verlag.

Baker, D. and Lanzerotti, L. (2008) 'A continuous L1 presence required for space weather', *Space Weather,* 6(November).

Baker, E., Fowlie, M., Lemoine, D. and Reynolds, S.S. (2013) The economics of solar electricity. The Energy Institute at Haas. Online. Available HTTP: <http://ei.haas.berkeley.edu/pdf/working_papers/WP240. pdf>.

Bamberger, R. (2009) *The strategic petroleum reserve—History, perspectives, and issues.* Tech. rept. Congressional Research Service, August 18.

Banerjee, A., Banerjee, J., Varsamopoulos, G., Abbasi, Z. and Gupta, S. (2013: May) 'Hybrid simulator for cyber-physical energy systems', in *IEEE Workshop on Modeling and Simulation of Cyber-Physical Energy Systems (MSCPES).* Berkeley, CA: Arizona State University, pp. 12–17.

Bank of America Corporation (2013) *10-K annual report.* Online. Available HTTP: <http://www.sec.gov/Archives/edgar/data/70858/00000708581 4000012/bac-12312013x10k.htm>.

Barabasi, L. and Albert, R. (1999) 'Emergence of scaling in random networks', *Science,* 286(October): 509–512.

Barbier, E. (1989) *Economics, natural-resource scarcity and development.* London: Earthscan.

Barnes, P.R., Rizy, D.T., McConnell, B.W., Tesche, F.M. and Taylor, E.R. (1991: November). *Electric utility experience with geomagnetic disturbances.* Tech. rept. ORNL-6665. Oak Ridge, TN: Oak Ridge National Laboratory.

Barnosky, A.D., Hadly, E., Bascompte, J., Berlow, E.L., Brown, J., Fortelius, M., Getz, W., Harte, J., Hastings, A., Marquet, P.A., et al. (2012) 'Approaching a state shift in Earth's biosphere', *Nature,* 486, 52–58.

Basu, P. and Gavin, W.T. (2011) 'What explains the growth in commodity derivatives?' *Federal Reserve Bank of St. Louis Review,* 93(1), 37–48.

Basulto, D. (2014) In the wake of Toledo, we need innovations to prevent a world water crisis. *Washington Post*, August 5.

Bauer, D. and Kramer, F. (2007) Risk and valuation of mortality contingent catastrophe bonds. *Journal of Banking and Finance*. Online. Available HTTP: <https://www.uni-ulm.de/fileadmin/website_uni_ulm/mawi.mort/pdf/Application/2007BauerKramer.pdf>. Accessed February 2015.

BCS, Incorporated (2008: March) *Waste heat recovery: technology and opportunities in the U.S. Industry*. Tech. rept. U.S. Department of Energy.

Beech, M. (2014) England's world cup opener underwhelms. *Utility Week*, June 16.

Behrens, C. and Glover, C. (2012: April). *U.S. energy: Overview and key statistics*. Tech. rept. Congressional Research Service.

Bencsath, B., Pek, G., Buttyan, L. and Felefyhazi, M. (2012) 'The cousins of Stuxnet: Duqu, Flame, and Gauss', *Future Internet*, 4(November), 971–1003.

Berges, M., Goldman, E., Matthews, H.S. and Soibelman, L. (2008) 'Training load monitoring algorithms on highly sub-metered home electricity consumption data'. *Tsinghua Science and Technology*, 13(S1), 406–411.

Bernabeu, E. and Katiraei, D. (2011) Aurora vulnerability: issues and solutions—Hardware mitigation devices, July 24. Online. Available HTTP: <https://www.smartgrid.gov/sites/default/files/doc/files/Aurora VulnerabilityIssuesSolutionHardwareMitigationDe201102.pdf >.

Bernstein, P.L. (2007) *Against the gods: The remarkable story of risk*. London: John Wiley and Sons.

Betz, A. (1966) *Introduction to the theory of flow machines*. Pergamon Press, Oxford.

Bhupathi, T. (2010) 'Technology's latest market manipulator? High frequency trading: The strategies, tools, risks, and responses', *North Carolina Journal of Law and Technology*, 11(2).

Biello, D. (2014) 'Fight over rooftop solar forecasts a bright future for cleaner energy'. *Scientific American*, March 25.

Bjørndal, M. and Jørnsten, K. (2007) 'Investment paradoxes in electricity networks', in *Pareto optimality, game theory and equilibria* (Springer Optimization and Its Applications series). New York: Springer, pp. 593–608.

Bjørndal, M. and Jørnsten, K. (2000) *Paradoxes in networks supporting competitive electricity markets*. Tech. rept. Norwegian School of Economics and Business Administration.

Blakley, B., McDermott, E. and Geer, D. (2001) 'Information security is information risk management', *Proceedings of Association of Computational Machinery Workshop on New Security Paradigms*, September: 97–104.

Bodeau, D. and Graubart, R. (2013: November) *Characterizing effects on the cyber adversary: A vocabulary for analysis and assessment*. Tech. rept. Project 51MSR615-DA. Bedford, MA: MITRE Corporation.

Boldrin, M. and Levine, D. (2008) *Against intellectual monopoly*. Cambridge: Cambridge University Press.

Bookstaber, R. (2009) *A demon of our own design: Markets, hedge funds, and the perils of financial innovation*. London: Wiley.

Booth, A., Greene, M. and Tai, H. (2010: Summer) *U.S. smart grid value at stake: The $130 billion question*. Tech. rept. McKinsey.

Born, P. and Viscusi, W. (2006) 'The catastrophic effects of natural disasters on insurance markets', *Journal of Risk Uncertainty*, 33: 55–72.

Bose, P.R. (2012) Coal India offers 70 mt pithead stock to power producers. *The Hindu Business Line*, July 5.

Boswell, A. (2014). Tres Amigas: Project 75 percent financed. *Portales News Tribune*, February 2.

Botterud, A., Bhattacharyya, A. and Ilic, M. (2002) 'Futures and spot prices—An analysis of the Scandinavian electricity market'. Proceedings of the 34th Annual North American Power Symposium (NAPS 2002), Tempe, AZ, October 2002.

Bouton, S., Creyts, J., Kiely, T., Livingston, J. and Naucier, T. (2010) *Energy efficiency—A compelling global resource*. Tech. rept. McKinsey & Company.

BPA (Bonneville Power Administration) (2012) *Northwest over generation: An assessment of potential magnitude and cost*. Tech. rept. Bonneville Power Administration.

Brager, G.S. and de Dear, R. (2001: April) 'Climate, comfort, & natural ventilation: A new adaptive comfort standard for ASHRAE Standard 55'. Proceedings: Moving Thermal Comfort Standards into the 21st Century. Windsor: Oxford Brookes University.

Branigan, S. and Cheswick, B. (1999) The effects of war on the Yugoslavian network. Online. Available HTTP: <http://www.cheswick.com/ches/map/yu/index.html>.

Broad, W.J., Marko, J. and Sanger, D.E. (2011) Israel tests on worm called crucial in Iran nuclear delay. *New York Times*, January 15.

Brock, H.W. (2002: May) *The transformation of risk: Main Street versus Wall Street*. Profile Report. Strategic Economic Decisions, Inc.

Brown, M.H. (2004) *Electricity transmission: A primer*. Tech. rept. National Conference of State Legislatures.

Bukhsh, W.A. (2012: November) Wide area blackouts: Causes and prevention. Lecture Slides, University of Edinburgh Energy and Optimization Group.

Burger, B. (2014) 'Electricity production from solar and wind in Germany in 2013'. Presentation at the Fraunhofer Institute for Solar Energy Systems ISE, January 9.

Burke, D. (1995: November) *All circuits are busy now: The 1990 AT&T long distance network collapse.* Tech. rept. CSC440-01. California Polytechnic State University.

Byres, E. (2013) 'Privacy and security—The air gap: SCADA's enduring security myth', *Communications of the ACM*, 56(8): 29–31.

CAISO (2013) California ISO: OASIS. Online. Available HTTP: <http://oasis.caiso.com>. Accessed August 2013.

CAISO (2014) CAISO fast facts: What the duck curve tells us about managing a green grid. Online. Available HTTP: <http://www.caiso.com/documents/exibleresourceshelprenewablesfastfacts.pdf>. Accessed August 2014.

Cambridge Systematics, Inc. (1999) Public transportation and the nation's economy: A quantitative analysis of public transportation's economic impact. Online. Available HTTP: <http://www.apta.com/resources/reportsandpublications/Documents/vary.pdf>.

Campbell, W.W. (1921) 'The aurora borealis of May 14, 1921', *Publications of the Astronomical Society of the Pacific*, 33: 164–167. Chicago: The University of Chicago Press.

Cannon, J.S. (2006) *Greening garbage trucks: Trends in alternative fuel use, 2002–2005.* New York: INFORM.

Cannon, P. (2013: February) *Extreme space weather: Impacts on engineered systems and infrastructure.* Tech. rept. London: Royal Academy of Engineering.

Cardenas, A., Amin, S., Sinopoli, B., Giani, A., Perrig, A. and Sastry, S. (2009). 'Challenges for securing cyber physical systems'. Online. Available HTTP: <http://chess.eecs.berkeley.edu/pubs/601/cps-security-challenges.pdf>. Accessed February 2015.

Carlson, W.B. (2006) 'Nikola Tesla and the business of invention, 1885–1905'. Paper presented at the Sixth International Symposium on Nikola Tesla, Belgrade, Serbia and Montenegro, October 18–20.

Carr, N. (2013) *The big switch: Rewiring the world, from Edison to Google.* New York: W.W. Norton & Company.

Carreras, B., Newman, D., Dobson, I. and Poole, A.B. (2001) 'Evidence for self-organized criticality in electric power system blackouts'. Paper presented at the Hawaii International Conference on System Sciences.

Carreras, B., Newman, D., Dobson, I. and Poole, A.B. (2004) (September). 'Evidence for self-organized criticality in a time series of electric power system blackouts', *IEEE Transactions on Circuits and Systems I: Regular Papers*, Vol. 51.

Carter, J. (1977) Televised speech by President Jimmy Carter on April 18, 1977. Online. Available HTTP: <http://www.pbs.org/wgbh/american experience/features/primaryresources/carter-energy/>.

Carter, N. (2013) *Energy-water nexus: The energy sector's water use*. Tech. rept. Congressional Research Service, August 30.

Cavoukian, A. (2013: July) *A primer on metadata: Separating fact from fiction*. Tech. rept. Information and Privacy Commissioner, Ontario, Canada.

Cazalet, E. (2010) 'TeMIX: a foundation for transactive energy in a smart grid world'. Grid Interoperability Forum. Online. Available HTTP: <http://www.pointview.com/data/files/2/1062/1878.pdf>.

CEA (2009) *17th electric power survey of India*. Tech. rept. Central Electricity Authority, Ministry of Power, Government of India.

CEA (2012: August) *Report of the enquiry committee on grid disturbance in Northern Region on 30th July 2012 and in Northern, Eastern, and North Eastern Region on 31st July 2012*. Tech. rept. Central Electricity Authority, New Delhi.

CEC (Commonwealth Edison Company) (1911) *Principal generating stations and transmission system*. Chicago: Commonwealth Edison.

CERC (2012) *Report on the grid disturbance on 30th July 2012 and grid disturbance on 31st July 2012*. Tech. rept. Central Electricity Regulatory Commission.

CERT-EU (2014: July) *DDoS overview and incident response guide*. Tech. rept. Computer Emergency Response Team for the EU Institutions, Bodies, and Agencies.

CEW (Confederation of European Waste-to-Energy Plants) (2011) Environmentally sound use of bottom ash. Online. Available HTTP: <http://www.cewep.eu/information/publicationsandstudies/statements/ceweppublications/m_722>.

CEW (2014) Heating and lighting the way to a sustainable future. Online. Available HTTP: <http://www.cewep.eu/m_1073>. Accessed: August 2014.

Chambers, L. and Bethea, K. (2013) 'Energy budget: Earth's most important and least appreciated planetary attribute'. *The Universe in the Classroom, Astronomical Society of the Pacific*, Summer.

Champ, A. (2010) 'The Lake Erie loop flow mystery', *Public Power*, January/February: 10–15.

Chen, J. (2005) *The physical foundation of economics: An analytical thermodynamic theory*. Singapore: World Scientific Publication Co.

Cheng, D., Onen, A., Arghandeh, R., Jung, J., Broadwater, R. and Scirbona, C. (2014: July). 'Model centric approach for Monte Carlo assessment of storm restoration and smart grid automation', *Proceedings of the ASME 2014 Power Conference*.

Cisco (2013) Cisco global cloud index: Forecast and methodology, 2011–2016. Online. Available HTTP: <http://tinyurl.com/7gnfeeb>. Accessed August 2013.

City of Seattle (2000) Solid waste by rail in the Pacific Northwest. Online. Available HTTP: <http://www.seattle.gov/util/groups/public/@spu/@garbage/documents/webcontent/cos_001786.pdf>. Accessed August 2014.

City of Seattle (2013) Seattle Climate Action Plan. Online. Available HTTP: <http://www.seattle.gov/Documents/Departments/OSE/2013_CAP_20130612.pdf>. Accessed February 2015.

Clark, C.E., Harto, C.B., Sullivan, J.L. and Wang, M.Q. (2011: January). *Water use in the development and operation of geothermal power plants*. Tech. rept. ANL/EVS/R-10/5, Argonne National Laboratory for the U.S. Dept. of Energy.

Clarke, A.C. (1964) Interview of Arthur C. Clarke, *Horizon*, BBC.

Clarke, A.C. (1968) '2001: A Space Odyssey', *New American Library*.

Clark, G. and Jacks, D. (2007) 'Coal and the industrial revolution, 1700–1869', *European Review of Economic History*, 11(1): 39–72.

Clarke, G., Reynders, D. and Wright, E. (2004) *Practical modern SCADA protocols: DNP3, 60870.5 and related systems*. London: Elsevier.

Clayton, B. (2012: September) 'Lessons learned from the 2011 strategic petroleum reserve release', Working Paper. Council on Foreign Relations.

Cleary, J. and Eades, D. (2014) 'SEC (Securities and Exchange Commission) issues cybersecurity risk alert', *The National Law Review*, April.

Clemente, D. (2013) *Cyber security and global interdependence: What is critical?* Tech. rept. London: Chatham House (The Royal Institute of International Affairs).

Cliver, E.W. and Dietrich, W.F. (2013) 'The 1859 space weather event revisited: limits of extreme activity'. *Journal of Space Weather and Space Climatology*, 3.

Cliver, E.W. and Svalgaard, L. (2004) 'The 1859 solar-terrestrial disturbance and the current limits of extreme space weather activity', *Solar Physics*, 224(October): 407–422.

Colarik, A. and Janczewski, L. (eds) (2007) *Cyber war and cyber terrorism*. Hershey, PA: The Idea Group.

Collins, J. (2001) *Good to great: Why some companies make the leap... and others don't*. New York: HarperCollins.

Columbia River Treaty Review (2015) Available HTTP: <http://blog.gov.bc. ca/columbiarivertreaty/faqs>. Accessed March 2015.

Comfort, L.K., Boin, A. and Demchak, C.C. (2010) *Designing resilience: Preparing for extreme events*. Pittsburgh, PA: University of Pittsburgh Press.

Concordia, C. (1990) 'Power system objectives' side effects: good and bad', *IEEE Power Engineering Review*, September, 12–13.

Cook, E. (1971) 'The flow of energy in an industrial society', *Scientific American*, 135–144.

Cook, E. (1976) *Man, energy, society*. W.H. Freeman Publishers.

Cooley, H. and Heberger, M. (2013: May) *Key issues for seawater desalination in California: energy and greenhouse gas emissions*. Tech. rept. Oakland, CA: Pacific Institute.

CBO (Congressional Budget Office) (2014) *The Highway Trust Fund and the treatment of surface transportation programs in the federal budget*.

Connecting Washington Task Force (2011a) Estimated 10-year transportation unfunded system needs. Online. Available HTTP: <http://www.digitalarchives.wa.gov/GovernorGregoire/priorities/trans portation/connect/20111129/meeting_notes.pdf>.

Connecting Washington Task Force (2011b) Initial presentation of statewide transportation needs. Online. Available HTTP: <http://www.digitalarchives.wa.gov/GovernorGregoire/priorities/trans portation/connect/20110719/transportation_needs.pdf>.

Connecting Washington Task Force (2012) Strategic transportation investments to strengthen Washington's economy and create jobs. Online. Available HTTP: <http://www.wsdot.wa.gov/NR/rdonlyres/0DD6F466 -6D52-4495-AAC6-78F2AA5B2332/0/ConnectingWashingtonfinal_ report.pdf.

Consolidated Edison Company of New York (n.d.) A brief history of conEdison. Online. Available HTTP: <http://www.coned.com/history/. Accessed January 2014.

Courtland, R. (2014) 'Ocean thermal energy: Back from the deep', *IEEE Spectrum*, September.

Courtney, H. (2001) *20/20 Foresight: Crafting strategy in an uncertain world*. Watertown, MA: Harvard Business Review Press.

Courtney, H., Kirkland, J. and Vigurie, P. (1997) Strategy under uncertainty. *Harvard Business Review*, November/December.

Covrig, C., Ardelean, M., Vasiljevska, J., Mengolini, A., Fulli, G. and Amoiralis, E. (2014) *Smart grid projects outlook*. Tech. rept. EUR 26651 EN. European Commission Joint Research Centre.

Crabtree, J., Rajan, P., Galvin, B., McCord, A. and Jia, J. (2010a: August) System and method for electric grid utilization and optimization. U.S. Patent Application, No. 20100217550.

Crabtree, J., McCord, A., Jia, J., Venn, O., Hoak, M. and Charneski, A. (2010b: December). System and method for participation in energy-related markets. U.S. Patent Application, No. 20100332373.

Crabtree, J., Rajan, P., Galvin, B., McCord, A. and Venn, O. (2010c: August). System and method for trading complex energy securities. U.S. Patent Application, No. 20100218108.

Crabtree, J., Rajan, P., Galvin, B., and McCord, A. (2011: February). Dynamic pricing system and method for complex energy securities. U.S. Patent Application, No. 20110040666.

Craddock, P. (1995) *Early metal mining and production.* Edinburgh: Edinburgh University Press.

Craft3 (n.d.) Home energy efficiency loans in Washington. Online. Available HTTP: <http://www.craft3.org/Borrow/CPW>. Accessed August 2013.

Creswell, J. and Gebelo, R. (2014) Traders profit as power grid is overworked. *The New York Times*, August 14.

Danish Energy Agency (2010) *Energy statistics report*, November 2009.

Danzig, R. (2014: July) *Surviving on a diet of poisoned fruit: Reducing national security risks of America's cyber dependencies.* Tech. rept. Center for New American Security.

Danzig, R. and Hosford, Z. (2012: November) Beyond Fukushima: A joint agenda for U.S.-Japanese disaster management. Policy Brief. Center for New American Security, Washington, DC.

Darby, S. (2006: April) *The effectiveness of feedback on energy consumption: A review for DEFRA of the literature on metering, billing and direct displays.* Tech. rept. Oxford: Environmental Change Institute, University of Oxford.

DARPA (2012: 23 February) High-assurance cyber military systems (HACMS). Broad Agency Announcement, DARPA-BAA-12-21.

Database of State Incentives for Renewable & Efficiency (2013) Renewable portfolio standard policies. Online. Available HTTP: <http://www.dsireusa.org/documents/summarymaps/RPS_map.pdf>. Accessed August 2013.

Davis, L., Gallman, R. and Gleiter, K. (1997) 'Technology', in: *In pursuit of Leviathan: Technology, institutions, productivity, and profits in American whaling, 1816–1906.* Chicago: University of Chicago Press, pp. 260–296.

de Avila, J. (2014) Temporary closing of Indian point power plant is considered: State proposal could save fish; operator is opposed. *Wall Street Journal*, July 21.

de Dear, R. and Brager, G. (2002) 'Thermal comfort in naturally ventilated buildings: revisions to ASHRAE Standard 55', *Energy and Buildings*, 34(6): 549–561.

Dehoust, G., Schüler, D., Vogt, R., and Giegrich, J. (2011: December). *Climate protection potential in the waste management sector examples: Municipal waste and waste wood*. Tech. rept. (UBA-FB) 001347/E. German Environmental Research of the Federal Ministry of the Environment, Nature Conservation and Nuclear Safety.

Deng, S.J. and Oren, S.S. (2006). 'Electricity derivatives and risk management', *Energy*, 31: 940–953.

Der Grüne Punkt (n.d.) Online. Available HTTP: <http://www.gruener-pun kt.de/>. Accessed September 2014.

DHS (Department of Homeland Security) (2008: December) *National power grid simulation capability: Needs and issues*. Tech. rept. Argonne, IL: U.S. Department of Homeland Security, Science and Technology Directorate.

DHS Risk Steering Committee (2010) *DHS risk lexicon*. Tech. rept. Department of Homeland Security, Washington, DC.

Dikötter, F. (2010) *Mao's great famine: The history of China's most devastating catastrophe, 1958–1962*. New York: Bloomsbury Publishing.

DOE (Department of Energy) (2006: December) *Energy demands on water resources. Report to Congress on the interdependency of energy and water*. Department of Energy.

DOE (Department of Energy/Office of Energy Efficiency and Renewable Energy) (2007) U.S. energy requirements for aluminum production: historical perspective, theoretical limits and current practices. Online. Available HTTP: <http://energy.gov/sites/prod/files/2013/11/f4/al_the oretical.pdf>.

DOE (2010) *Data access and privacy issues related to smart grid technologies*. Tech. rept. Department of Energy.

DOE (2012) *U.S. Department of Energy smart grid privacy workshop summary report*. Tech. rept. Department of Energy, January 31.

DOE (2014) Monthly energy review: March 2014. U.S. Energy Information Administration, March. DOE/EIA-0035(2014/03). Online. Available HTTP: <http://www.eia.gov/totalenergy/data/monthly>.

Doucette, R. and McCullough, M. (2011) 'Modeling the CO_2 emissions from battery electric vehicles given the power generation mixes of different countries', *Energy Policy*, 39(November): 803–811.

Drower, M.S. (1954) 'Water-supply, irrigation, and agriculture', in: *A History of Technology*. New York: Oxford University Press.

Duggan, D.P., Thomas, S.R., Veitch, C.K.K. and Woodard, L. (2007) *Categorizing threat: building and using a generic threat matrix*. Tech. rept. Albuquerque, NM: Sandia National Laboratories.

Duke Energy Carolinas, LLC (2013) Schedule SGS (NC), small general service. Rate schedule. Online. Available HTTP: <http://www. duke-energy.com/pdfs/NCScheduleSGS.pdf>. Accessed February 2013.

Durumeric, Z., Wustrow, E. and Halderman, J. (2013) 'ZMap: Fast internet-wide scanning and its security applications', *Proceedings of the 22nd USENIX Security Symposium*.

Eaton, J. and Eckstein, Z. (1984) 'The U.S. strategic petroleum reserve: An analytic framework', in: *The Structure and Evolution of Recent U.S. Trade Policy*. Chicago: University of Chicago Press, pp. 237–276.

Ecomaine (2011) Process diagram for Waste-to-Energy. Online. Available HTTP: <http://www.ecomaine.org/our-facility/waste-to-energy-plant/>.

Economist (2004) Asia's tsunami: the cruel sea. *The Economist*, December 29.

Economist (2005) Hurricane Katrina: After the flood. *The Economist*, September 1.

Economist (2012) Hurricane Sandy: Wild is the wind. *The Economist*, November 3.

Economist (2013). Natural gas—Difference engine: Fuel for the future? *The Economist*, May 27.

Economist (2014a) Brazil's oil auction—Cheap at the price. *The Economist*, October 26.

Economist (2014b) Money for nothing: Financing energy efficiency. *The Economist*, April 26.

EIA (U.S. Energy Information Administration) (2003) Commercial building energy consumption survey: data. Online. Available HTTP: <http://www.eia.gov/consumption/commercial/>. Accessed August 2014.

EIA (2009) Residential energy consumption survey: data. Online. Available HTTP: <http://www.eia.gov/consumption/residential/>. Accessed August 2014.

EIA (2011a) Electricity storage technologies can be used for energy management and power quality. Online. Available HTTP: <http://1.usa.gov/uUvMSh>. Accessed August 2013.

EIA (2011b) *Annual energy review*. Table 8.10: average retail prices of electricity, 1960–2012. Available HTTP: <www.eia.gov/totalenergy/data. annual>.

EIA (2012a) *2012 Brief: Average wholesale electricity prices down compared to last year.* Tech. rept. U.S. Energy Information Administration. Online. Available HTTP: <http://www.eia.gov/todayinenergy/detail.cfm ?id=9510>. Accessed August 2013.

EIA (2012b: September). *Annual energy review 2011.* Tech. rept. DOE/EIA-0384(2011). U.S. Energy Information Administration. Online. Available HTTP: <http://www.eia.gov/totalenergy/data/annual/ pdf/aer.pdf>.

EIA (2012c) California energy consumption estimates. Online. Available HTTP: <http://www.eia.gov/forecasts/aeo/pdf/0383%282012%29.pdf>. Accessed August 2014.

EIA (2013a) Feed-in tariffs and similar programs: State policies as of May 2013. U.S. Energy Information Administration. Online. Available HTTP: <http://www.eia.gov/electricity/policies/provider_programs.cfm>.

EIA (2013b) Electric power sales, revenue, and energy efficiency. Form EIA-861 detailed data files. Online. Available HTTP: <http://www.eia.gov/electricity/data/eia861/>. Accessed August 2013.

EIA (2013c: July) *International energy outlook 2013.* Tech. rept. DOE/EIA-0484(2013). U.S. Energy Information Administration. Online. Available HTTP: <http://www.eia.gov/forecasts/ieo/pdf/0484% 282013%29.pdf>.

EIA (2013d: April) *Updated capital cost estimates for utility scale electricity generating plants.* Tech. rept. U.S. Energy Information Administration. Online. Available HTTP: <http://www.eia.gov/forecasts/capitalcost/pdf/ updated_capcost.pdf>.

EIA (2013e) *Lower residential energy use reduces home energy expenditures as a share of household income.* US Energy Information Administration.

EIA (2014a: September) *International energy outlook 2014.* Tech. rept. DOE/EIA-0484(2014). U.S. Energy Information Administration. Online. Available HTTP: <http://www.eia.gov/forecasts/ieo/>.

EIA (2014b: July) Japan: country analysis brief. U.S. Energy Information Administration. Online. Available HTTP: <http://www.eia.gov/countri es/country-data.cfm?fips=JA>.

EIA (2014c: April) Levelized cost and levelized avoided cost of new generation resources in the Annual Energy Outlook 2014. White Paper. U.S. Energy Information Administration.

EIA (2014d: July) *Monthly biodiesel production report with data for May 2014.* Tech. rept. U.S. Energy Information Administration. Online. Available HTTP: <www.eia.gov>.

EIA (2014e) *Monthly energy review: petroleum overview.* Tech. rept. U.S. Energy Information Administration, June 25. Online. Available HTTP: <www.eia.gov>.

EIA (2014f) *Annual energy outlook 2014.* U.S. Energy Information Administration.

Elcock, D. (2008) *Baseline and projected water demand data for energy and competing water use sectors.* Tech. rept. ANL/EVS/TM/08-8. Argonne National Laboratory for the U.S. Dept. of Energy, November.

Elgie, S. and McClay, J. (2013) 'BC's carbon tax shift is working well after four years (attention Ottawa)', *Canadian Public Policy*, 39(s2): 1–10.

Elimelech, M. (2012) 'Seawater desalination', Presentation at the 2012 NWRI Clarke Prize Conference, Newport Beach, CA, November.

Elmer-Dewitt, P. (1990) Ghost in the machine. *Time Magazine*, January 29: 58.

Enerdata (2012: January) *India energy report 2012.* Tech. rept. Enerdata.

EPA (U.S. Environmental Protection Agency) (2006) Municipal solid waste generation, recycling, and disposal in the United States: facts and figures for 2005. Online. Available HTTP: <http://www.epa.gov/solidwaste/nonhaz/municipal/pubs/mswchar05.pdf>.

EPA (U.S. Environmental Protection Agency) (2007) The role of distributed generation and combined heat and power (CHP) systems in Data Centers. Combined Heat and Power Partnership. Online. Available HTTP: <http://www.epa.gov/chp/documents/datactr_whitepaper.pdf>.

EPA (U.S. Environmental Protection Agency) (2009a: September). Clean air markets: Emissions, compliance and market data. Online. Available HTTP: <http://www.epa.gov/airmarkets/progress/historical-reports.html>. Accessed August 2013.

EPA (2009b) Clean water act, Chapter 3, Section 316b. U.S. Environmental Protection Agency. Online. Available HTTP: <http://www.nmfs.noaa.gov/pr/consultation/opinions/biop_epa_cwa316b_2014.pdf>.

EPA (U.S. Environmental Protection Agency) (2011) Solid waste and emergency response: municipal solid waste generation, recycling, and disposal in the United States: Facts and figures for 2010. Online. Available HTTP: <http://www.epa.gov/waste/nonhaz/municipal/pubs/msw_2010_factsheet.pdf.

EPA (2012) *Light-duty automotive technology, carbon dioxide emissions, and fuel economy trends: 1975 through 2011.* Tech. rept. EPA-420-R-12-001a. U.S. Environmental Protection Agency.

EPA (U.S. Environmental Protection Agency) (2014) Wastes - non-hazardous waste - municipal solid waste. Online. Available HTTP: <http://www.epa.gov/epawaste/nonhaz/municipal/>.

EPRI (Electric Power Research Institute) (2001) The cost of power disturbances to industrial and digital economy companies. Consortium for electric infrastructure to support a digital society. Online. Available HTTP: <http://www.epri.com/abstracts/Pages/ProductAbstract.aspx? ProductId=000000003002000476>.

EPRI (2005). *Security, quality, reliability, and availability: metrics definition: Progress Report 1008569*, Palo Alto, CA: EPRI. Online. Available HTTP: <http://www.epri.com/abstracts/Pages/ProductAbstract.aspx? ProductId=000000000001008568>.

EPRI (2010: May) *Accuracy of digital electricity meters*. Tech. rept. Electric Power Research Institute.

Eto, J. and La Commare, K. (2008: October) *Tracking the reliability of the U.S. electric power system: An assessment of publicly available Information reported to State Public Utility Commissions*. Tech. rept. Lawrence Berkeley National Laboratory.

Eurostat (2014) Electricity prices for domestic consumers, from 2007 onwards—bi-annual data. Online. Available HTTP: <http://appsso. eurostat.ec.europa.eu/nui/show.do?dataset=nrg_pc_204&lang=en>. Accessed February 2015.

Evans, D., Hagiu, A. and Schmalensee, R. (2008) *Invisible Engines: How Software Platforms Drive Innovation and Transform Industries*. Cambridge, MA: MIT Press.

Ewart, D.N. (1975) 'Automatic generation control—performance under normal conditions', in: *Systems Engineering for Power: Status and Prospects*, US Government Document, pp. 1–14.

ExxonMobil (2014) *The outlook for energy: a view to 2040.* Tech. rept. ExxonMobil.

Fabbri, A., Roman, San Tomas, G., Abbad, J.R. and Quezada, V.M. (2005) 'Assessment of the cost associated with wind generation prediction errors in a liberalized electricity market', *IEEE Transactions on Power Systems*, 20(3), 1440–1446.

Faeth, P., Sovacool, B., Thorkildsen, Z., Rao, A., Purcell, D., Eidsness, J., Johnson, K., Thompson, B., Imperiale, S. and Gilbert, A. (2014: July) *A clash of competing necessities: water adequacy and electric reliability in China, India, France, and Texas*. Tech. rept. CNA Research.

Falkenmark, M. (1989) 'The massive water scarcity threatening Africa—why it isn't being addressed', *Ambio*, 18(2), 112–118.

Falliere, N., Murchu, L. and Chien, E. (2011) W32: Stuxnet dossier. White Paper, Security Response. Symantec Corporation.

Faturechi, R. (2014) Snowden leaks prompt firms to focus cyber security on insider threats. *Los Angeles Times*, August 10.

Faulkner, R. (2013) 'AC vs DC powerlines and the electric grid'. The Energy Collective. Available HTTP: <http://theenergycollective.com/rogerrethinker/204396/ac-versus-dc-powerlines>. Accessed March 2015.

Fawzi, H., Tabuada, P. and Diggavi, S. (2011: September) 'Secure state-estimation for dynamical systems under active adversaries', *2011 Allerton Conference on Communication, Control, and Computing*: 337–344.

FDA (U.S. Food and Drug Administration) (2006) Guidance for industry: Use of recycled plastics in food packaging: chemistry considerations. Online. Available HTTP: <http://www.fda.gov/Food/GuidanceRegulation/GuidanceDocumentsRegulatoryInformation/IngredientsAdditivesGRASPackaging/ucm120762.htm>. Accessed August 2014.

Federal Highway Administration of Transportation, U.S. Department (2014: April). Moving 12-month total vehicle miles traveled, January 1971—April 2014. FRED Economic Data, Federal Reserve Bank of St. Louis.

FERC (Federal Energy Regulatory Commission) (2009) RTO/ISO Map. Online. Available HTTP: <http://1.usa.gov/141u94Y>. Accessed August 2013.

FERC (2010: March 18) Order on application for authorization to sell transmission services at negotiated rates, FERC Docket ER10-396, March 18.

FERC (2012) FERC letter to public service company of New Mexico re: Docket ER12-2424, November 2.

FERC (2013a) Accepting Tres Amigas, SPS and SPP interconnection agreement. FERC Docket ER13-912, April 9.

FERC (2013b) FERC Order 791, Version 5 critical infrastructure protection reliability standards, November 22.

FERC/Swidler, J.C. (1965) *Report to the President on the power failure in the Northeastern United States and the Province of Ontario on November 9–10, 1965*. Tech. rept. FERC (December).

Ferguson, E. (1962) *Kinematics of Mechanisms From the Time of Watt*. Washington, DC: Smithsonian Institute.

Field, C.B., Barros, V., Stocker, T.F. and Dahe, Q. (2012) *Managing the Risks of Extreme Events and Disasters to Advance Climate Change Adaptation*. Cambridge: Cambridge University Press.

Fisher, G. and Babbar, S. (2007) *Private financing of toll roads*. Tech. rept. RCM Discussion Paper Series 117. The World Bank.

Flynn, S. (2012: July) The New Homeland Security Imperative: The case for building greater societal and infrastructure resilience. Written Testimony to the Committee on Homeland Security and Governmental

Affairs U.S. Senate on The Future of Homeland Security: Evolving and Emerging Threats.

Foran, J. (n.d.) The day they turned The Falls on: The invention of the Universal Electric Power System. Online. Available HTTP: <http://library.buffalo.edu/libraries/projects/cases/niagara.htm>. Accessed August 2013.

Forbes, K. and St. Cyr, O.C. (2011: June) 'The smart grid and the imperative for improved modeling of electricity transmission flows: The case of Denmark and its trading partners'. Presentation at the 34rd IAEE International Conference Stockholm, Sweden.

Foster, J., Gjelde, E., Graham, W., Hermann, R., Kluepfel, H., Lawson, R., Soper Jr., G., Wood, L. and Woodard, J. (2004) *Executive report: Report of the Commission to Assess the Threat to the United States from Electromagnetic Pulse (EMP) Attack.* Tech. rept. Commission to Assess the Threat to the United States from Electromagnetic Pulse (EMP) Attack.

Foster, J., Gjelde, E., Graham, W., Hermann, R., Kluepfel, H., Lawson, R., Soper, G., Wood, L., and Woodard, J. (2008: April). *Critical national infrastructures: Report of the Commission to Assess the Threat to the United States from Electromagnetic Pulse (EMP) Attack.* Tech. rept. Commission to Assess the Threat to the United States from Electromagnetic Pulse (EMP) Attack.

Foster, R. (1986) *Innovation: The Attacker's Advantage.* New York: Summit Books.

FPC (1965) *Investigative report into November 9–10 Northeastern power outage.* Tech. rept. Federal Power Commission, Washington, DC: December 6.

FPC (1967) *Prevention of power failures Volumes I–III.* Tech. rept. U.S. Federal Power Commission, Washington, DC.

Fremdling, R. (2000) 'Transfer patterns of British technology to the continent: The case of the iron industry', *European Review of Economic History,* 4(2): 195–222.

Fresco, L. (2011) *Michael Pollan's Misguided Food Nostalgia.* ZesterDaily, February 21.

Fresco, L. (2014) 'The future of food and agriculture', *South African Journal of Science,* 110: 5–6.

Fried, I. (2009) Bill Gates offers the world a physics lesson. CNET, July.

Fried, I. (2010) Bill Gates: Better software modeling is a key. CNET, August 6.

Friederich, R., Jaron, A. and Schultz, J. (2011: July). *Closed-loop waste management: Recovering wastes—conserving resources.* Tech. rept.

German Federal Ministry for the Environment, Nature Conservation, Building and Nuclear Safety.

Funabashi, Y. and Kitazawa, K. (2012) 'Fukushima in review: A complex disaster, a disastrous response', *Bulletin of the Atomic Scientists*, March 5.

Galbraith, K. (2012) Proudly isolated Texas power grid reaches out a bit. *New York Times*, March 29.

Gallagher, P. (2013) 'Revealed: The science behind Usain Bolt power burst used to break the 100m world record'. *The Independent*, July 26, 2013.

Gauthier, D., Cavalcante, de S., Hugo L.D., Oria, M., Sornette, D. and Ott, E. (2013) 'Predictability and suppression of extreme events in a chaotic system'. *Physical Review Letters*, 111: 198701, November 4, 2013.

Geer, D. (2008) March 13 Keynote Speech. Source Boston Conference.

Geer, D. (2012) 'The future of security: Criticality, rejectionists, risk tolerance', *login*, 37(4).

Geer, D. (2014) 'Heartbleed as metaphor', *Lawfare*, April.

Geer, D. and Aucsmith, D. (2003) 'Monopoly considered harmful', IEEE Computer Society, *IEEE Security and Privacy*, 1(6): 14–19.

Geer, D., Bace, R., Gutmann, P., Metzger, P., Pfleeger, C., Quarterman, J. and Schneier, B. (2003: September). *CyberInsecurity: The cost of monopoly—how the dominance of Microsoft's products poses a risk to security*. Tech. rept. Introduced by Computer and Communications Industry Association.

Gellings, C. (2011: March) *Estimating the costs and benefits of the smart grid*. Tech. rept. Electric Power Research Institute.

Gent, S. (2014: January). CEO Seattle Steam. Interview.

Georgescu-Roegen, N. (1971) *The Entropy Law and the Economic Process*. Cambridge, MA: Harvard University Press.

Georgescu-Roegen, N. (1986) 'The entropy law and the economic process in retrospect', *Eastern Economic Journal*, 12(1).

Gerland, P., Raftery, A., Sevcikova, H., Li, N., Gu, D., Spoorenberg, T., Alkema, L., Fosdick, B., Chunn, J., Lalic, N., Bay, G., Buettner, T., Heilig, G. and Wilmoth, J. (2014) 'World population stabilization unlikely this century', *Science*, September.

German Federal Ministry for the Environment, Nature Conservation, Building and Nuclear Safety (2006) Municipal solid waste management in 2006. TASi one year on: no wastes landfilled without pretreatment in Germany since 1 June 2005. Online. Available HTTP: <http://www.bmub.bund.de/fileadmin/bmu-import/files/pdfs/allgemein/application/pdf/bericht_siedlungsabfallentsorgung_2006_engl.pdf>. Accessed August 2014.

GGBHTD (Golden Gate Bridge Highway and Transportation District) (2014) Bridge design and construction statistics. Online. Available HTTP: <http://goldengatebridge.org/research/factsGGBDesign.php>. Accessed July 2014.

Gleick, P.H. (1993a) 'Water and conflict: Resources and international security', *International Security*, 18(1): 79–112.

Gleick, P.H. (ed) (1993b) *Water in Crisis: A Guide to the World's Fresh Water Resources*. New York: Oxford University Press.

Gleick, P.H. (1994) 'Water and energy', *Annual Reviews in Energy and the Environment*, 19, 267–299.

Gleick, P.H. (1998) 'Water in crisis: paths to sustainable water use', *Ecological Applications*, 8(3): 571–579.

Gleick, P.H. (2014) *The World's Water: The Biennial Report on Freshwater Resources*. Vol. 8. London: Pacific Institute.

Goldemberg, J. (2012) *Energy: What Everyone Needs to Know*. Oxford: Oxford University Press.

Goldman Sachs Group (2013) *10-K annual report 2013*. Online. Available HTTP: <http://www.sec.gov/Archives/edgar/data/886982/00011931251 4073792/d655877d10k.htm>.

Gomber, P. and Haferkorn, M. (2013) 'High-frequency trading', *Business and Information Systems Engineering*, 5(2): 97–99.

Gorton, G. and Rouwenhourst, K.G. (2004: June) *NBER Working Paper 10595*. Cambridge, MA: National Bureau of Economic Research.

Graham, R. (2014) MASSCAN: Mass IP port scanner. Online. Available HTTP: <https://github.com/robertdavidgraham/masscan>.

Graham, R. and Johnson, P. (2014) 'Finite state machine parsing for internet protocols: Faster than you think'. Hanover, NH: Dartmouth College IEEE Security and Privacy Workshops.

Greater London Authority (2012) Carbon dioxide emissions, borough. Online. Available HTTP: <http://data.london.gov.uk/dataset/carbon-dioxide-emissions-borough>. Accessed August 2014.

Green, M., Emery, K., Hishikawa, Y. and Warta, W. (2011) 'Solar cell efficiency tables (version 37)', *Progress in Photovoltaics: Research and Applications*, 19, 84–92.

GWAC (2013) *GridWise transactive energy framework—DRAFT version*. Tech. rept. GridWise Architecture Council.

GWAC (2014) *Transactive energy (TE) principles (v1.0)*. Tech. rept. GridWise Architecture Council.

Halff, A. Sovacool, B. and Rozhon, J. eds. (2014) *Energy poverty: global and local solutions*. Oxford University Press.

Hall, C. and Klitgaard, K. (2011) *Energy and the Wealth of Nations*. New York: Springer.

Hammerstrom, D.J., Ambrosio, R., Carlon, T.A., DeSteese, J.G., Horst, G.R., Kajfasz, R., Kiesling, L., Michie, P., Pratt, R.G., Yao, M. et al. (2007) *Pacific Northwest GridWiseTM testbed demonstration projects.* Tech. rept. PNNL-17167. Pacific Northwest National Laboratory.

Hammond, P. (2011) *Connecting Washington: Forces affecting the state's transportation system.* Connecting Washington Task Force.

Hampson, A., Bourgeois, T., Dillingham, G. and Panzarella, I. (2013: March). *Combined heat and power: Enabling resilient energy infrastructure for critical facilities.* Tech. rept. ORNL/TM-2013/100. Oak Ridge National Laboratory.

Hansen, J., Sato, M., Kharecha, P. and von Schuckmann, K. (2011) 'Earth's energy imbalance and implications', *Atmospheric Chemistry and Physics*, 11, 13421–13449.

Harlan, C. (2011). Japan's Prime Minister Kan announces resignation. *Washington Post*, August 26, 2014.

Harris, E., Perlroth, N., Popper, N. and Stout, H. (2014) A sneaky path into target customers' wallets. *The New York Times*, January 17.

Hassan, F. (1997) 'The synamics of a riverine civilization: A geoarchaeological perspective on the Nile Valley, Egypt', *World Archaeology*, 29(1), 51–74.

Haub, C. (2002) How many people have ever lived on earth? *Population Today*, November/December.

Hawking, S. (2000) Interview, *San Jose Mercury News*, Morning Final Edition, January 23.

Herring, H. (2006) 'Energy efficiency—A critical view', *Energy*, 31: 10–20.

Herzog, S. (2011) 'Revisiting the Estonian cyber attacks: Digital threats and multinational responses', *Journal of Strategic Security*, 4(2), 49–60.

Hills, R. (1996) *Power from Wind: A History of Windmill Technology.* Cambridge: Cambridge University Press.

Hilt, D.W. (2005) Northeast blackout impacts and actions and the energy policy act of 2005. Paper presented at ISPE Annual Conference, August 14, 2003.

Hines, P., Apt, J., Liao, H. and Talukdar, S. (2006) The frequency of large blackouts in the United States electrical transmission system: an empirical study. Paper presented at Carnegie Mellon University Electricity Conference.

Hines, P., Apt, J. and Talukdar, S. (2009) 'Trends in the history of large blackouts in the United States', *Energy Policy*, 27, 5249–5259.

Hodge, B.-M., and Milligan, M. (2011: March) *Wind power forecasting error distributions over multiple timescales.* Tech. rept. NREL/CP-5500-50614. National Bureau of Economic Research.

Hodge, B.-M., Lew, D., Milligan, M., Holttinen, H., Sillanpää, S., Gomez-Lazaro, E., Larsen, X.G., Giebel, G., Flynn, D. and Dobschinski, J. (2012: September) *Wind power forecasting error distributions: An international comparison*. Tech. rept. NREL/CP-5500-56130. National Renewable Energy Laboratory.

Hoekstra, A. and Mekonnen, M. (2012) 'The water footprint of humanity', *Proceedings of the National Academy of Sciences*, 109(9), 3232–3237.

Hogan, W. (2012: July) Financial transmission rights, up to congestion transactions and multi-settlement systems. Online. Available HTTP: <http://www.hks.harvard.edu/fs/whogan/Hogan_UTC_071612.pdf>. Accessed February 2015.

Holder, T. (1882) *A complete record of Niagara Falls and Vicinage*. Niagara Falls, published for the author, p. 53.

Hooper, D.U., Adair, E.C., Cardinale, B.J., Byrnes, J.E., Hungate, B.A., Matulich, K.L., Gonzalez, A., Duffy, J.E., Gamfeldt, L. and O'Connor, M.I. (2012) 'A global synthesis reveals biodiversity loss as a major driver of ecosystem change', *Nature*, 486: 105–108.

Hoornweg, D. and Bhada-Tata, P. (2012: March) *What a waste: A global review of solid waste management*. Tech. rept. No. 15. The World Bank.

Hoste, G., Dvorak, M. and Jacobson, M. (2009) *Matching hourly and peak demand by combining different renewable energy sources: a case study for California in 2020*. Tech. rept. Stanford University.

Hottenroth, S., Albrecht, T., Raer, D., Cantner, J. and Rommel, W. (2004: July) *Transfer of national experiences with waste prevention and minimisation to a regional or international level*. Tech. rept. Bavarian Institute for Applied Environmental Research and Technology.

Houser, T. and Mohan, S. (2011: September). Energy poverty American style. Online. Available HTTP: <http://rhg.com/notes/energy-poverty-american-style>. Accessed February 2015.

Houston-Galveston Area Council (2007) *Bridging our Communities 2035: The 2030 Houston-Galveston Regional Transportation Plan*. Tech. rept.

Huang, Y.-F., Werner, S., Huang, J., Kashyap, N. and Gupta, V. (2012) 'State estimation in electric power grids', *IEEE Signal Processing Magazine*, September, 33–43.

Hughes, J.A. (2013) Utilities right to seek most bang for their solar buck. Online. Available HTTP: <http://www.azcentral.com/opinions/articles/20130531solar-powerguidelines-should-fair-all.html>. Accessed June 2013.

Hughes, T.P. (1988). *Networks of Power: Electrification in Western Society, 1880–1930*. Baltimore, MD: The Johns Hopkins University Press.

Hun, G., Kandel, R. and Mecherikunnel, A. (1986) 'A history of presatellite investigations of the Earth's radiation budget', *Reviews of Geophysics*, 24(2), 351–356.

Hurricane Sandy Rebuilding Task Force 2013 (August). *Hurricane Sandy rebuilding strategy: Stronger communities, a resilient region*. Report. Accessed August 2013.

Hutchins, E., Clopper, M. and Amin, R. (2011) 'Intelligence-driven computer network defense informed by analysis of adversary campaigns and intrusion kill chains', *Leading Issues in Information Warfare & Security Research*, 1.

IEA (International Energy Agency) (1999). *World energy outlook*. Tech. rept. International Energy Agency.

IEA (International Energy Agency) (2008) Combined heat and power: Evaluating the benefits of greater global investment. Online. Available HTTP: <http://www.iea.org/publications/freepublications/publication/chp_report.pdf>.

IEA (International Energy Agency) (2011a) Co-generation and renewables: Solutions for a low-carbon energy future. Online. Available HTTP: <http://www.iea.org/publications/freepublications/publication/CoGeneration_RenewablesSolutionsforaLowCarbonEnergyFuture.pdf>.

IEA (International Energy Agency) (2011b) *World energy outlook 2011*. Tech. rept. Online. Available HTTP: <http://www.worldenergyoutlook.org/publications/weo-2011/>.

IEA (International Energy Agency) (2012) *World energy outlook 2012*. Tech. rept. Online. Available HTTP: <http://www.worldenergyoutlook.org/publications/weo-2012/>.

IEA (2013) *Key world energy statistics*. Tech. rept. International Energy Agency.

IEEE (2003: June) IEEE 1547 standard for interconnecting distributed resources with electric power systems. IEEE Standards Board.

Ifrah, G. (2001) *The Universal History of Computing: From the Abacus to the Quantum Computer*. London: John Wiley and Sons.

Igure, V., Laughter, S. and Williams, R. (2006) 'Security Issues in SCADA Networks', *Computers and Security*, 25(7), 498–506.

IISD (International Institute for Sustainable Development) (2012) *A citizen's guide to energy subsidies in India*. Tech. rept. Online. Available HTTP: <http://www.iisd.org/gsi/sites/default/files/ffs_india_czguide.pdf>.

Ilic, M. (2007) 'From hierarchical to open access electric power systems', in: S. Haykin and E. Moulines (eds) *IEEE Special Issue on Large-Scale Dynamical Systems*, Vol. 95.

Ilic, M. (2010a) *Securing Electricity Supply in the Cyber Age: Exploring the Risks of Information in Tomorrow's Electricity Infrastructure*, Chapter 2. London: Springer, pp. 13–32.

Ilic, M. (2010b) Unit commitment for sustainable integration of large scale wind power and responsive demand. Paper presented at FERC Technical Conference on Unit Commitment Software, Docket AD10-12.

Ilic, M. (2012: June) A method for assessing relative value of central generation and distributed energy resources (DERs) in distribution feeders. Paper presented at Staff Technical Conference on Increasing Real-Time and Day-Ahead Market Efficiency through Improved Software (Docket No. AD10?12?003).

Ilic, M. (2013: September) 'Possible framework for integrating new technologies at value in electric energy systems', in: *Future Green Technologies: Challenges and Opportunities*, Lehigh University.

Ilic, M., Joo, J.Y., Xie, L., Prica, M. and Rotering, N. (2011). 'A decision-making framework and simulator for sustainable electric energy systems', *IEEE Transactions on Sustainable Energy*, Vol. 2.

IMF (International Monetary Fund) (2013) Energy subsidy reform: Lessons and implications. Online. Available HTTP: <http://www.imf.org/external/np/pp/eng/2013/012813.pdf>.

Indexmundi (2014) DAP fertilizer monthly price: US dollars per metric ton. Online. Available HTTP: <http://www.indexmundi.com/commodities/?commodity=dap-fertilizer>. Accessed August 2013.

Insull, S. (1914) Centralization of energy supply. In: Central-Station Electric Service. Online. Available HTTP: <http://archive.org/stream/centralstationel00insurich/centralstationel00insurich_djvu.txt>.

International District Energy Association (2009) U.S. district energy systems map. Online. Available HTTP: <http://www.districtenergy.org/u-s-district-energy-systems-map>. Accessed August 2014.

International Institute for Sustainable Development; Global Subsidies Initiative (2012) A citizen's guide to energy subsidies in India. Online. Available HTTP: <http://www.iisd.org/gsi/sites/default/files/ffs_india_czguide.pdf>.

IRENA (2014: September) *A path to prosperity: Renewable energy for islands*. Tech. rept. International Renewable Energy Agency.

IRG (International Risk Governance Council) (2010) Severe space weather: solar storms. Online. Available HTTP: <http://irgc.org/wp-content/uploads/2012/04/Emerging_risks_solar_storms.pdf>.

Ives, A.R. and Carpenter, S.R. (2007) 'Stability and diversity of ecosystems', *Science*, 317, 58–62.

Jaffe, M. (2014) A battle over energy efficiency waged at Colorado PUC. *The Denver Post*, April 25.

Jaffee, D. and Russell, T. (1996: May) Catastrophe insurance, capital markets, and uninsurable risks. Working Paper 96–12. Wharton Financial Institutions Center.

Japan Cabinet Office (2011: June) *Report of Japanese Government to the IAEA Ministerial Conference on Nuclear Safety—The accident at TEPCO's Fukushima nuclear power stations.*

Jappenman, J. (2010: January). *Low-frequency protection concepts for the electric power grid: Geomagnetically Induced Current (GIC) and E3 HEMP mitigation.* Tech. rept. Meta-R-322. Metatech Corporation for Oak Ridge National Laboratory, Goleta, CA.

Jaron, A. and Flaschentreher, N. (2012: December) *Waste management in Germany 2013: Facts, data, diagrams.* Tech. rept. German Federal Ministry for the Environment, Nature Conservation, Building and Nuclear Safety.

Jaron, A., Rothkirch, U.G., and Schultz, J. (2006: December) *Waste Management in Germany: A driving force for jobs and innovation.* Tech. rept. German Federal Ministry for the Environment, Nature Conservation, Building and Nuclear Safety.

Jevons, W. (1865) *The Coal Question.* Second edition. London: Macmillan and Co.

Jha, P. and Peth, R. (2014) 'Global effects of smoking, of quitting, and of taxing tobacco'. *New England Journal of Medicine*, January.

John Fritz Medal Board (1926) Presentation of the John Fritz Gold Medal to Edward Dean Adams. Online. Available HTTP: <http://catalog.hathitrust.org/Record/002011472>. Accessed August 2013.

Johnson, B. and Barz, G. (2005) *Energy Modeling: Advances in the Management of Uncertainty.* Second edition. Chapter 1, London: Riskbooks, pp. 9–57.

Johnson, E. and Paulson, L. (2011: September) *Public port districts: connecting us to the future.* Washington Public Ports Association. Available HTTP: www.digitalarchives.wa.gov/governorgregoire/priorities/transportation/connect/20110909/03_ports.pdf. Accessed March 2015.

Johnson, L. (2012) 'Catastrophe bonds and financial risk: Securing capital and rule through contingency', *Geoforum*, 45(July): 30–40.

Johnson, K. (2014) 'Exploding world, cheap oil', *Foreign Policy*, September 11, 2014.

Jolly, D. (2010) Energy access seen as vital to abolishing worst poverty. *The New York Times*, September 21.

Jones, W. (2008) 'Lessons from the $1 billion Intel trade-secret theft', *IEEE Spectrum*, November 1.

Jonnes, J. (2003) *Empires of Light: Edison, Tesla, Westinghouse and the Race to Electrify the World*. New York: Random House Digital Inc.

Joo, J.-Y. and Ilic, M. (2011) 'Multi-temporal risk minimization of adaptive load management in electricity spot markets', *IEEE Power and Energy Systems Innovative Smart Grid Technologies*, December.

Joo, J.-Y. and Ilic, M. (2012) Distributed multi-temporal risk management approach to designing dynamic pricing. IEEE Power and Energy Society General Meeting, July.

Joo, J.-Y. and Ilic, M. (2010) A Multi-Layered Adaptive Load Management (ALM) system. Paper presented at the IEEE PES Transmission and Distribution Conference, April.

Joosten, I. (2004) *Technology of Early Historical Iron Production in the Netherlands*. Ph.D. Thesis, Vrije Universiteit, Amsterdam.

Jordan, G. and Venkataraman, S. (2012 June) Analysis of Cycling Costs in Western Wind and Solar Integration Study. Subcontract Report NREL/SR-5500-54864. National Renewable Energy Laboratory.

Joskow, P. (2012) 'Creating a smarter US electricity grid'. *Journal of Economic Perspectives*, 26, 29–48.

Joskow, P. and Tirole, J. (2000) 'Transmission rights and market power on electric power networks', *RAND Journal of Economics*, 31(3), 450–487.

JP Morgan Chase & Co. (2013) *10-K annual report 2013*. Online. Available HTTP: <http://www.sec.gov/Archives/edgar/data/19617/000001961714 000289/corp10k2013.htm>.

Juris, A. (1998) *Development of natural gas and pipeline capacity markets in the United States*. Tech. rept. World Bank.

Kaminski, V. (2012) *Energy Markets*. London: Risk Books.

Kanter, R.M. (1991) Championing change: An interview with Bell Atlantic's CEO Raymond Smith. *Harvard Business Review*, January.

Kapp, G. (1886) Electric Transmission of Energy and its Transformation, Subdivision, and Distribution: A Practical Handbook. Chiswick Press.

Kappenman, J. (2010: January) *Geomagnetic storms and their impacts on the U.S. power grid*. Tech. rept. Meta-R-319. Metatech Corporation for Oak Ridge National Laboratory, Goleta, CA.

Kardashev, N. (1964) 'Transmission of Information by extraterrestrial civilizations', *Soviet Astronomy*, 8(217).

Karkanas, P., Shahack-Gross, R., Ayalon, A., Bar-Matthews, M., Barkai, R., Frumkin, A., Gopher, A. and Stiner, M. (2007) 'Evidence for habitual use of fire at the end of the Lower Paleolithic', *Journal of Human Evolution*, 53(April): 197–212.

Karnouskos, S. (2011: November) Stuxnet worm impact on industrial cyber-physical system security. Paper presented at 37th Annual Conference of the IEEE Industrial Electronics Society IECON.

Kaspersky, E. (2011) The man who found Stuxnet—Sergey Ulasen in the spotlight. Kaspersky Labs Blog, November 2.

Kataoka, R. (2013) 'Probability of occurrence of extreme magnetic storms', *Space Weather*, 11(March), 1–5.

Kaufmann, D. (1960) *Sodium Chloride: The Production and Properties of Salt and Brine*. American Chemical Society Monograph Series. Reinhold Publishing, New York.

Kaufmann, D. and Penciakova, V. (2011) Japan's triple disaster: Governance and the earthquake, tsunami and nuclear crises. *Brookings Opinion*, March 16.

Kenny, J.F., Barber, N.L., Hutson, S.S., Linsey, K.S., Lovelace, J.K. and Maupin, M.A. (2009) 'Estimated use of water in the United States in 2005. Circular 1344'. U.S. Geological Survey.

Keogh, M. and Cody, C. (2012: June) Cybersecurity for state regulators with sample questions for regulators to ask utilities. Online: energy.gov.

Keohane, G. (2014) Preparing for disaster by betting against it. *The New York Times*, February 12.

Kerr, P., Rollins, J. and Theohary, C. (2010: December) *The Stuxnet computer worm: harbinger of an emerging warfare capability*. Tech. rept. Congressional Research Service.

Kersting, W. (2007) *Distribution System Modeling and Analysis*. Second edition. Florida: CRC Press.

Kesmodel, D. (2011) Boeing examines supply chain for weak links. *Wall Street Journal*, December 30.

Kessler, R. (2014) Interview: Tres Amigas' Phil Harris. *REcharge*, August 27.

Keys, D. (2010) The first industrial revolution. *BBC History Magazine*, March.

Kiehl, J.T. and Trenberth, K.E. (1997) 'Earth's annual global mean energy budget', *Bulletin of the American Meteorological Society*, 78(2), 197–208.

King, C. (2001: June). *The economics of real-time and time-of-use pricing for residential consumers*. Tech. rept. American Energy Institute.

King, T. (2014: August) Visualizing energy resources dynamically on earth—VERDE: A real-time national visualization resource. PDF

Brochure. Online. Available HTTP: <http://web.ornl.gov/sci/electricdeli very/pdfs/VERDEFactSheet.pdf>.

Kinsey, G. (2014) Tracking the duck. Solar Curator, May 6. Online. Available HTTP: <http://www.solarcurator.com/2014/05/06/tracking-duck-curve-solar-pvtracking-systems/>.

Kirby, B., Hirst, E. and Vancoevering, J. (1995: May). *Identification and defintion of unbundled electric generation and transmission services.* Tech. rept. ORNL/CON-415. Oak Ridge National Laboratory, Oak Ridge, TN.

Knapp, E. (2011). Industrial Network Security: Securing Critical Infrastructure Networks for Smart Grid, SCADA, and Other Industrial Control Systems. Syngress.

Knight, M., Widergren, S., Mater, J. and Montgomery, A. (2013) 'Maturity Model for Advancing Smart Grid Interoperability', Innovative Smart Grid Technologies, *IEEE PES*, February, 1–6.

Koonce, A.M., Apostolakis, G.E. and Cook, B.K. (2006: September) *Bulk power grid risk analysis: ranking infrastructure elements according to their risk significance.* Tech. rept. ESD-WP-2006-19. MIT, Boston, MA.

Kotze, A.A. (2011) *Foreign exchange derivatives: Effective theoretical and practical techniques for trading, hedging and managing FX derivatives.* Tech. rept. Financial Chaos Theory Pty. Ltd., March.

Kravets, D. (2009) Feds: Hacker disabled offshore oil platforms' lead-detection system. *WIRED Magazine*, March 19.

Kumar, N., Besuner, P., Lefton, S., Agan, D. and Hilleman, D. (2012: July). Power Plant Cycling Costs. Subcontract Report NREL/SR-5500-55433. National Renewable Energy Laboratory.

Kümmel, R. (2011) *The Second Law of Economics.* The Frontiers Collection. New York: Springer.

Kushner, D. (2013) 'The real story of Stuxnet: How Kaspersky Lab tracked down the malware that stymied Iran's nuclear-fuel enrichment program', *IEEE Spectrum*, February 26.

Kwoka, John Jr. (1996) *Power Structure: Ownership, Integration, and Competition in the U.S. Electricity Industry.* New York: Kluwer Academic Publishers.

Langer, R. (2013: November). *To kill a centrifuge: A technical analysis of what Stuxnet's creators tried to achieve.* Tech. rept. The Langer Group.

Langton, C.G. (1990) 'Computation at the edge of chaos: Phase transitions and emergent computation', *PhysicaD*, 42: 12–37.

Lawrence Livermore National Labs (n.d.) Energy flow. Online. Available HTTP: <https://flowcharts.llnl.gov>. Accessed August 2014.

Lee, F.N. (1989: November) 'A new approach for determining thermal unit priority commitment order', *IEEE Conference Proceedings on Systems, Man, and Cybernetics*, Vol. 3. pp. 1166–1172.

Lee, K.K. (1992) *Lectures on Dynamical Systems, Structural Stability and the Applications.* Singapore: World Scientific.

Lemay, A. (2013) *Defending the SCADA Network Controlling the Electrical Grid from Advanced Persistent Threats.* Ph.D. thesis, Ecole Polytechnique de Montreal, Montreal, QC.

Leverett, E. (2011: June) *Quantitatively Assessing and Visualizing Industrial System Attack Surfaces.* M.Phil. thesis, University of Cambridge.

Lew, D., Brinkman, G., Kumar, N., Besuner, P., Agan, D. and Lefton, S. (2012: July) 'Impacts of wind and solar on fossil-fuel generators', Presented at IEEE Power and Energy Society General Meeting. National Renewable Energy Laboratory and Intertek APTECH, San Diego, CA.

Lewis, J. (2010: March). *The electrical grid as a target for cyber attack.* Tech. rept. Center for Strategic and International Studies.

Li, Z., Moreau, L. and Arking, A. (1997) 'On solar energy disposition: A perspective from surface observation, satellite estimation and GCM simulation', *Bulletin of the American Meteorological Society*, 78(1), 53–70.

Lidberg, W. (2012: December) *Moshyttan: Sweden's Oldest Known Blast Furnace?* M.Phil. thesis, UMEA Universitet.

Lieber, D. (2013) In the world of Texas electricity, free is not always free. *Dallas News*, October 12.

Linden, D. and Reddy, T.B. (eds) (2002) *Handbook of Batteries.* London: McGraw-Hill.

Lindner, F. (2006) 'Software security is software reliability', *Communications of the ACM*, 49(6), 56–61.

Lipscy, P., Kushida, K. and Incerti, T. (2013) 'The Fukushima disaster and Japan's nuclear plant vulnerability in comparative perspective', *Environmental Science and Technology, American Chemical Society*, 47(May), 6082–6088.

Little, L. (2014) TXU energy gives customers even more hours to use free electricity. TXU Energy Press Release, May 7.

Livingston, P., Cole, P., Elliot, D. and Bartlett, R. (2014: March). Building energy codes program: National Benefits Assessment. Tech. rept. PNNL-22610. Pacific Northwest National Laboratory.

LLNL (Lawrence Livermore National Labs) (2013) Energy flow. Online. Available HTTP: <https://flowcharts.llnl.gov/>. Accessed August 2014.

Long, T. (2008) Edison fries an elephant to prove his point. *WIRED Magazine*, January 4.

Loomis, E. (1859) 'The great auroral exhibition of August 28 to September, 1859'. *American Journal of Science*, 28(November), 385–408.

Loreau, M. (2010) *From Populations to Ecosystems: Theoretical Foundations for a New Ecological Synthesis*. Princeton, NJ: Princeton University Press.

Lotka, A.J. (1956) *Elements of Mathematical Biology*. London: Dover Publications.

Love, J. (2012) 'Credible occurrence probabilities for extreme geophysical events: Earth-quakes, volcanic eruptions, magnetic storms', *Geophysical Research Letters*, 39(May).

Low Income Home Energy Assistance Program (LIHEAP) (2014) LIHEAP Service Eligibility FY 2014. Online. Available HTTP: <http://www.acf.hhs.gov/programs/ocs/resource/liheap-eligibility-criteria>. Accessed May 2015.

Lunde, S. (2012) 'Energy storage for the grid: Watchful waiting or the perfect storm?' Companion Paper from the MIT Enterprise Forum Northwest.

Luo, G., Xie, L., Zhou, Q. and Angelidaki, I. (2011) 'Enhancement of bioenergy production from organic wastes by two-stage anaerobic hydrogen and methane production process', *Bioresource Technology*, Vol. 102.

Macaulay, T. and Singer, B. (2011) *Cybersecurity for Industrial Control Systems: SCADA, DCS, PLC, HMI, and SIS*. Florida: CRC Press.

MacDougall, A.S., McCann, K.S., Gellner, G., and Turkington, R. (2013) 'Diversity loss with persistent human disturbance increases vulnerability to ecosystem collapse', *Nature*, 494(December): 86–89.

Macknick, J., Newmark, R., Heath, G. and Hallett, K.C. (2011) *A review of operational water consumption and withdrawal factors for electricity generating technologies*. Tech. rept. NREL/TP-6A20-50900. National Renewable Energy Laboratory.

Malas, N. (2014) Mosul Dam's takeover by ISIS raises risk of flooding. *Wall Street Journal*, August 11.

Maloney, S. (1991) *To Secure Command of the Sea: NATO Command Organization and Naval Planning for the Cold War at Sea, 1945–54*. M.Phil. thesis, University of New Brunswick.

Management Information Services (2011: October) 60 years of energy incentives: Analysis of federal expenditures for energy development. Report.

Margolies, D. and von Rosenvinge, T. (1998: July). *Advanced Composition Explorer (ACE): lessons learned and final report*. Tech. rept. NASA/Goddard Space Flight Center, Greenbelt, MD.

Marketline (2014) Global tobacco 2014. Online. Available HTTP: <http://store.marketline.com/Product/global_tobacco?productid=MLIP 1376-0012>. Accessed August 2014.

Markey, E. and Waxman, H. (2013) *Electric grid vulnerability: Industry responses reveal security gaps.* Tech. rept. Congressional offices of Edward J. Markey (D-MA) and Henry A. Waxman (D-CA).

Marsh, G. (2009) 'Boeing's 787: Trials, tribulations, and restoring the dream', *Reinforced Plastics*, November/December.

Mateski, M., Trevino, C., Veitch, C., Michalski, J., Harris, J., Maruoka, S., and Frye, J. (2012) (March). *Cyber threat metrics.* Tech. rept. SAND2012-2427. Sandia National Laboratories, Albuquerque, NM.

Matrosov, A., Rodionov, E., Harley, D. and Malcho, J. (2010: September). *Stuxnet under the microscope.* Tech. rept. ESET LLC.

Matsuyama, K. (2013) Shutdown of Japan's last nuclear reactor raises power concerns. Bloomberg, September 15.

Maw, W.H. and Dredge, J., eds. (1890a) 'The utilization of Niagara no. II', *Engineering: An Illustrated Weekly Journal*, 50: 449.

Maw, W.H. and Dredge, J., eds. (1890b) 'The utilization of Niagara no. IV', *Engineering: An Illustrated Weekly Journal*, 50: 505.

Maw, W.H. and Dredge, J., eds. (1890c) 'The utilization of Niagara no. V', *Engineering: An Illustrated Weekly Journal*, 50: 535.

McCaffrey, S.C. (1993) 'Water politics and international law', in: *Water in Crisis: A Guide to the World's Fresh Water Resources.* New York: Oxford University Press, pp. 92–104.

McDonald, G., Murchu, L., Doherty, S. and Chien, E. (2013) *Stuxnet 0.5: The missing link.* Tech. rept. Symantec Corporation.

McEntee, K. (2005) National compost prices. Online. Available HTTP: <http://www.recycle.cc/compostprices.pdf>. Accessed October 2010.

McGee, D. and Nussbaum, P. (2006: June) *Cogeneration in Louisiana: an updated (2005) tabulation of independent power producer (IPP) and cogeneration facilities.* Report.

McGurk, S. (2011) *Cybersecurity: Assessing the immediate threat to the United States.* Tech. rept. 112-55. Subcommittee on National Security, Homeland Defense and Foreign Operations, Committee on Oversight and Government Reform, U.S. House of Representatives, 112th Congress, May 25.

Melford, C. (2012) *Washington state employment estimates and forecast for transportation investment planning.* Community Attributes.

Mendes, V.M.F., Mariano, S.J.P.S., Catalao, J.P.S. and Ferreira, L.A.F.M. (2004) 'Emissions contraints on short-term schedule of thermal units', 39th International Universities Power Engineering Conference, Vol. 3: 1068–1072.

Mercatus Energy Advisors (2013) An overview of energy basis, basis risk and basis hedging. Mercatus Energy Blog. Online. Available HTTP: <http://www.mercatusenergy.com/blog/bid/38368/An-Overview-of-Energy-Basis-Basis-Risk-and-Basis-Hedging>.

Meridian Corp. (1989) *Energy system emissions and materials requirements*. Tech. rept. Washington, DC: U.S. Department of Energy.

METI (Ministry of Economy, Trade and Industry) (2014: April) *4th Basic Energy Plan*. Tech. rept. Ministry of Economy, Trade and Industry, Government of Japan.

Midwest ISO and PJM (2007) *Investigation of loop flows across combined midwest ISO and PJM footprint*. Tech. rept.

Miller, J.H. and Page, S.E. (2007) *Complex Adaptive Systems: An Introduction to Computational Models of Social Life*. Princeton, NJ: Princeton University Press.

Milligan, M. (2011: January) Costs of integration for wind and solar energy: Large-scale studies and implications. Presentation at the MIT Wind Integration Workshop.

MIT (Massachusetts Institute of Technology) (2003) The future of nuclear power: An interdisciplinary MIT study. Massachusetts Institute of Technology, July. Online. Available HTTP: <http://web.mit.edu/nuclearpower/>.

Mo, Y., Kim, T.H., Brancik, K. and Dickinson, D. (2012) 'Cyber-physical security of a smart grid infrastructure', *Proceedings of the IEEE*, 100(January), 195–209.

Mohanty, B. (2001) 'Standby power losses in household electrical appliances and office equipment'. Presented at: Regional Symposium on Energy Efficiency Standards and Labelling, April.

Moritz, L.A. (1956) 'Vitruvius' water mill', *The Classical Review*, 6(3/4): 193–196.

Morris, C. (2005) *The Tycoons: How Andrew Carnegie, John D. Rockefeller, Jay Gould, and J.P. Morgan Invented the American Supereconomy*. New York: Owl Books.

Morrow, D. (2011: November) *Impacts of severe space weather on the electric grid*. Tech. rept. JSR-11-320. McLean, VA: The MITRE Corporation.

Moslehi, K., Kumar, A.B., and Hirsch, P. (2006) 'Valuating infrastructure for a self-healing grid', presented at Carnegie Mellon University Electricity Conference.

Moss, D. and Kintgen, E. (2010) *The Dojima rice market and the origins of futures trading*. Tech. rept. HBS Case No 9-709-044. Harvard Business School, November 10.

Mosse, J. (1967) 'The Albion Mills 1784–1791', *Transactions of the Newcomen Society*, 40(1), 47–60.

Müller-Sievers, H. (2012) *The Cylinder: Kinematics of the Nineteenth Century*. Oakland, CA: University of California Press.

Mulrine, A. (2014) Is US vulnerable to EMP attack? A doomsday warning, and its skeptics. *The Christian Science Monitor*, August 1.

Mumford, L. (1934) *Technics and Civilization*. London: Harcourt, Brace and Company.

Munson, R. (2005) *From Edison to Enron: The Business of Power and What it Means for the Future of Electricity*. Santa Barbara, CA: Praeger.

Myers, J. (2014) Minnesota Power offers off-peak rates in test program. *Twin Cities, Pioneer Press*, August 21.

NAS (2008) *Severe space weather events - understanding societal and economic impacts*. Tech. rept. National Research Council.

National Center for Charitable Statistics (2013) NCCS all registered nonprofits Table Wizard. Online. Available HTTP: <http://urbn.is/WzG2NY>. Accessed August 2013.

National Grid (2014) National Grid up for the World Cup. Press release, May 30.

National Opinion Research Center (1966) *Public response to the northeastern power blackout*. Tech. rept. National Opinion Research Center, Commissioned by the U.S. Office of Civil Defense.

Natural Resource Defense Council (2014) 'Scaling up energy efficiency across the data center industry: evaluating key drivers and barriers'. *Data Center Efficiency Assessment*, August 2014.

Nazari, M. (2012) *Making the Most out of Distributed Generation Without Endangering Normal Operation: A Model-Based Technical-Policy Approach*. Ph.D. thesis, Carnegie Mellon University.

NCS (2004) Supervisory Control and Data Acquisition (SCADA) Systems. Technical Information Bulletin 04-1, October. National Communications System, Arlington, VA.

NEPCC (Northeast Power Coordinating Council) (1966: September). *A study of the effect of possible but improbable events on the Northeast Power Interconnected Systems under 1968 conditions*. Tech. rept. New York: Northeast Power Coordinating Council.

NERC (2004: July) *Technical analysis of the August 14, 2003 blackout: What happened, why, and what did we learn*. Tech. rept. Report to the NERC Board of Trustees by the NERC Steering Group.

NERC (2012a: May) *Cyber attack task force (final report)*. Tech. rept. Cyber Attack Task Force, North American Electric Reliability Corporation.

NERC (2012b) *History of NERC*. Tech. rept. North American Electric Reliability Council.

Neumann, P.G. (1990) Cause of AT&T network failure. *Telephony*, 22 January: 11.

Newsweek (2008: May) Manhole covers stolen for scrap metal. *Newsweek*.

New York Times (1921a) Cables damaged by sunspot aurora, *New York Times*, May 17.

New York Times (1921b) Sunspot aurora paralyzes wires, *New York Times*, May 15.

Ngo, C. and Natowitz, J. (2009) *Our Energy Future: Resources, Alternatives and the Environment*. John Wiley and Sons.

Nipon P., Chananchida N. and Sirisukpoka, U. (2014) 'Biological hydrogen and methane production from food waste in two-stage CSTR', *Energy Procedia*, Vol. 50.

NIST (2010a: August) *Guidelines for smart grid cyber security: Vol. 1, Smart grid cyber security strategy, architecture, and high-level requirements*. Tech. rept. NIST IR-7628. National Institute of Standards and Technology.

NIST (2010b: January). *NIST framework and roadmap for smart grid interoperability standards, release 1.0*. NIST Special Publication 1108. National Institute of Standards and Technology.

NIST (2014: July) *High integrity software system assurance*. Tech. rept. National Institute of Standards and Technology.

Nixon, R. (1974) Presidential radio address about the national energy crisis, January 17. The American Presidency Project. Online. Available HTTP: <http://www.presidency.ucsb.edu/ws/?pid=4208>.

NMAP (2014) Online. Available HTTP: <http://nmap.orgNmap2014>. Accessed September 2014.

Nott, J. (2006) *Extreme Events: A Physical Reconstruction and Risk Assessment*. Cambridge University Press.

NREL (National Renewable Energy Laboratory) (2012) Worth the Trouble: New Market Tax Credits. Online. Available HTTP: <https://financere.nrel.gov/finance/content/new-market-tax-credits-rene wable-energy>.

NREL (National Renewable Energy Laboratory) (2013) *Best research-cell efficiencies*. Tech. rept. National Renewable Energy Laboratory.

NWC (The Northwest Power and Conservation Council (2011) The effects of an increasing surplus of energy generating capability in the Pacific Northwest NW Council. Online. Available HTTP: <https://www.nwcouncil.org/media/29943/2011_01.pdf>.

Nyberg, M. (2012) *Thermal efficiency of gas-fired generation in California: 2012 update.* Staff Paper CEC-200-2013-002. Electricity Analysis Office, California Energy Commission.

Odenwald, S. and Green, J. (2008) 'Bracing for a solar superstorm', *Scientific American*, August.

OECD (Organization for Economic Co-operation and Development) (2011) *Policy roundtables: Competition in ports and port services.* Tech. rept. of the Directorate for Financial and Enterprise Affairs. Online. Available HTTP: <http://www.oecd.org/regreform/sectors/48837794.pdf>.

Ostrower, J. (2014) Boeing assessing damage to aircraft components after rail accident. *Wall Street Journal*, July 6.

Owen, D. (2010) The efficiency dilemma. *The New Yorker*, December 20.

Packard, N.H. (1988) 'Adaptation toward the edge of chaos', in: J.A.S. Kelso, A.J. Mandell and M.F. Shlesinger (eds), *Dynamic Patterns in Complex Systems*. Singapore: World Scientific, pp. 293–301.

Pagels, H.R. (1988) *The Dreams of Reason: The Computer and The Rise of the Sciences of Complexity*. New York: Simon and Schuster.

Palensky, P. and Dietrich, D. (2011: August) 'Demand side management: demand response, intelligent energy systems, and smart loads', *IEEE Transactions on Industrial Informatics*, 7: 381–388.

Parfomak, P. (2014) *Physical security of the U.S. power grid: High voltage transformer substations.* Tech. rept. Congressional Research Service, June 17.

Pargal, S. and Banerjee, S. (2014) *More power to India: The challenge of electricity distribution.* Tech. rept. 88906. World Bank.

Parsons, R.H. (1939) *The Early Days of the Power Station Industry*. Cambridge: Cambridge University Press.

Perry, W.J., Schlesinger, J.R., Cartland, H., Ikle, F., Foster, J., Payne, K., Glenn, J., Tarter, B., Halperin, M., Williams, E. et al. (2009) *America's strategic posture: The final report of the Congressional Commission on the Strategic Posture of the United States.* Tech. rept. DTIC Document.

Petersen, M., Glenn, F. Cada, M., Stale, M. and Eddlemon, G. (2003). Regulatory approaches for addressing dissolved oxygen concerns at hydropower facilities. *US DOE Energy Efficiency and Renewable Energy 2003*.

Phetteplace, G. (1995: September) *Efficiency of steam and hot water heat distribution systems.* Tech. rept. CRREL Report 95-18. U.S. Army Corps of Engineers.

Piggott, G. (2007: February). *Commuting in London.* Tech. rept. DMAG Briefing 2007-3. Greater London Authority.

Platts (2013) *Megawatt daily*. Tech. rept. McGraw Hill Financial. Online. Available HTTP: <www.platts.com>. Accessed August 29.

Polimeni, J., Mayumi, K. and Giampietro, M. (2007) *Jevons' Paradox: The Myth of Resource Efficiency Improvements*. London: Earthscan.

Pollan, M. (2007) *The Omnivore's Dilemma*. London: Penguin.

Poterba, J. (1989) 'Lifetime incidence and the distributional burden of excise taxes', *American Economic Review*, May 29.

Poumadere, M., Mays, C., Le Mer, S. and Blong, R. (2005) 'The 2003 heat wave in france: dangerous climate change here and now', *Risk Analysis*, 25(6): 1483–1494.

Preservation Green Lab, National Trust for Historic Preservation (2011) *The role of district energy in greening existing neighborhoods: A primer for policy makers and local government officials*. Report, September.

PTI (2013) 6 gas-based power plants sitting idle in Andhra Pradesh. *The Economic Times, India Times*, March 13.

PWC (2012) *The Indian coal sector: Challenges and future outlook*. Tech. rept. The Indian Chamber of Commerce.

Randazzo, S. and Fitzgerald, P. (2014) Indiana toll road operator files for bankruptcy. *The Wall Street Journal*, September.

Rawsthorn, A. (2009) The demise of 'form follows function'. *The New York Times*, 1 June.

Rayner, S. (1995) 'A conceptual map of human values for climate change decision making', in: A. Katama (ed.) *Equity and social considerations related to climate change*. ICIPE Science Press, Nairobi, Kenya.

Red Owl Analytics, LLC (2013: July) Red Owl Analytics announces strategic partnership and technology development agreement with In-Q-Tel. Press release.

Revkin, A. (2014) Indian Point's tritium problem and the N.R.C.'s regulatory problem. *The New York Times*, June 12.

Reynolds, T. (2002) *Stronger Than a Hundred Men: A History of the Vertical Water Wheel*. Baltimore, MD: Johns Hopkins University Press.

Rhodes, P.S. (1962) *Guide to the Ballycopeland Windmill*. Tech. rept. H.M.S.O., Belfast.

Rietkerk, M., Dekker, S.C., de Ruiter, P.C., and van de Koppel, J. 2004. 'Self-organized patchiness and catastrophic shifts in ecosystems', *Science*, 305, 1926–1929.

Rittel, H. and Webber, M. (1973) 'Dilemmas in a general theory of planning', *Policy Sciences* 4: 155–169.

Roberts, N. (2000) 'Wicked problems and network approaches to resolution', *The International Public Management Review*, 1(1).

Robinson, W.D. (ed.) (1986) *The Solid Waste Handbook: A Practical Guide*. New York: Wiley-Interscience.

RMI (Rocky Mountain Institute) (2013a) Simple solar balance of systems. Online. Available HTTP: <http://www.rmi.org/simple>. Accessed August 2013.

Rogers, H. (2014) Don't drink the water: West Virginia after the chemical spill, *Rolling Stone*, March 12.

Ronneau, C. (2004) *Energie, Pollution de L'Air et Developpement Durable*. Louvain-la-Neuve, Belgium: Presses Universitaires de Louvain.

Rosenthal, A.M. (ed.) (1965) *The Night the Lights Went Out*. New York: The New York Times Company.

Rotem-Gal-Oz, A. (2006) Fallacies of distributed computing explained. Online. Available HTTP: <http://www.rgoarchitects.com/Files/fallacies.pdf>.

Rothleder, M. (2013) Flexible supply and renewable energy: Solar and the impact on load curves. CAISO and EPRI Summer Seminar on Innovating for Flexibility, August 5.

Sagan, C. (2000) *The Cosmic Connection*, 2nd edition. Cambridge: Cambridge University Press.

Sagara, J. and Ishiwatari, M. (2012). *Knowledge Note 4-1 Cluster 4: Recovery planning infrastructure rehabilitation*. Report, September.

Saul, J. and Vukmanovic, O. (2014) Ship glut burdens LNG tanker market, slashes profits. Reuters, March 11, 2014.

Schaller, B. (2010) 'New York City's congestion pricing experience and implications for road pricing acceptance in the United States', *Transport Policy*, 17(August): 266–273.

Schiermeier, Q. (2008) 'Water: Purification with a pinch of salt', *Nature*, 452(March): 260–261.

Schrödinger, E. (1944) *What is Life?* Cambridge: Cambridge University Press.

Schweppe, F.C. and Wildes, J. (1970) 'Power system static-state estimation, Part I, II, III', *IEEE Transactions on Power Systems*, 89(1), 120–135.

Seattle City Light (2012) *2012 Integrated Resource Plan*. Seattle City Light, Seattle, WA.

Seattle City Light (2013) Electric Rates and Provisions: Schedule MDC. Medium Standard General Service: City. Rate Schedule.

SEC (2011) *CF disclosure guidance: Topic No. 2: Cybersecurity*. Tech. rept. Division of Corporation Finance, Securities and Exchange Commission, October.

Serning, I. (1979) 'Prehistoric iron production', in: *Iron and Man in Prehistoric Sweden*. Stockholm, pp. 50–98.

Shakarian, P. (2011) 'Stuxnet: cyberwar revolution in military affairs', *Small Wars Journal*, April.

Shannon, C. (1948) 'A mathematical theory of communication', *Bell System Technical Journal*, 27(July and October): 379–423; 623–656.

Shapiro, J. (2001) *Mao's War Against Nature: Politics and the Environment in Revolutionary China*. Cambridge: Cambridge University Press.

Sharples, B. and Inajima, T. (2014) Japanese public seen as biggest obstacle to nuke restart. Bloomberg, August 2014.

Shearer, M. (2010: August) SHODAN for penetration testers. Online. Available HTTP: <https://www.defcon.org/images/defcon-18/dc-18-presentations/Schearer/DEFCON-18-Schearer-SHODAN.pdf>.

Shepherd, D. (1994) 'Historical development of the windmill', in: *Wind Turbine Technology, Fundamental Concepts of Wind Turbine Engineering*. New York: ASME Press, pp. 1–46.

Silver, N. (2012) *The Signal and The Noise: Why So Many Predictions Fail—But Some Don't*. London: Penguin.

Silverman, S.M. and Cliver, E.W. (2001) 'Low-latitude auroras: the magnetic storm of 14–15 May 1921', *Journal of Atmospherical and Solar Terrestrial Physics*, 63: 523–535.

Singer, P.W. and Friedman, A. (2014) *Cybersecurity and Cyberwar: What Everyone Needs to Know*. Oxford: Oxford University Press.

Singh, A. (2006) 'Power sector reform in India: current issues and prospects', *Energy Policy*, 34(16): 2480–2490.

Singh, M. (2013: July) *Growth of electricity sector in India from 1947–2013*. Tech. rept. Central Electricity Authority, Ministry of Power, Government of India, New Delhi.

Singh, R.K., and Katakey, R. (2012) Ambani, Tata 'Islands' shrug off grid collapse: Corporate India. Bloomberg, August 14.

Slay, J. and Miller, M. (2007) 'Lessons learned from the Maroochy water breach', in: *Critical Infrastructure Protection*, Vol. 253. London: Springer, pp. 73–82.

Smil, V. (2003) Energy at the crossroads: Global perspectives and uncertainties. Cambridge, MA: MIT Press.

Smil, V. (2009) The iron age and coal-based coke: A neglected case of fossil-fuel dependence. MasterResource: A Free-Market Energy Blog, September, 17.

Smil, V. (2010a) *Energy Myths and Realities*. Washington, DC: American Enterprise Institute Press.

Smil, V. (2010b) *Energy Transitions: History, Requirements, Prospects*. Santa Barbara, CA: Praeger.

Smith, A. (1970) *An Inquiry into the Nature and Causes of the Wealth of Nations*. New York: Pelican Books.

Smith, A.H.V. (1997) 'Provenance of coals from Roman sites in England and Wales', *Britannia*, 28, 297–324.

Smith, S. (2014) Congestion pricing in New York may come back from the dead. *Next City*, March 25.

Smith, T. (2004) 'Electricity theft: a comparitive analysis', *Energy Policy*, 32: 2067–2076.

Solley, W., Pierce, R. and Perlman, H. (1998) *Estimated use of water in the United States in 1995*. Tech. rept. Circular 1200. U.S. Geological Survey.

Solove, D. (2011) *Nothing to Hide: The False Trade Between Privacy and Security*. Newhaven, CT: Yale University Press.

Southern California Edison (2013) Schedule TOU-GS-2, rate schedule. Rosemead, California: Southern California Edison. Online. Available HTTP: <https://www.sce.com/NR/sc3/tm2/pdf/CE281.pdf>.

Spiegel Online (2006: 18 July) The war over resources: "Energy Security Will Be one of the Main Challenges of Foreign Policy". Spiegel Interview of Daniel Yergin. Online. Available HTTP: <http://www.spiegel.de/international/spiegel/the-war-over-resources-energysecurity-will-be-one-of-the-main-challenges-of-foreign-policy-a-427350.html>.

Spiegel Online (2008) Theresienstadt robbery: concentration camp hit by scrap metal thieves. Online. Available HTTP: <http://www.spiegel.de/international/europe/theresienstadt-robbery-concentration-camp-hit-by-scrap-metal-thieves-a-548295.html2010>.

SRM (2013) 'Alban Lakata wins UCI European Marathon Championships'. Available HTTP: <www.srm.de/news/mountain-bike/alban-lakata-wins-uci-european-marathon-championships>. Accessed March 2015.

Stein, C. (1995) 'Heat flow of the earth', in: T. Ahrens (ed), *Global Earth Physics*. London: Wiley.

Steinbeck, J. (1962) *Travels with Charley in Search of America*. New York: Viking Press.

Stephens, G., Li, J., Wild, M., Clayson, C.A., Loeb, N., Kato, S., L'Ecuyer, T., Stackhouse, P., Lebsock, M. and Andrews, T. (2012) 'An update on Earth's energy balance in light of the latest global observations', *Nature Geoscience*, September.

Stewart, B. (1861) 'On the great magnetic disturbance which extended from August 28 to September 7, 1859 as recorded by photography at Kew Observatory', *Philosophical Transactions of the Royal Society of London*, 151: 423–430.

Stiles, M. (2014) Heating a skyscraper with a data center: Amazon's one-of-a-kind solution. *Puget Sound Business Journal*, September 18.

Stockholm International Peace Research Institute (2013). Military expenditure: Data by country. Online. Available HTTP: <http://www.sipri.org/research/armaments/milex/milex_database>. Accessed August 2013.

Strickland, E. (2013) Japan prepares to restart nuclear plants. *IEEE Spectrum*, July 16.

Sugie, M. (1985) Theory of speculation and futures trading. Tokyo: Chikura Shobo.

Sukumar, S.R., Shankar, M., Olama, M., Hadley, S., Protopopescu, V., Malinchik, S. and Ives, B. (2010) *Quantifying state-policy incentives for the renewable energy investor*. Tech. rept. Oak Ridge National Laboratory and Lockheed Martin Corporation.

Sumner, J., Bird, L. and Smith, H. (2009: December). *Carbon taxes: A review of experience and policy design considerations*. Tech. rept. NREL/TP-6A2-47312. National Renewable Energy Laboratory.

Sundberg, R.E. and Nyman, H.O. (1979: October). *District heating/cogeneration application studies for the Minneaplis-St. Paul area: Methods and cost estimates for converting existing buildings to hot water district heating*. Tech. rept. ORNL/TM-6820/P4. Oak Ridge National Laboratory.

Sutterlin, P. and Downey, W. (1999) 'A power line communication tutorial: Challenges and technologies', in: *International Symposium on Power-Line Communications and its Applications*.

Swanstrom, T. (2008: April) Regional resilience: A critical examination of the ecological framework. Working Paper 2008-07. Saint Louis University.

Tabors, R.D. (2009: October) Demand response and smart grid. Charles River Associates, APEX Conference Presentation.

Taleb, N.N. (2001) *Fooled by Randomness: The Hidden Role of Chance in Life and in the Markets*. London: Penguin.

Taleb, N.N. (2007) *The Black Swan: The Impact of the Highly Improbable*. London: Penguin.

Taneja, J., Katz, R. and Culler, D. (2012) Defining CPS challenges in a sustainable electricity grid. Proceedings of the 2012 IEEE/ACM Third International Conference on Cyber-Physical Systems, IEEE Computer Society.

Tang, C. and Zimmerman, J. (2009) 'Managing new product development and supply chain risks: The Boeing 787 case', *Supply Chain Forum*, 10(2): 74–86.

Tate, Z. (2011: April) 'Estimating and visualizing the impact of forecast errors on system operations', in: *Smarter Grids Conference at Los Alamos National Laboratory*. University of Toronto.

Tax Policy Center (2012) The tax policy briefing book: A citizens' guide for the election, and beyond. Online. Available HTTP: <http://www.taxpolicycenter.org/briefing-book/>. Accessed August 2013.

Taylor, C.H. (ed.) (1917) History of the board of trade of the City of Chicago. Vol. I-III. Robert O. Law Company.

Tesla, N. (1900) The problem of increasing human energy with special references to the harnessing of the sun's energy. *Century Illustrated Magazine*, June.

Texas House of Representatives (2009–2010) Texas House Bill 1831 relating to disaster preparedness and emergency management and to certain vehicles used in emergencies; providing a penalty. Online. Available HTTP: <http://www.tdi.texas.gov/reports/leg/81bills/billsummary.html>. Accessed February 2015.

Thermal, NRG. (2014) The benefits of district energy. Website. Accessed August 2014.

Thornton, C.P., Golden, J.M., Killick, D.J., Pigott, V.C., Rehren, T.H. and Roberts, B.W. (2010) 'A chalcolithic error: Rebuttal to amzallag 2009', *American Journal of Archaeology*, 114: 305–315.

Thronton, R.P. (2009) Copenhagen's district heating system recycling waste heat reduces carbon emissions and delivers energy security. Report.

Tidball, R., Bluestein, J., Rodriguez, N. and Knoke, S. (2010: November). *Cost and performance assumptions for modeling electricity generation technologies*. Tech. rept. NREL/SR-6A20-48595. National Renewable Energy Laboratory.

Tiley, J. (2011) *Studies in the History of Tax Law*, Vol. 5. Oxford: Hart Publishing.

Tolk, A. and Muguira, J. (2003) 'The levels of conceptual interoperability model', in: Proceedings of the 2003 Fall Simulation Interoperability Workshop, Vol. 7, IEEE.

Tolk, A., Bair, L. and Diallo, S. (2013) 'Supporting network enabled capability by extending the levels of conceptual interoperability model to and interoperability maturity model', *The Journal of Defense Modeling and Simulation: Applications, Methodology, Technology*, 10(2): 145–160.

Transportation Research Board of The National Academies (2006) The Fuel Tax and alternatives for transportation funding. Online. Available HTTP: <http://onlinepubs.trb.org/onlinepubs/sr/sr285.pdf>.

Trautmann, N. (1996) Cornell Waste Management Institute. Online. Available HTTP: <http://compost.css.cornell.edu/physics.html>.

Trenberth, K.E., Fasullo, J. and Kiehl, J. (2009) 'Earth's global energy budget', *Bulletin of the American Meteorological Society*, 90(3): 311–323.

Tversky, A. and Kahneman, D. (1974) 'Judgment under uncertainty: Heuristics and biases', *Science*, 185(4157): 1124–1131.

Tylecote, R.F., Austen, J.N. and Wraith, A.E. (1971) 'The mechanisms of the Bloomery process in shaft furnaces', *Journal of the Iron and Steel Institute*, 209: 342–363.

Union of Concerned Scientists (2012) UCS EW3 Energy-water database. Online. Available HTTP: <http://www.ucsusa.org/ew3database>.

United Nations Environment Programme (2008) *Reforming energy subsidies: Opportunities to contribute to the climate change agenda.* Tech. rept. United Nations.

Upton, C. (2013) John Inshaw's steam clock. Online. Available HTTP: <http://www.search.digital-ladywood.org.uk/engine/resource/exhibition/sequential/default.asp?theme=&originator=/engine/theme/default.asp&page=1&records=&direction=1&pointer=4659&text=0&resource=7666>. Accessed August 2013.

Urquhart, A. and Thompson, M. (2014) 'Impacts of demand data time resolution on estimates of distribution system energy losses', *IEEE Transactions on Power Systems*, August.

USACE (2014: June) Columbia River Treaty: History and 2014/2024 Review. U.S. Army Corps of Engineers and Bonneville Power Administration Fact Sheet.

UK Department of Energy & Climate Change (2014) Annual Fuel Poverty Statistics Report, 2014 URN 14D/146.

United Kingdom Department for Transport (2011) Transport statistics Great Britain, 2011. Online. Available HTTP: <https://www.gov.uk/government/statistics/transport-statistics-great-britain-2011>

U.S. Attorney's Office, District of Maryland (2010: 17 December). Fannie Mae corporate intruder sentenced to over three years in prison for attempting to wipe out Fannie Mae financial data malicious code would have destroyed mortgage information. Press release.

U.S. Attorney (2007a: November) United States Attorney, California—Sacramento man pleads guilty to attempting to shut down California's power grid. Online. Available HTTP: <http://www.usdoj.gov/usao/cae/pressreleases/docs/2007/12-14-07DenisonPlea.pdf>.

U.S. Attorney (2007b: November) United States Attorney, Eastern District of California, Willows man arrested for hacking into Tehama Colusa Canal Authority computer system. Online. Available HTTP: <http://www.usdoj.gov/usao/cae/pressreleases/docs/2007/11-28-07KeehnInd.pdf>.

U.S. Census Bureau (2012) American Community Survey. Online. Available HTTP: <http://www.census.gov/acs/www/>. Accessed August 2013.

U.S. Census Bureau (2013) State and county quick-facts. Data derived from population estimates, American Community Survey, census of population and housing, state and county housing unit estimates, county business patterns, nonemployer statistics, economic census, survey of business owners, building permits. Online. Available HTTP: <http://quickfacts.census.gov/qfd/states/00000.html>.

USDA (2013) The rural energy for America Program. Online. Available HTTP: <http://www.rurdev.usda.gov/BCP_Reap.html.> Accessed January 2013.

U.S. Department of Transportation; Federal Highway Administration (2010) Public-private partnership concessions for highway projects: A primer. Online. Available HTTP: <http://www.fhwa.dot.gov/ipd/pdfs/p3/p3_concession_primer.pdf>.

U.S. Federal Highway Administration (2000). Addendum to the 1997 Federal Highway Cost Allocation Study Final Report. Online. Available HTTP: <http://www.fhwa.dot.gov/policy/hcas/addendum.htm>.

Valia, H. (1994) *Coke production for blast furnace ironmaking.* Tech. rept. American Iron and Steel Institute.

van der Linden, P. (1994) *Expert C Programming: Deep C Secrets.* Aarhus, Denmark: SunSoft Press.

VanSlyck, N., Jaleeli, L.S., Ewart, D.N., Fink, L.H. and Hoffmann, A.G. (1992) 'Understanding automatic generation control', *IEEE Transactions on Power Systems*, 7(3), 1106–1122.

Varian, H.R. (2001) Economic scene: Catastrophe bonds could fill the gaps in reinsurance. *The New York Times*, October 25.

Veldhuijzen, H. and Rehren, T. (2007) 'Slags and the city: early iron production at Tell Hammeh, Jordan, and Tel Beth-Shemesh, Israel', in: *Metals and Mines—Studies in Archeaometallurgy.* London: Archetype Books.

Verizon Enterprise Solutions (2014) *2014 data breach investigations report.* Tech. rept. Verizon.

Verweij, M. and Thompson, M. (2006) *Clumsy solutions for a complex world: governance, politics and plural perceptions.* Palgrave Macmillan, Basingstoke, UK.

Vyas, U.K. and Gentilcore, E.B. (2010) 'Growing demand for green construction requires legal evolution', *The Construction Lawyer* 30(3).

Wachs, M. (2011) *Transportation, Jobs and Economic Growth.* ACCESS.

Wadha, V. (2014) The coming era of unlimited—and free—clean energy. *The Washington Post*, September 19.

Wakita, S. (2001) 'Efficiency of the Dojima rice futures market in Tokugawaperiod, Japan', *Journal of Banking and Finance*, 25(3), 535–554.

Wald, M. (2014) Agency urges quake study for Indian point. *The New York Times*, May 10.

Wall, J.S. (2007) 'From waste to profit: Modeling program evaluates energy, use alternatives', *Energy, Environment & Sustainability*.

Washington State Department of Transportation (2009) Highway Performance Monitoring System. Online. Available HTTP: <http://www.wsdot.wa.gov/mapsdata/travel/hpms/hpms.htm>. Accessed August 2013.

Washington State Legislature (n.d.) RCW 47.04.280; Transportation System Policy Goals. Online. Available HTTP: <http://apps.leg.wa.gov/rcw/default.aspx?cite=47.04.280>. Accessed August 2013.

Waste Market Overview and Outlook (2012) Waste Business Journal, San Diego, CA.

Watkins, K. (2006) *Human Development Report 2006—Beyond Scarcity: Power, Poverty and The Global Water Crisis.* London: Palgrave Macmillian.

Watson, W., Paduano, N., Raghuveer, T. and Thapa, S. (2011) *US coal supply and demand: 2010 year in review.* US Energy Information Administration, June 2011.

Weber, B., Depew, D., Dyke, C., Salthe, S., Schneider, E., Ulanowicz, R. and Wicken, J. (1989) 'Evolution in thermodynamic perspective: An ecological approach', *Biology and Philosophy*, 4: 373–405.

White, N., Ingham, K., von Bechtolsheim, M. Haischer, M., and Francis, R. (2013) 'The future of energy utilities: How utilities can survive the "perfect storm"', Boston, MA: Arthur D. Little, p. 2.

White House (1998) *The Clinton Administration's policy on critical infrastructure protection.* Tech. rept. Presidential Decision Directive 63, May 22. Online. Available HTTP: <https://www.fas.org/irp/offdocs/paper598.htm>.

Whitmore, C. (2008: July) Electric Power Markets in the West and Southeast. Presentation by the Office of Enforcement, Division of Energy Market Oversight, U.S. Federal Energy Regulatory Commission.

Wilford, J. (1994) Enduring mystery solved as tin is found in Turkey. *The New York Times*, January 4.

Williams, J. (2007: October) *Paying at the pump: Gasoline taxes in America.* Tech. rept. Background Paper No. 56. Tax Foundation.

Williamson, R. (2014) $1.65B deal could make Clovis, N.M., hub of nation's power. *The Bond Buyer*, November 26, 2012.

Wilson, C., Grubler, A., Gallagher, K.S. and Nemt, G.F. (2012) 'Marginalization of end-use technologies in energy innovation for climate protection', *Nature Climate Change* (2).

Wilson, I. (2014) Japan regulator's safety probe heralds nuclear revival: timeline. *Washington Post*, July 15.

Wines, M. (2007) Toiling in the dark: Africa's power crisis. *The New York Times*, July 29.

Winn, C. and Cavoukian, A. (2012: March) *Applying privacy by design: Best practices to SDG&E's smart pricing program*. Tech. rept. SDG&E and Privacy-By-Design.

Wissner-Gross, A.D. and Freer, C.E. (2010) 'Relativistic statistical arbitrage', *Physical Review Letters*, 82 (056104).

Wissner-Gross, A.D. and Freer, C.E. (2013) 'Causal entropic forces', *Physical Review Letters*, 110 (168702).

Witthaut, D. and Timme, M. (2012) 'Braess's paradox in oscillator networks, desynchronization and power outage', *New Journal of Physics*, 14(August).

Wolchover, N. (2012) Why it took so long to invent the wheel. *Scientific American*, March 6.

Woodcroft, B. (1851) *Translation from the Greek of The Pneumatics of Hero of Alexandria*. London: Taylor Walton and Maberly.

Woolsey, R.J. and Pry, P.V. (2014) The growing threat from an EMP attack: A nuclear device detonated above the U.S. could kill millions, and we've done almost nothing to prepare. *Wall Street Journal*, August 12.

Wright, S. (2012) Tipping fees vary across the U.S., *Waste & Recycling News*, Vol. 18, July.

WWAP (World Water Assessment Programme) (2012: March). *United Nations World Water Development Report 4*. Tech. rept. Paris: UNESCO.

Xie, L. and Ilic, M. (2009) Model predictive economic/environmental dispatch of power systems with intermittent resources. IEEE Power and Energy Society General Meeting.

Yan, Y., Quan, T., Sharif, H. and Tipper, D. (2012) A survey on smart grid communication infrastructures: Motivations, requirements, and challenges. IEEE Communications Surveys and Tutorials.

Zetter, K. (2013) Power plants and other vital systems are totally exposed on the internet. *WIRED Magazine*, November 8.

Zhang, Z., Wang, W. and Qiu, B. (2014) 'Oceanic mass transport by mesoscale eddies', *Science*, 345(6194): 322–324.

Zurich (2014) *Risk Nexus: Global interconnections of cyber risk: impact on small and medium-sized enterprises*. Report, April.

Index